SAE International's Dictionary of Testing, Verification, and Validation

In Remembrance of Linda Ayres-DeMasi

In the quiet solitude of the written word, you found your place of grace. Your keen eye, your unwavering commitment to clarity, and your unending pursuit of excellence breathed life into the pages of countless manuscripts.

Your gentle guidance and your tireless encouragement propelled authors forward when doubt and uncertainty loomed large. Without you, many books would have remained unpublished.

You may have left this world but your spirit lives on in the books that bear your mark, in the hearts of the authors whose journeys you guided and in the memory of everyone inspired by your professionalism.

With heartfelt gratitude,

Sherry Nigam

Sherry Nigam
Publisher, SAE Books
SAE International

SAE International's Dictionary of Testing, Verification, and Validation

BY
JON M. QUIGLEY, PMP, CTFL

INTERNATIONAL®

Warrendale, Pennsylvania, USA

400 Commonwealth Drive
Warrendale, PA 15096-0001 USA
E-mail: CustomerService@sae.org
Phone: 877-606-7323 (inside USA and Canada)
724-776-4970 (outside USA)
FAX: 724-776-0790

Library of Congress Catalog Number 2023948007
http://dx.doi.org/10.4271/9781468605914

ISBN-Print 978-1-4686-0590-7
ISBN-PDF 978-1-4686-0591-4
ISBN-epub 978-1-4686-0592-1

To purchase bulk quantities, please contact: SAE Customer Service

E-mail: CustomerService@sae.org
Phone: 877-606-7323 (inside USA and Canada)
724-776-4970 (outside USA)
Fax: 724-776-0790

Visit the SAE International Bookstore at books.sae.org

Publisher
Sherry Dickinson Nigam

Development Editor
Publishers Solutions, LLC
Albany, NY

Director of Content Management
Kelli Zilko

Production and Manufacturing Associate
Michelle Silberman

Contents

Preface

I have long thought about communication within groups. As an engineer, I understood the need for clear articulation of things and intentions. It is not as easy as blurting a collection of words; words mean something, and what they mean depends upon the interpretation of the person receiving the information. We might be surprised at how often we think we have communicated with one another but we actually have not. Understanding this has led me to actions best told via a story.

> *At one time in my career, I was selected to develop and manage a test and verification group. I found people in and outside the company for these critical roles, from environmental engineers to software testers, HIL rig creating and maintaining, Systems integration, and many other test domains. The company had been performing verification and testing on its product line. Still, this effort was dispersed among departments, with gaps between departments, insufficient tools, and a general lack of focus.*
>
> *A lexicon refers to a collection or a list of words and their meanings within a specific language or domain. It serves as a comprehensive repository of the vocabulary used by speakers of that language or* ***professionals*** *within a particular field of study. Lexicons can vary in size and scope, ranging from simple dictionaries that cover everyday words to extensive specialized lexicons for technical jargon in specific industries.*
>
> *In linguistics, a lexicon is considered one of the key components of a language's structure, along with grammar and phonology. It is an essential resource for understanding and using a language effectively, as it provides the definitions, pronunciations, and sometimes additional information about words, such as part of speech, usage examples, and etymology.*

Upon getting sufficient talent to perform the task, I decided our team would attend a software testing certification class. I did this even though not all of the team would be testing the software, and I did not care if the team member took the certification test; it was more important to sit for the training for the certification. I hold the PMP certification and teach the course from time to time. Many times in those classes, a student would approach me about answering the test questions as just semantics. I would tell them that words have specific meanings and tell them my favorite story about a project manager that would approach me with a particular item on the "critical path." They had no Gantt chart or demonstration of this fact; we find a blocked task upon exploration. Both of these present project difficulties and have different ways to resolve them. How the project reacts depends on knowledge of the specific situation and circumstances. The more technical and complex the work environment, the more precise our language requirements. A common lexicon goes a long way to common understanding.

Coordinated work requires clear communication for efficiency and effectiveness. We need not maintain this lexicon forever as is, but this is a starting point for the team. The team language can change over time but is controlled with input from the group and influenced by new technology and emerging circumstances. We hope this dictionary provides such structure to build the language base for your test and verification teams.

A

"Ask people's advice, but decide for yourself."

—Ukrainian Proverb

Accelerated Testing

Accelerated testing subjects a product or system to conditions that simulate real-world use over a shorter period than it would take under normal circumstances. For example, the product's exposure to anticipated temperature extremes over its eight-year life, 200 hours of exposure. The product is tested at 200 hours of extreme temperature. This continuous exposure is over the course of approximately nine days.

Accelerated testing aims to identify potential issues or defects in a product or system more quickly and efficiently to address them before the product is released to the public—for example, this method tests durability or reliability. In addition, it can be beneficial for detecting latent issues, which may only become apparent after long periods of use. For more information, see **J2020_202210**, **J2100_202101**, **J3014_201805**, and **J2017-01-0389** [1, 2, 3, 4, 5, 6].

Acceleration Factor

The acceleration factor in reliability engineering allows estimates of the stimulus's effect on a product's reliability under normal use conditions. The idea is to speed up the aging process and accelerate failures by exposing the product to higher-than-normal stress or operating conditions, allowing engineers to collect data and evaluate the product's reliability more quickly.

The acceleration factor is a ratio of the time it would take for the product to fail under normal use conditions (or "useful life") to the time it takes to fail under accelerated conditions. For example, if a product has a useful life of ten years under normal conditions and fails after three months under accelerated conditions, the acceleration factor would be 40. This result of the calculation means that every three months of accelerated testing is equivalent to ten years of regular use. The acceleration factor can then estimate the product's reliability under normal conditions based on the observed failures under accelerated conditions.

It's important to note that the acceleration factor is based on assumptions and models, and the validity of the accelerated testing depends on how well these models represent the real-world behavior of the product [1].

Acceptance Criteria

Acceptance criteria of stress-test-driven approaches are typically "test to pass," which means that the value of the qualification statement depends entirely on the validity of the model parameters because quality and reliability are not measured. Therefore, a product's robustness is unknown after performing this qualification. The resulting evaluation is qualitative, as the relationship between the applied stress during the stress-test-driven qualification conditions and lifetime at conditions of use is usually not established. The sensitivity of stress-test-driven methods concerning new or changed materials or technologies is insufficient to demonstrate the robustness of a component in the harsh automotive environment [7, 8].

Acceptance Testing

Acceptance testing determines whether a system or application meets specified requirements and is ready for delivery to the end user. It is usually the final testing stage before a product is released. The system or application is tested in a real-world environment during acceptance testing, simulating how the end users would use it. Customers or end users typically do this testing and provide feedback on the system's performance and functionality.

Acceptance testing aims to ensure that the system or application meets required specifications and functions as expected. It is an essential step in the development process because it helps identify issues or defects before releasing the product to the market.

Acceptance testing verifies whether a software system meets the requirements of the business or user and whether it is fit for its intended use. Acceptance testing typically occurs after functional testing and before the software deployment to production. It is often performed by the software's customers or end users, although a third-party testing company can also test it. Acceptance testing aims to ensure that the software meets the user's needs and is ready for use in a real-world environment [7, 9].

Accessibility Testing

Accessibility testing evaluates how well people with disabilities can use a website, application, or digital product. Accessibility testing can explore low vision, hearing, mobility, and cognitive impairments. Accessibility testing is essential because it ensures that all users can use the product effectively and efficiently regardless of their abilities. Some standard tools used in accessibility testing include screen readers, magnifiers, and color contrast checkers. For more information, see **EPR202224**, **J2678_201609**, and **J2364_201506** [10, 11].

Accuracy

Accuracy measures how well a system, process, or test precisely produces correct results (see Figure A.1). It is often expressed as a percentage and is calculated by dividing the correct results by the total number of results. A high accuracy rate indicates that the system, process, or test consistently produces accurate results. In contrast, a low accuracy rate suggests that there may be issues with the system or method that require adjustment to improve the accuracy of the results. Accuracy is essential in many fields, including science, medicine, and engineering, as it ensures that the results of experiments or tests are reliable and trustworthy.

FIGURE A.1 An illustration of accuracy and precision.

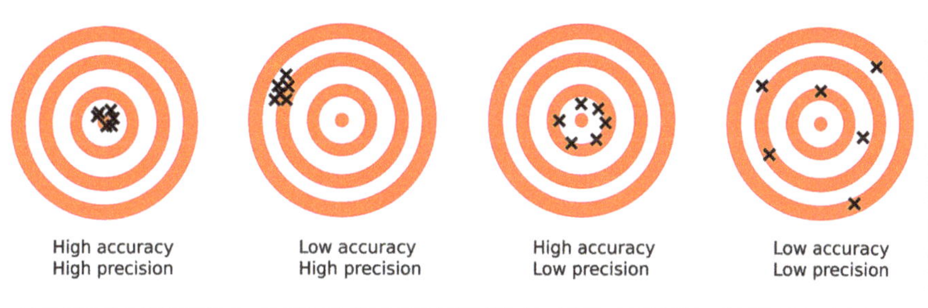

pOrbital.com/Shutterstock.com.

Active Design of Experiments (ADoE)

Active design of experiments (ADoE) is a systematic approach to designing and conducting experiments that focuses on actively manipulating variables and maximizing the information gained. It involves carefully planning the experimental conditions and testing them in a controlled manner to identify the factors that significantly impact the outcome.

ADoE typically involves statistical analysis to determine the most effective design for an experiment, including the number of variables and the sample size needed to measure the effects of those variables accurately. It also involves using various techniques, such as randomization and blocking, to reduce the impact of extraneous variables on the results.

ADoE is used in various fields, including engineering, biology, and social science, to improve products, processes, or understanding of a particular phenomenon. It is a powerful tool that allows researchers to identify the key factors that drive outcomes and optimize them for maximum impact.

Actor-Observer Bias

Actor-observer bias is a psychological phenomenon whereby individuals attribute their behavior to situational factors while attributing the behavior of others to their dispositions or personality traits. This bias manifests in how people explain events and view their actions as influenced by external circumstances. In contrast, they see others' actions as indicative of their inherent characteristics. For example, if someone is late to a meeting, they might explain it by saying they got stuck in traffic, but if someone else is late, they might attribute it to their lack of punctuality.

This bias can affect how people perceive and interact with others, leading to misunderstandings and misinterpretations. It can also impact decision-making in various fields, such as psychology, law, and management. Understanding the actor-observer bias and recognizing when it's affecting one's perceptions can help to reduce its influence and promote more accurate understanding.

Actor-observer bias and fundamental attribution error are related but distinct biases in social cognition. Actor-observer bias refers to individuals' tendency to attribute their own behavior to situational factors while attributing the behavior of others to dispositional or personality traits. By comparison, fundamental attribution error is the tendency to overstate dispositional explanations for other people's behavior while ignoring situational factors. In other words, it is the tendency to attribute the behavior of others primarily to their characteristics, such as their personality or motives, rather than to situational factors.

While actor-observer bias and fundamental attribution error involve behavior attribution, the former is specific to how people explain their own behavior versus others' behavior. The latter is about overemphasizing dispositional explanations for others' behavior.

Actual Result

Testing actual results refers to evaluating the performance of a product, system, or program by comparing its observed outcomes (actual) with its expected outcomes. The purpose of testing actual results is to assess the accuracy, reliability, and effectiveness of the product or system tested and to identify and address issues or problems.

Testing actual results is a crucial step in developing and deploying any product or system, as it helps ensure that the product meets the requirements and expectations of its intended users. Testing actual results can involve various methods, including manual, automated, and real-world testing. It may be performed by internal teams or by external third-party testers.

Ad Hoc Review

An ad hoc review typically refers to an inspection done on a case-by-case basis rather than following a formal, standardized process. Use ad hoc reviews when there is an urgent or unexpected need for a review or when a more detailed or formal review is not required or feasible. For example, an ad hoc review might be done by a supervisor or manager to quickly evaluate the performance of an employee who has repeatedly been absent from work or by a group of colleagues to assess the submitted proposal.

Ad hoc reviews can be helpful when a quick decision or evaluation is needed. Still, they may not always be as thorough or rigorous as a more formal process. Therefore, it is essential to consider the context and purpose of the review before deciding whether an ad hoc approach is appropriate [12, 13, 14].

Ad Hoc Testing

Informally conducted testing is referred to as ad hoc testing. There is no formal test preparation, no recognized test design technique, few expectations for the findings, and the test execution activity is random [15].

Ad hoc testing is product testing conducted "as needed." It is a flexible and informal testing method that does not follow specific guidelines or test cases. Instead, it explores the system and finds defects or issues through unstructured and random testing. Ad hoc testing supplements more formal testing methods performed by any testing team member without specialized training or knowledge. It helps identify defects not identified by more structured testing methods and can quickly and effectively test new features or system changes.

Adaptability

Adaptability is a product's (software and hardware) capacity for customization for various defined situations without using methods other than those made available by the software under consideration.

Adaptation

Vehicle adaptations are accessories that you can have installed on your car to give you a trip mobility solution. These can also be original equipment manufacturer (OEM) variations. For example, commercial heavy truck manufacturers can have individual customer adaptations. Many adaptations of the product increase testing complexity and likely duration, which often have the impact of reducing the total test coverage.

Advanced Driver Assistance System (ADAS)

Advanced driver assistance systems (ADAS) are designed to help drivers in various ways, from providing warnings about potential hazards to automatically controlling certain vehicle functions. These systems use multiple technologies, including cameras, radar, and lidar, to gather information about the vehicle's surroundings and the driver's actions. As a result, ADAS can help drivers avoid collisions, maintain safe speeds, and stay in their lane (see Figure A.2), among other things. Some examples of ADAS features include adaptive cruise control, lane-keeping assist, and automatic emergency braking.

FIGURE A.2 Illustration of an advanced driver assistance system.

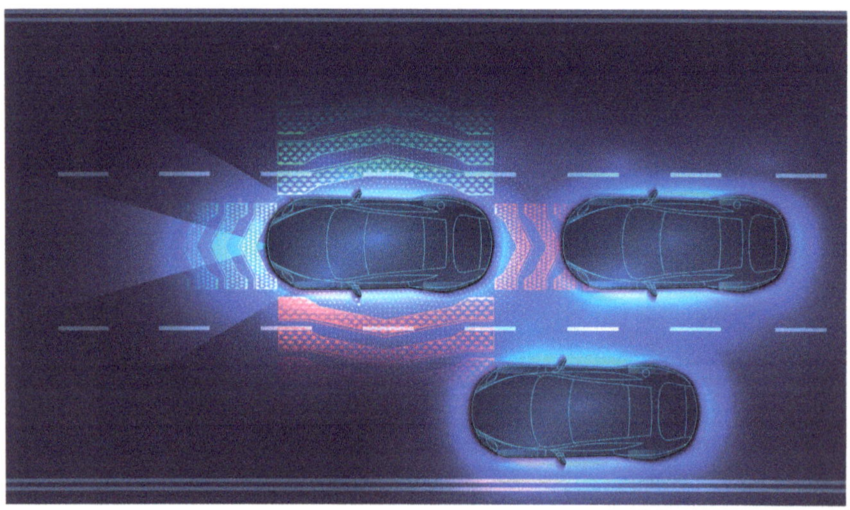

ZinetroN/Shutterstock.com.

Agile

Agile describes a method of project management that emphasizes flexibility, collaboration, and customer satisfaction. The Agile Manifesto outlines software development values and principles that prioritize individuals and interactions, working software, and customer collaboration over processes and tools. Agile methodologies, such as Scrum and Kanban, are widely used in software development and have been applied to other fields. Agile processes prioritize adaptability and involve quick incremental and iterative deliveries of the developed product (see Figure A.3).

FIGURE A.3 An example of the Agile approach with integrated testing.

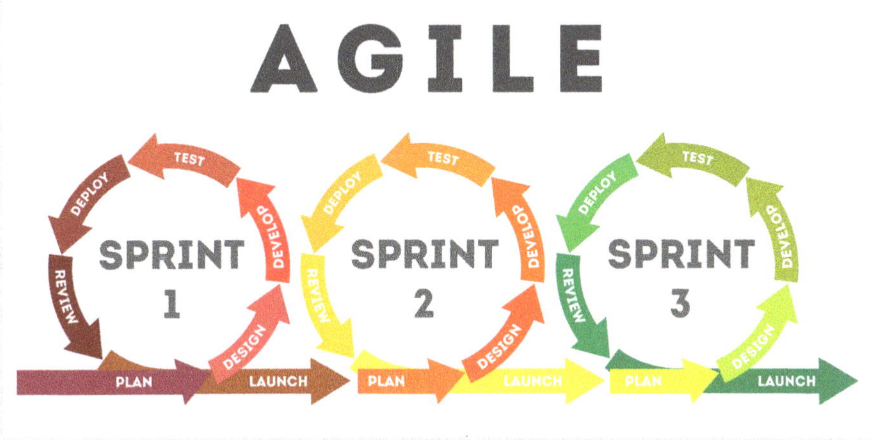

300 librarians/Shutterstock.com.

Agile Testing

Agile testing is a method that follows Agile software development principles. It involves continuous testing and integration of new features and improvements throughout the development process rather than waiting until the end to perform comprehensive testing. This approach enables faster delivery of high-quality software, as it allows for early detection and correction of defects. Agile testing also emphasizes collaboration between developers, testers, and other stakeholders to ensure that software meets users' needs.

Aging Tests

An aging test, or reliability testing, is performed on a product to determine its ability to withstand the effects of time and wear. Typically this testing is performed on products with a long life span, such as vehicles and appliances. It involves simulating conditions the product will likely encounter over its lifetime. For example, aging tests may expose the product to high or low temperatures, moisture, vibration, or other environmental conditions that cause wear and tear. Aging testing aims to identify potential failures or weaknesses in the product. These weaknesses can be understood and addressed before the product goes on sale. For more information, see **J1211_201211** [16].

Alpha Testing

Alpha testing is software testing that occurs early in the development process after the completion of the initial testing but before the released software goes to crucial customers or the general public. A small group of internal testers or developers familiar with the software perform Alpha testing to provide feedback on its functionality and performance.

Alpha testing identifies and fixes any significant issues or bugs in the software before releasing it to a broader audience. Alpha testing is conducted in a controlled environment, such as a lab, test bed, or test track. It allows developers to gather valuable feedback and make necessary improvements before moving on to the next testing stage. The goal of Alpha testing is to ensure that the software is of high quality and ready for use by the public.

Anomaly

A test anomaly is a deviation or unexpected result encountered while testing a system or product. It is a behavior or outcome inconsistent with the expected or intended behavior and may indicate a defect or issue in the tested system [17]. Different types of testing, such as functional, performance, or security, detect different types of anomalies. They can result from various factors, such as incorrect assumptions, coding errors, data inconsistencies, or environmental factors.

A test anomaly is carefully analyzed to determine the root cause and potential impact on the system being tested. Then, depending on the severity and effects of the abnormality, further testing or corrective action may be necessary to ensure the system functions correctly. Overall, detecting and addressing test anomalies is an essential part of the testing process, as it helps improve the quality and reliability of the tested system [18, 19].

Altitude Test

Except for air shipment of unenergized controls, operation in a vehicle should follow the anticipated operating limits. Altitude testing stresses a product over these limits of absolute pressure. Product altitude testing refers to testing a product's performance and functionality at different altitudes. This type of testing is used in aerospace and aviation industries, where products such as aircraft, spacecraft, and missiles must perform reliably and safely at various altitudes. However, it applies to ground vehicle components traveling mountain roads or undergoing air shipping.

The testing process typically involves exposing the product to different states (operating/nonoperating), and other altitude conditions, such as high, low, and rapid altitude changes, and measuring its performance and behavior (see Figure A.4). Various parameters may be tested during altitude testing, including the product's speed, acceleration, stability, power, fuel efficiency, and the performance of its multiple components and systems.

FIGURE A.4 An example of altitude exposure for a product [20].

Condition	Altitude	Atmospheric Pressure
Operating	3.6 km (12000 feet)	62.0 kPa absolute pressure (9 lbf/in^2 absolute)
Nonoperating	12.2 km (40000 feet)	18.6 kPa absolute pressure (2.7 lbf/in^2 absolute)

Altitude testing is essential because atmospheric conditions can significantly affect the performance of many products, particularly those that rely on air pressure, temperature, and density to operate. For example, air pressure and temperature decrease at high altitudes, impacting the performance of engines, hydraulics, and other systems.

By testing a product's performance at different altitudes, engineers and designers can identify and address any performance or functionality issues that may arise and ensure that the product meets the necessary safety and performance standards. This type of testing is often conducted in specialized facilities, such as altitude chambers or wind tunnels, that can simulate various altitude and atmospheric conditions. For more information, see **J1455** [20].

Amp Meter

Also referred to as an ammeter, an ampere meter or amp meter is an instrument used to measure the electric current in a circuit. It measures the flow of electric charge, typically measured in amperes (amps), the unit of electric current.

An ammeter is connected in series with the circuit component or device measured. When electric current flows through the circuit, it also flows through the ammeter, and the ammeter measures the amount of current flowing through it.

Ammeters can be either analog or digital. Analog ammeters display the current reading using a needle and a graduated scale, while digital ammeters display the current value numerically.

Analyzability

Analyzability and testing are closely related concepts in software development. Analyzability refers to the ease with which an exploration of a system is possible or broken down into its constituent parts. At the same time, testing involves evaluating the performance and functionality of a software system to identify and correct defects.

Analyzability plays a critical role in testing because the easier it is to break down a system or problem into smaller components, the easier it becomes to test those components individually. In addition, by testing each component individually, developers can identify and fix any defects or errors, improving the software system's overall quality.

Analyzability can also help developers to design effective test cases that cover all the critical aspects of the software system. For example, analyzing a system's architecture, data flow, and processing logic can help developers identify potential problem areas and prioritize their testing efforts accordingly.

Analyzability is an essential factor to consider when designing and testing software systems. By focusing on analyzability and breaking down systems into smaller, more manageable components, developers can create more effective and reliable software systems that meet the needs of their users.

Anechoic Chamber

An anechoic chamber is a room designed to be acoustically "dead," meaning it has minimal sound reflection. The walls, floor, and ceiling are lined with materials that absorb sound waves, and the room is designed to be isolated from outside noise sources. Anechoic chambers are used for various purposes, including acoustics testing, audio recordings, and scientific experiments that require a quiet environment.

An anechoic chamber is used for several purposes:

1. Acoustics Testing: An anechoic chamber is used to test and evaluate the acoustic properties of various devices and systems, such as speakers, headphones, microphones, and other audio equipment.
2. Scientific Research: Anechoic chambers are used for various scientific experiments that require a quiet environment, such as studying human hearing, animal behavior, and electromagnetic compatibility.
3. Shielding and EMI Testing: Anechoic chambers can also be used to test electronic devices' electromagnetic compatibility (EMC) and evaluate their shielding effectiveness against electromagnetic interference (see Figure A.5).

Overall, the unique acoustic properties of an anechoic chamber make it a valuable tool in a range of applications where a controlled and quiet environment is required. For more information, see **J2883_202003**, **J1400_201707**, **J1503_200409** and **J377_200712** [21, 22, 23, 24].

FIGURE A.5 An anechoic chamber is used for product EMC testing.

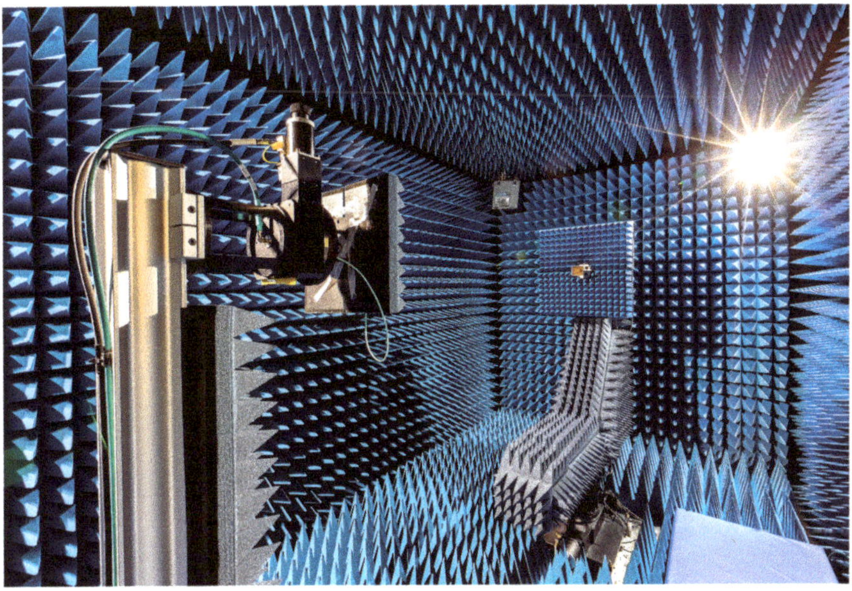

emre topdemir/Shutterstock.com.

API Testing

These tests evaluate application programming interfaces (APIs) to ensure they function correctly and meet specified requirements. This type of testing is typically done at the integration level and involves sending requests to an API and verifying the response received. In addition, it ensures that the API can handle various input types and produce the expected output. API testing is essential because it helps identify any API issues that could impact the functionality of the system integration [25, 26].

Application Binary Interface

An application binary interface (ABI) bridges two separate binary components in computer software. The OS or a library may provide one of these modules, while the other is user-created software.

An ABI test verifies that a software application is compatible with another specific operating system or platform's ABI. The ABI is a set of standards that define how different software components and libraries interact with each other and the operating system.

ABI testing ensures the application performs without issues on a specific operating system or platform. ABI testing is essential because different operating systems and platforms have different ABIs. An application not designed to be compatible with a particular ABI may not function properly or run at all.

ABI testing can involve verifying that the application is compatible with the system's ABI at the binary level and testing the application's performance and functionality on the target operating system or platform.

Architectures

Vehicle architecture refers to the overall design and layout of a vehicle's components, systems, and subsystems, including mechanical and electrical aspects (see Figure A.6).

Mechanical architecture refers to the physical components of the vehicle, such as the engine, transmission, suspension, and steering systems, as well as the overall chassis, body, and exterior design. These mechanical systems work together to support the vehicle's weight, provide power and torque to the wheels, and enable the vehicle to move and handle in a desired manner.

FIGURE A.6 An example of a vehicle architecture.

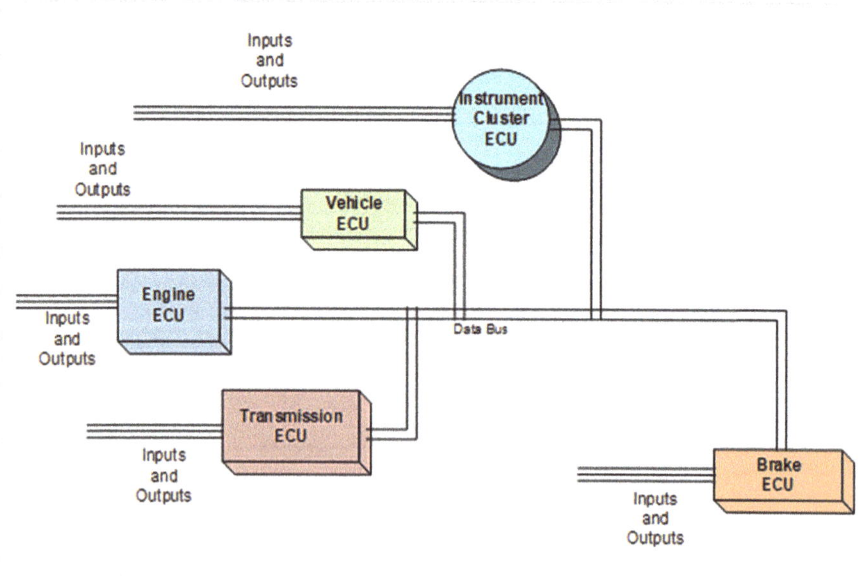

Electrical architecture refers to the electronic components and systems within the vehicle, such as the battery, wiring, sensors, and control modules.

These systems allow the vehicle to control lighting, climate control, entertainment, safety features, and critical components such as the engine and transmission.

In modern vehicles, the mechanical and electrical architectures are tightly integrated, with electronic systems controlling and communicating with mechanical systems to optimize performance, fuel efficiency, and safety. This integration has led to the development of advanced techniques like hybrid and electric drivetrains, adaptive cruise control, lane departure warning, and other driver assistance systems. For more information, see **J2057/4_202212**, and **J3131_202203** [27, 28, 29, 29].

Artifacts

Test artifacts are documents or other materials that are created and used during the testing process. These include test plans, test cases, test processes, test data, test scripts, test results, and other documentation related to the testing process. They may also have tools or software for executing tests, such as test management or automation tools. Test artifacts help to provide a clear and structured approach to testing. In addition, artifacts communicate the testing process and attributes to stakeholders and ensure that tests are conducted effectively and efficiently.

Artificial Intelligence (AI)

Artificial intelligence (AI) is a field of computer science that focuses on creating intelligent machines that can think and act like humans. These intelligent systems are designed to learn from their experiences and improve their performance over time. Some examples of AI include voice recognition systems, autonomous vehicles, and machine learning algorithms. AI can potentially transform many industries and has already begun to impact fields such as healthcare, finance, and education. In addition, AI can help generate test cases for the product or at least the high-level descriptions of the test cases required.

Attack Vectors

Product testing involves identifying potential vulnerabilities or weaknesses in a product and trying to exploit them to determine its security. Penetration testing includes testing the product's resistance to various types of attacks, such as denial of service attacks, injection attacks, and cross-site scripting attacks.

Attack vectors refer to attackers' methods or channels to access a system or product. These include exploiting vulnerabilities in software, using social engineering tactics to trick users into divulging sensitive information, or physically tampering with hardware (see Figure A.7):

FIGURE A.7 An example of the range of domains in information security management.

d3verro/Shutterstock.com.

1. Physical tampering or manipulation of the product.
2. Network vulnerabilities or exploits within the product's software or firmware.
3. Social engineering tactics to gain access or exploit vulnerabilities.
4. Supply chain attacks, where the product is compromised during the manufacturing or distribution process.
5. Physical or electronic access to the product's data or user information.
6. The exploitation of vulnerabilities within the product's hardware or components.
7. Unsecured data storage or transmission of sensitive information through the product.
8. Lack of proper security measures or product updates leads to the exploitation of vulnerabilities.

Audit

Auditing in product development is a necessary process that involves independently evaluating and verifying the effectiveness of the product development process. In addition, auditing in product development aims to ensure that the developed products meet required quality standards and comply with relevant regulations [30].

Product development auditing typically involves reviewing the product development process from start to finish. Therefore, the audit may cover various stages of product development, such as design, prototyping, testing, and final production. It may also include a review of the quality control measures and procedures that are in place to ensure that the products meet all required specifications.

During an audit in product development, auditors use various techniques and tools to assess the effectiveness of the process. For example, the audit may include conducting interviews with key personnel involved in product development, reviewing documentation and records, and performing product tests and inspections.

The audit results identify areas of improvement in the product development process and develop recommendations for corrective action. The audit may also help to identify potential risks and issues that could affect the quality and safety of the product.

Overall, auditing in product development is a critical process that can help to ensure that the products developed meet required quality standards and comply with relevant regulations. In addition, by identifying and addressing potential issues early in the product development process, organizations can improve the quality of their products, reduce the risk of defects and recalls, and enhance customer satisfaction [12].

Augmented Reality

Augmented reality (AR) is a technology that overlays digital information in the real world, allowing users to interact with virtual objects as if they were real. AR provides various benefits to testing, from exploring test case alternatives to improving efficiency, accuracy, and safety.

One application of AR in testing is assembling and maintaining complex machinery or equipment. AR can provide workers with visual guides and instructions that help them perform complex tasks more efficiently and accurately. For example, an AR system could overlay a 3D model of a piece of equipment onto the real-world view, showing workers how to assemble or disassemble the equipment in real time.

AR can be used to explore prototypes and product iterations. In addition, AR can provide virtual environments to explore and develop specific test cases. For example, an AR system could simulate a car's or other vehicle's actions in various driving conditions, allowing engineers to test the vehicle's performance in a safe and controlled environment. AR can also be used in training and education, preparing the test personnel with an understanding of the product and application beyond reading the myriad of documents. For testing and manufacturing exploration, the development of specific tools for practical testing.

Authority Bias

Authority bias occurs when people tend to give greater weight to the opinions or beliefs of authorities or experts, even if those beliefs are not supported by evidence or facts. This cognitive bias can lead people to accept information without critically evaluating it, resulting in people making decisions that are not in their best interest.

This deference to authority can make devising test approaches and reporting testing results difficult for the testing group.

Automated Testing

Automated testing is a method of testing software where the testing process is executed automatically rather than being performed manually by a human tester. Automated testing can validate that a piece of software or hardware functions correctly and meets the specified requirements. It can also identify software defects and bugs and ensure the software is reliable and performs consistently.

There are many different types of automated testing, including unit, integration, functional, and regression. Unit testing involves testing individual units or components of the software to ensure they are working correctly. Integration testing involves testing how different components or parts of the software work together. Functional testing involves testing the applicable requirements of the software to ensure that it meets the users' needs. Finally, regression testing involves re-running previously conducted tests to ensure that changes to the software have not introduced any new defects.

Automated testing can help ensure software quality. Automation often runs quickly and accurately and is repeatable. It can also reduce the time and resources required for testing performed by a machine rather than a human tester. However, automated testing is not a replacement for manual testing. Manual software testing may be required to identify any issues automated testing may miss [31, 32].

Automated Testware

Automated testware is software designed to perform automated testing on a system or application. Automated testware includes testing tools such as test scripts, test cases, and test frameworks programmed to run tests on a system or application and report the results. Automated testware is often used in software development and quality assurance processes to ensure that a system or application functions correctly and meets specific standards or requirements. It is used for various purposes, such as unit testing, integration testing, system testing, acceptance testing, and testing multiple platforms and devices [31].

Automotive Safety Integrity Level

See Functional Safety.

Autosar

AUTOSAR (Automotive Open System Architecture) is a standardized software architecture for automotive electronic control units (ECUs) that aims to improve the development and integration of software in the automotive industry. AUTOSAR defines a set of standards, guidelines, and software modules that automotive manufacturers and suppliers can use to build complex software systems for vehicles.

Testing is an integral part of the development process in the AUTOSAR framework. Therefore, several critical aspects of testing are relevant to AUTOSAR:

1. Integration Testing: AUTOSAR provides a standardized architecture for integrating software modules from different suppliers. Integration testing is necessary to ensure that the various software components work together effectively and without conflicts.
2. Functional Testing: AUTOSAR provides a standardized approach to defining software interfaces and behavior. Functional testing is necessary to ensure that the software modules perform the intended functions as specified.
3. Safety Testing: In the automotive industry, safety is of critical importance. AUTOSAR includes a range of safety features and mechanisms used to improve the safety of software systems. Safety testing is necessary to ensure that the safety features are working correctly and the software systems are safe.
4. Performance Testing: AUTOSAR software systems can be complex and resource-intensive. Performance testing is necessary to ensure that software systems operate efficiently without excessive resource usage.

In addition to these testing aspects, AUTOSAR also provides a range of tools and frameworks that support testing activities. For example, AUTOSAR provides a standardized test framework that can be used to automate testing and improve the efficiency of the testing process. Overall, testing is an essential part of the development process in the AUTOSAR framework and plays a vital role in ensuring the quality, safety, and reliability of automotive software systems [9, 33, 34, 35].

Availability

Availability refers to the ability of a system or service to be accessed by users at any given time. Availability includes the system's uptime, reliability, and performance, usually written as a percentage.

B

"Bureaucracy is the layer, or layers, of management that lie between the person who has decision-making authority on a project and the highest-level person who is working on it full-time."

—Herbert Rees, Eastman Technology

B10

In reliability engineering, B10 refers to the probability of a component or system failing within its useful life. Specifically, B10 represents the point at which 10% of components or systems are expected to fail due to regular use or wear and tear.

Reliability is an essential factor to consider in the design and development of software systems, as it affects their overall quality, performance, and usability. By ensuring that a software system is reliable, developers can minimize the risk of system failures or downtime and improve the user experience.

In the context of B10, reliability engineering seeks to identify the factors that can impact a component or system's useful life, such as environmental conditions, manufacturing processes, and usage patterns. By analyzing these factors and conducting tests and simulations, engineers can estimate the B10 value for a given component or system and design it accordingly.

Back-to-Back Testing

Comparison testing involves running many versions of a given component or system with the same inputs, then looking for differences in the results and determining why they occurred.

Another version of back-to-back testing evaluates the performance of a system by comparing it to a similar system. It involves running both systems simultaneously and comparing the results to determine which performs better. This type of testing is often used in industries such as manufacturing, telecommunications, and software development to compare different models or versions of a product or system. Back-to-back testing allows for a more comprehensive evaluation of the systems compared and a direct comparison of their performance under similar conditions.

Basis Path Testing

Basis path testing, also known as control flow testing, is a white-box testing technique used in software testing to test the different execution paths of a program systematically. It is based on the concept that the number of independent paths through a program's source code determines the minimum number of test cases needed to achieve a certain level of coverage.

The objective of basis path testing is to ensure that all possible paths within a program have been tested at least once. In addition, it aims to uncover errors or faults that may occur due to incorrect control flow, loop iterations, or conditional statements.

The basis path testing technique follows a structured approach that involves the following steps:

1. Control Flow Graph (CFG) Creation: The first step is to create a control flow graph representing the program's control flow structure. The CFG consists of nodes representing individual program statements and edges representing the possible flow of control between statements.
2. Cyclomatic Complexity Calculation: The cyclomatic complexity of the program is determined based on the CFG. Cyclomatic complexity is a measure of the number of independent paths within a program and serves as an indicator of the number of test cases required for basis path testing.
3. Test Case Generation: Test cases are designed to cover all independent paths within the program. Each path is traversed at least once, ensuring all statements, branches, and loops are exercised during testing. This can be achieved using techniques such as control flow testing, decision testing, or loop testing.
4. Test Case Execution: The generated test cases are executed, and the program's behavior is observed and compared against the expected results. Any discrepancies or failures are recorded as potential defects.

Basis path testing helps identify and target specific areas of a program's control flow with a higher risk of errors or faults. It provides a systematic approach to achieve a higher level of code coverage and can be combined with other testing techniques to achieve comprehensive testing.

It's important to note that basis path testing focuses on a program's internal logic and structure and is typically performed by developers or testers with knowledge of the program's source code. It complements other testing techniques such as functional testing, unit testing, and integration testing to ensure thorough testing coverage.

Basis Set

The basis set is the rationale for product testing and refers to the criteria or standards for evaluating the product via test. Some common elements of a basis set for product testing include the following:

1. Product Specifications: These include the product's design specifications, performance requirements, and any other requirements that the manufacturer or regulatory bodies have set.
2. Safety: Testing the product to ensure that it is safe for use by consumers and meets any relevant safety standards or regulations.
3. Reliability: Testing the product to ensure it performs consistently and reliably over time.
4. Quality: Testing the product to ensure that it meets the required level of quality and that it is free from defects or issues.
5. User Experience: Testing the product to ensure that it is easy to use and meets the needs and expectations of the target user group.
6. Environmental Impact: Testing the product to ensure that it meets defined relevant environmental standards and has a minimal environmental impact.
7. Legal Compliance: Testing the product to ensure it conforms to relevant legal requirements or regulations.

Behavior

The behavior of a product refers to how it performs and responds to different inputs and conditions. The requirement defines the expected behavior of the product. Testing is a critical part of understanding and verifying a product's actual behavior and can be used to identify defects, weaknesses, and areas for improvement.

There are several types of testing to evaluate the behavior of a product:

1. Functional Testing: This testing ensures that the product actually performs the functions it is designed to perform. Functional testing typically involves creating test cases that simulate different usage scenarios and inputs and verifying that the product responds correctly.
2. Performance Testing: This type of testing evaluates how well the product performs under different workloads and conditions. Performance testing can help identify performance bottlenecks and areas for optimization.

3. Compatibility Testing: This testing ensures the product is compatible with different hardware, software, and network environments. Compatibility testing typically involves testing the product on various devices and configurations.
4. Security Testing: This type of testing evaluates the product's resistance to attacks and vulnerabilities. Security testing typically involves exploiting the product to identify weaknesses and vulnerabilities.
5. Usability Testing: This type of testing evaluates how straightforward and intuitive the product is. Usability testing typically involves creating user scenarios and observing how users interact with the product.

By using a combination of these testing types, it is possible to gain a comprehensive understanding of the behavior of a product. Testing can help identify defects and issues before they become major problems and ensure that the product meets the required quality and performance standards. Overall, testing plays a critical role in ensuring that a product behaves optimally and meets the needs of its intended users.

Belief Bias

The cognitive bias trait known as belief bias leads people to rely excessively on preexisting beliefs and knowledge when evaluating an argument's conclusions, rather than properly considering the argument's content and structure. Belief bias states that people frequently accept arguments that support their preexisting ideas, even if those arguments are weak, illogical, or logically flawed. They often reject views contradicting their preexisting beliefs, even if those arguments are powerful and good.

Belief Revision

Belief revision refers to updating or modifying one's beliefs or knowledge in response to new evidence or information. It involves reevaluating previously held beliefs or assumptions and adjusting them based on the latest evidence.

Product technical testing involves subjecting a product to various tests to evaluate its performance, reliability, and safety. These test results can provide new information or evidence that may require a revision of one's beliefs about the product's performance or quality.

For example, suppose a product is believed to be highly durable and reliable based on previous testing, yet a new test reveals it is prone to failure under certain conditions. In that case, this new evidence may require a revision of one's beliefs about the product's performance and reliability.

In some cases, belief revision leads to changes in a product's design, manufacturing process, or marketing strategy, as when new evidence indicates the need for improvements or changes to ensure the product meets customer expectations.

Benchmark Test

A benchmark test is a standardized test to measure the performance or effectiveness of a system, device, or process. It typically compares the performance of different designs or assesses a system's progress or improvement over time. Benchmark tests are often used in the technology industry to compare the performance of different computer systems, software programs, or hardware components. They may also be used in other industries, such as manufacturing, to measure the efficiency or productivity of processes or equipment. Benchmark test areas include:

1. A benchmark that measures or compares one product performed against another.
2. A procedure used to compare systems or parts to one another or a standard, such as (1).

Berkson's Paradox

Berkson's paradox is a statistical phenomenon that can occur when studying the relationship between two variables. It was named after the American mathematician Joseph Berkson.

In the context of testing automotive products, Berkson's paradox may arise when evaluating the relationship between two factors, such as the quality of a product and the likelihood of it undergoing testing. Suppose there are two groups of automotive products: those that have undergone rigorous testing and those that have not. Suppose a tester or manager only examines the tested products. In that case, they may mistakenly conclude that there is a negative relationship between product quality and testing, that is, that tested products are of lower quality. However, this conclusion is misleading.

Berkson's paradox arises because the selection process for testing can introduce an artificial correlation. In this case, products of higher quality may be more likely to be selected for testing because they already exhibit good performance, while products of lower quality may not be selected because they are prone to failure and do not meet the testing criteria. Consequently, if one only focuses on the tested products, quality and testing may be negatively correlated, even though in reality, they are independent or possibly positively correlated.

To avoid falling into Berkson's paradox, it is crucial to consider the larger population of products, including both tested and untested ones. Analyzing the relationship between quality and testing based on the entire population can provide a more accurate understanding of the true association between these variables.

When conducting statistical analyses or making decisions based on automotive product testing, it is important to be aware of potential biases and paradoxes like Berkson's paradox. Proper study design, random sampling, and comprehensive data collection can help mitigate these issues and provide more reliable insights into the relationship between variables.

Beta Testing

Beta testing occurs after alpha testing and before the final release of a product. A group of users who are not part of the development team is selected to conduct the product Beta testing and provide feedback. Beta testing is essential because it allows developers to gather real-world data and insights about a product's use and what issues or bugs may need addressing. Beta testing typically lasts a few weeks to a few months and is usually followed by a final product release to the general public.

Bias

Bias can significantly impact testing and influence test designs, execution, and interpretation, leading to inaccurate or unreliable results. Bias refers to a systematic error in the testing process that skews results in a particular direction. This skewing can occur at any phase of the testing process.

Bias is introduced in many ways, such as through the selection of test subjects, the testing environment's design, the test case wording, or the interpretation of the results. In addition, various factors, including personal beliefs, cultural values, and social norms, can intentionally or unintentionally influence bias.

The impact of bias on testing can be significant, leading to incorrect conclusions or recommendations based on flawed data. Bias can be particularly problematic in fields such as engineering or product testing, where inaccurate results can have severe consequences for individuals or society.

To minimize the impact of bias on testing, carefully design tests that are free from discrimination and use objective and rigorous methods to collect and analyze data. It is also essential to recognize and address potential sources of bias and to be transparent about any limitations or uncertainties in the testing process.

Big Bang Testing

Big bang testing is a testing approach in which all system components are evaluated together in a single testing cycle. This approach involves testing an entire system rather than breaking it down into smaller parts (subassemblies) and testing them individually before testing the whole system. Big bang testing is selected to respond to a tight deadline or when the system is so integrated that components cannot be tested effectively in isolation. The main advantage of big bang testing is its ability to identify and resolve issues early in the development process, reducing the risk of later problems. However, it can also be risky and require more resources and time [7].

Binary Portability Testing

Binary portability testing checks whether a software application can run on different operating systems and hardware platforms without any issues. This type of testing is significant for software that needs to be compatible with a wide range of devices and systems.

Binary portability testing starts with the software application installed on various operating systems and hardware platforms. Then the functionality of each platform is tested. Testing may include installing and uninstalling the software, launching the application, and performing various tasks to ensure the software works as expected on each platform.

Binary portability testing aims to ensure that the software application is compatible with various devices and systems and can be easily deployed and operated by users on different platforms. Binary portability testing helps to ensure that the software is widely accessible and can be employed by a wide range of users (vehicle hardware platforms), regardless of their specific hardware and software configurations.

Black-Box Testing

A black-box test is software testing in which the tester does not know the system's internal workings. The tester only has access to the inputs and outputs of the system and cannot see or modify the code or internal logic of the system (see Figure B.1). A black-box test focuses on verifying that the system functions correctly and meets the specified requirements rather than finding and fixing coding errors. Black-box testing tests a system's functionality and usability rather than its internal structure or implementation [7, 36].

FIGURE B.1 In black-box testing, the tester does not know the product's inner workings but the inputs and anticipated outputs.

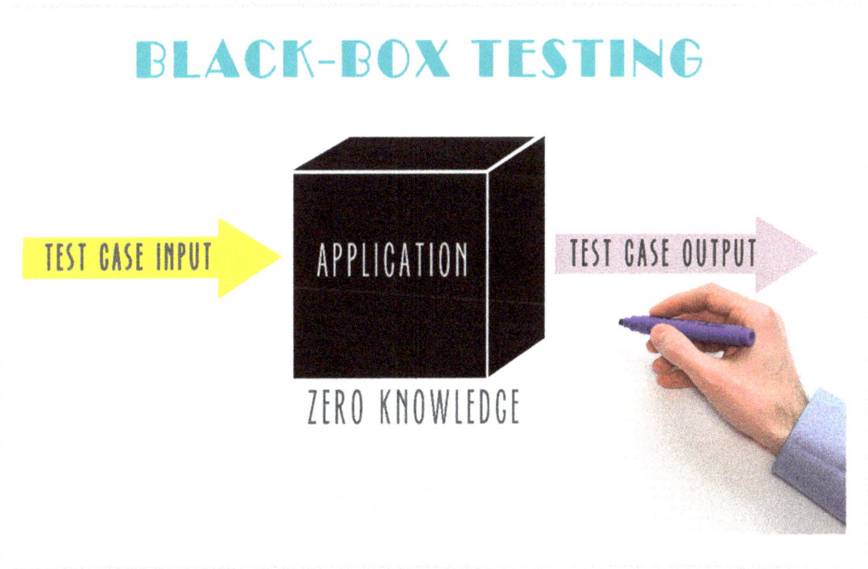

Blocked Test Case

A blocked test case is unexecutable during a testing phase. It could be because the test case is irrelevant to the current testing goals or because of a problem with the test case itself (e.g., it is not formatted correctly, is missing test data, or has incomplete information). On the other hand, it may be due to system availability or supporting test infrastructure. A blocked test case cannot verify a system's functionality or performance under test [7].

Bottom-Up Testing

Bottom-up testing is a software testing approach that first tests a system's smallest units or components and gradually builds up to test the higher-level integration of those components. This approach is the opposite of top-down testing, which starts with the overall system and then drills down to the individual components.

In bottom-up testing, the tester starts with the lowest level of the system and works up to the highest level. This approach is employed when the lower-level components are already stable and well-defined, and the focus is on testing the integration of these components. It is also helpful in identifying issues that may not be immediately visible when testing the overall system, as it allows the tester to isolate and identify problems at the component level.

Some advantages of bottom-up testing include the following:

1. It allows for early detection and resolution of issues at the component level, preventing larger problems from arising later in the testing process.
2. Testing individual components separately before integrating them can be easier, as it allows for a more focused testing approach.
3. It allows for testing components before full integration into the overall system.

Overall, bottom-up testing can help ensure the stability and functionality of a system by testing components individually before integrating them into the larger system [7].

Boundary Condition

Boundary conditions refer to the specific requirements at the edges or boundaries of a function, problem, or simulation. These conditions may involve constraints on the values of variables, typical behavior or behavior changes at the borders, or other requirements to be met when analyzing or solving the system.

Regarding boundary conditions and testing, it is essential to consider the boundary conditions when designing and conducting tests, as they can significantly impact the accuracy and reliability of the results. For example, if a boundary condition specifies that the system must operate within certain temperature limits, it would be necessary to ensure that it is tested under these conditions. This way, boundary conditions and testing work together to ensure that a system is functional, reliable, and meets the specified requirements [7].

Boundary Testing

Software engineers frequently employ the black box testing technique known as boundary testing to look for mistakes at a function's boundaries or extreme ends. All potential inputs made available by a software application are included in an input domain.

Boundary testing focuses on the limits or boundaries of a system or software application. This includes testing input limits, testing the system's performance under extreme conditions, or testing the system's behavior at the edges of its operating range (see Figure B.2). Boundary testing is important because it helps identify any issues or bugs that may occur when the system is operating at or near its limits, which can help improve the overall stability and reliability of the system [7].

FIGURE B.2 An illustration of the boundary conditions for detecting sensor performance.

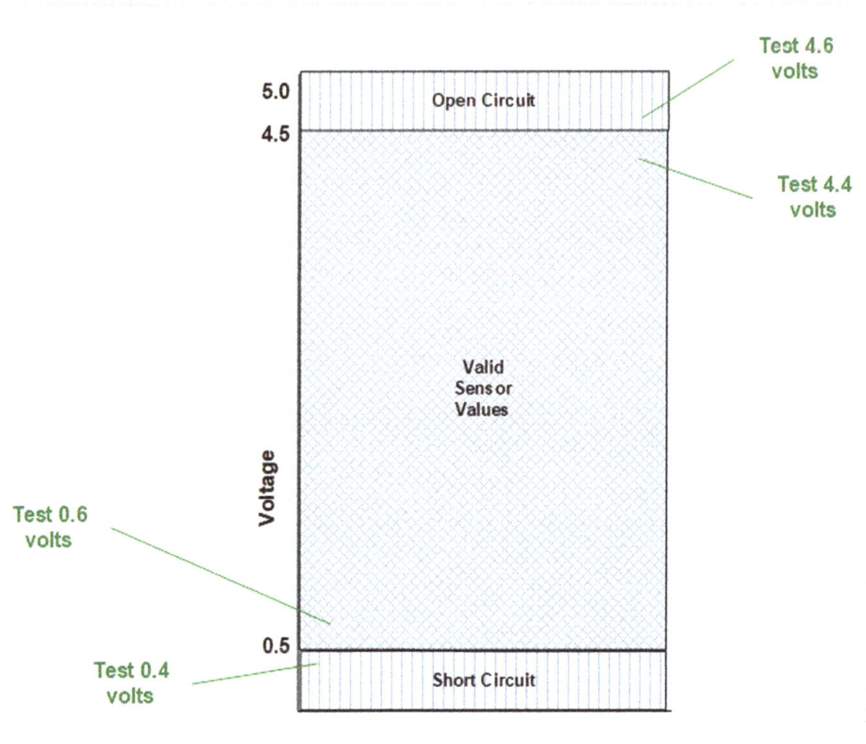

Boundary Testing—Three Point

A software boundary-value testing variation that involves assessing three data points on each side of the boundary. This testing procedure chooses boundary values depending on inputs at various ends of the testing range after equivalence class partitioning, which initially divides classes before dividing at the borders [7].

Boundary Testing—Two Point

A software boundary-value testing variation that involves assessing two data points on each side of the boundary. This testing procedure chooses boundary values depending on inputs at various ends of the testing range after equivalence class partitioning, which initially divides classes before dividing at the borders [7].

Boundary Value

For example, a range's minimum or maximum value is an edge value because it lies at the smallest incremental distance from the range's boundary where a state change happens in the product.

Boundary Value Analysis

Boundary value analysis is a technique used in software testing to evaluate the behavior of a system or software application at the limits or boundaries of an input domain. It involves testing the system with input values at the minimum or maximum of the defined range and values just inside and outside the range (see Figure B.3). This can help identify defects or errors in the system that may only occur when the input values are at the edges of the defined range.

FIGURE B.3 Part B impact and microphone setup [36].

Boundary Value Coverage

Boundary value coverage is a software testing evaluation technique that estimates the percentage of boundaries evaluated. Tests are designed to verify the behavior of a system at the border or edge. This requires testing the limits of the functional transitions and performance. This technique evaluates the level of completeness of testing the product's various functional boundaries.

$$Boundary\ Value\ Coverage = \frac{Boundaries_{tested}}{Boundaries_{total}}$$

Brake Rotor and Drum Modal Frequencies Verification

This document describes the standard test, analysis, and reporting methods for measuring the resonant modes of automotive disc brake rotors and drums for design/development and production verification of these components. Part A of this procedure may be used to determine the brake rotor or drum's resonant frequencies and mode shapes during the design and development phase. Part B of this procedure may be used to verify the production capability of a plant to produce brake rotors or drums that are consistent with the production part approval process (PPAP; see Figure B.4). In addition, the procedure may be used to fingerprint parts used for the design validation process [36].

FIGURE B.4 An example of boundary value analysis.

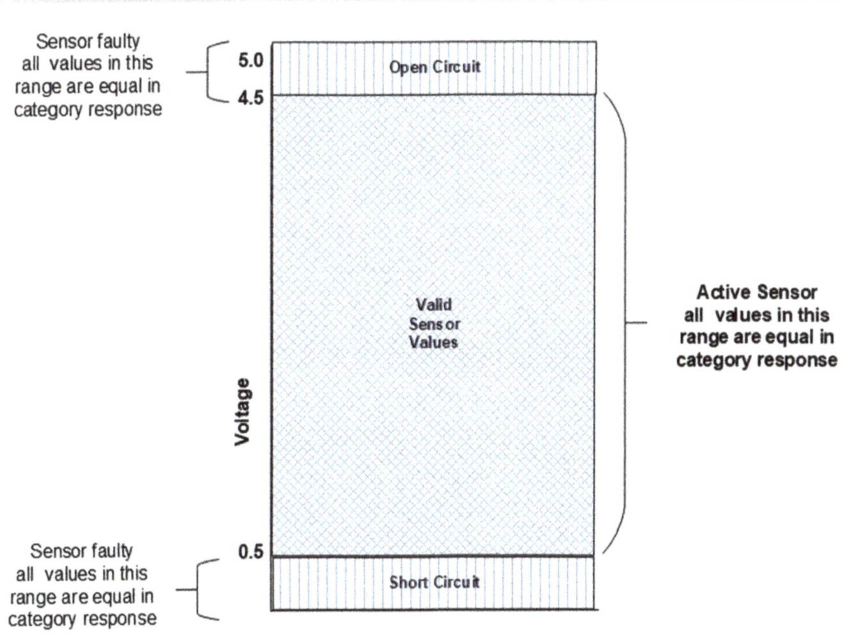

Branch Coverage

Branch coverage is a software testing technique that evaluates the percentage of program branches executed at least once during testing. Branch coverage includes all possible outcomes of a decision or conditional statement, such as "if," "else," or similar statements. Branch coverage measurement assesses the number of branches or code paths tested and measures testing completeness.

$$Branch\ Coverage = \frac{Branches_{tested}}{Branches_{total}}$$

Branch Testing

Branch testing focuses on the various branches or paths within a software program (see Figure B.5). It involves testing each possible branch or path to ensure that the program functions correctly in all cases. This type of testing is significant in programs with many branching points, as it helps ensure that the program will behave correctly regardless of the path taken.

FIGURE B.5 A code excerpt example of code branches.

```
#include <stdio.h>

int main() {
    int num;
    printf("Enter a number: ");
    scanf("%d", &num);

    switch (num) {
        case 1:
            printf("You entered 1\n");
            break;
        case 2:
            printf("You entered 2\n");
            break;
        case 3:
            printf("You entered 3\n");
            break;
        default:
            printf("You entered something else\n");
    }

    return 0;
}
```

Branch testing covers a wide range of product or software decisions in a short amount of time. The development team (white-box testing) typically executes this testing to identify decision issues or defects in a system or product.

Branch testing ensures all possible paths through the code are exercised, or at least the prioritized paths are covered, and that the software behaves correctly under different conditions and inputs. As a result, branch testing can help identify software defects or errors that may not be visible through other testing techniques.

Branch testing typically involves creating test cases that exercise different branches or decision points in the code. Test cases are designed and performed to ensure that each component in the software functions correctly and that all possible software outcomes of the decision points are exercised.

Effective branch testing requires an excellent understanding of the code and the decision points and the identification of all possible outcomes of each decision point. This can be challenging in complex software systems and may require specialized tools or techniques to help identify and test all possible branches.

Developers often perform branch testing. Branch testing is one approach; practical testing approaches should include other testing techniques, such as unit, integration, and system testing. However, it is noteworthy that achieving 100% branch coverage does not guarantee that a software system is entirely error-free or that it will behave correctly in all possible scenarios. Therefore, other types of coverage, such as path coverage and function coverage, should also be considered to ensure a comprehensive testing process [7].

Bug

A bug in testing is a flaw or error in a piece of software, hardware, or system that is discovered during testing and may require fixing before the software or system can be considered fully functional. Various factors, such as coding errors, design flaws, or compatibility issues with other systems or software, can cause these bugs. It is essential to identify and address these bugs to ensure that the software or system is reliable and performs as intended. In addition, bugs can have a wide range of undesirable effects, from improperly rendered user interface elements to complete program termination [18, 19].

Bulk Current Injection (BCI)

Bulk current injection (BCI) is a testing technique used to evaluate the susceptibility of electrical and electronic systems to conducted electromagnetic interference (EMI). It involves injecting a controlled current directly into the system being tested to simulate the effects of conducted EMI.

The purpose of BCI testing is to assess the ability of the system to withstand conducted interference and maintain its proper operation. In addition, engineers can evaluate a system's immunity and identify potential vulnerabilities by subjecting it to control currents that simulate real-world EMI conditions.

BCI testing is typically performed in a laboratory setting using specialized equipment. The test setup includes a power amplifier, a coupling device, and a current probe (see Figure B.6). The power amplifier generates a controlled current waveform, which is then injected into the system under test through the coupling device. The current probe monitors the injected current and ensures that the desired waveform is achieved.

FIGURE B.6 Differential bulk current injection (DBCI) test setup [37].

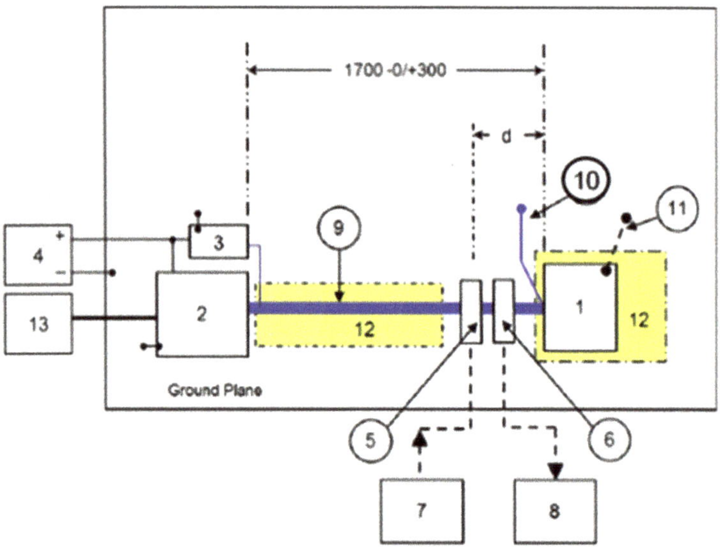

Key:

1. DUT (and DUT connector)
2. Load simulator
3. Artificial network
4. Power supply
5. Injection probe
6. Monitor probe (unless otherwise specified by the manufacturers, refer to ISO 11452-4)
7. RF generation equipment
8. Current monitoring equipment
9. DUT wire harness
10. DUT power return removed from wire harness and connected directly to sheet metal; wire length is 200 mm ± 50 mm
11. DUT metal case ground (if applicable, see manufacturer's specifications)
12. Dielectric support ($\varepsilon_r \leq 1.4$)
13. Support/monitoring equipment

d. Injection probe distance from DUT (see 4.3)

The injected current waveform used in BCI testing is designed to simulate the characteristics of conducted EMI that the system may encounter in its intended operating environment. The waveform includes specific frequency components and amplitudes based on the requirements and standards applicable to the system being tested.

The system's response to the injected current is observed and analyzed during BCI testing. This may involve monitoring its functional operation, performance metrics, or any deviations from expected behavior. The goal is to determine if the system remains operational and meets the required performance criteria despite the injected current.

BCI testing is commonly employed in various industries, including automotive, aerospace, telecommunications, and consumer electronics. It helps manufacturers ensure their products comply with electromagnetic compatibility (EMC) standards and regulations. It also allows engineers to make design improvements to enhance a system's immunity to conducted EMI.

SAE J1113-4 outlines the specific setup, equipment, and procedures required for BCI testing in the automotive industry. It specifies the waveform characteristics, frequency range, current levels, and duration of the injected currents for testing. The standard also guides measuring and evaluating the system's response to the injected currents. For more information, see **J1113-4** [37].

Burndown Chart

A burndown chart is a visual representation of the progress of a project over time. It is commonly used in Agile software development methodologies such as Scrum, but it can be applied to other projects as well. It can also be used to track testing efforts.

The chart displays the amount of work remaining in a project on the vertical axis, typically regarding story points, tasks, or hours. The horizontal axis represents time, typically in sprints or iterations. (See Figure B.7.)

FIGURE B.7 An example of an Agile burndown chart.

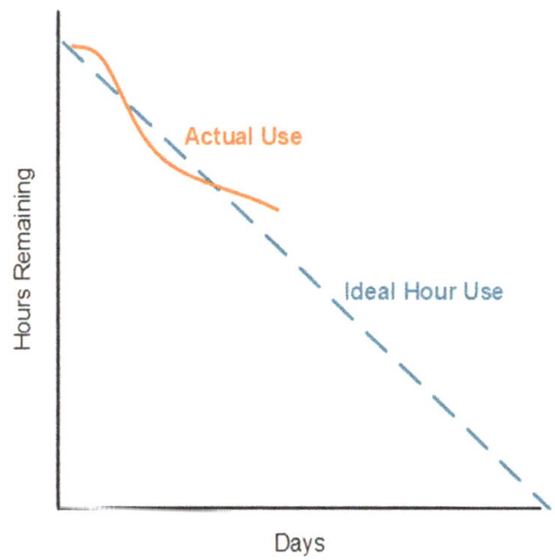

The chart starts with a line representing the total amount of work remaining at the beginning of the project. As the project progresses, the line should trend downward toward zero, indicating that the remaining work is being completed.

There are two types of burndown charts: sprint burndown and release burndown. Sprint burndown charts track the progress of a single sprint, while release burndown charts track the entire project's progress across multiple sprints.

Burndown charts can be used to track progress and identify potential issues or delays. For example, if the graph shows that progress is slower than expected, the project team can investigate the reasons for the delay and take corrective action. Burndown charts can also be used to forecast the project's completion date based on the current rate of progress.

Business Process-Based Testing

Business process-based testing focuses on evaluating the processes and systems within an organization to ensure that they are functioning as intended. This testing approach is used to verify that business processes are working effectively and efficiently to meet an organization's needs.

The main goal of business process-based testing is to identify and resolve any issues or defects that may impact business operations. This can involve testing the entire process from start to finish, including all steps, inputs, outputs, and interactions with other systems and processes.

To conduct business process-based testing, testers need to thoroughly understand the business processes and how they work within the organization. They also need to clearly understand the business goals and objectives and how the strategies contribute to meeting those goals.

Some standard techniques used in business process-based testing include the following:

- Process flow testing involves verifying that the process flows smoothly from one step to the next and that all steps are performed correctly.
- Data flow testing involves verifying that data is accurately captured and passed between different steps in the process.
- Integration testing involves verifying that the process integrates seemlessly with other systems and processes within the organization.
- Performance testing involves evaluating the performance and efficiency of the process to ensure that it meets the organization's needs.

Business process-based testing is essential to quality assurance and can help organizations identify and fix issues before they impact the business. It is often used with other testing approaches, such as functional testing, to provide a comprehensive view of the quality and effectiveness of an organization's processes and systems.

C

"If an organization is to work effectively, the communication should be through the most effective channel regardless of the organization chart."

—David Packard, founder, Hewlett-Packard

Calibration

Calibration is adjusting a device or system to ensure that it is operating at its optimal performance level. Calibration can involve changing the settings or parameters of the device to match a specific standard or reference point, or it may include testing and adjusting the device to ensure that it accurately measures or functions as intended. Calibration is often done on scientific instruments, measurement devices, and other types of equipment to ensure that they operate reliably and accurately. Calibration is vital to maintaining the quality and precision of devices and systems, and regulatory agencies and industry standards often require it.

Caliper

A caliper is a tool used to measure the distance between two opposite sides of an object. It is commonly used in engineering, machining, and metalworking to take precise measurements of the dimensions of a workpiece (see Figure C.1). Calipers come in various forms, including digital, dial, and vernier calipers. They can measure internal and external sizes, depths, and steps and are used with micrometers for greater accuracy. Calipers can measure various objects, from small parts to large structures, and are an essential tool in many industries.

FIGURE C.1 Calipers test dimensions of mechanical parts before and after testing.

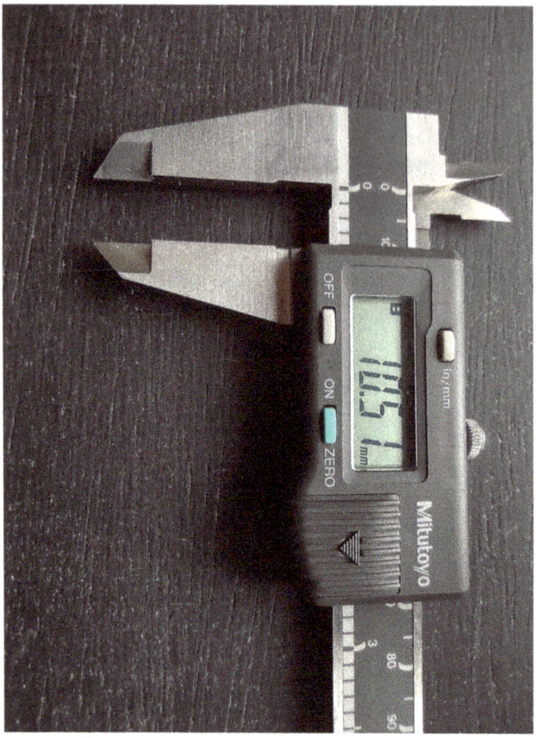

Capability Maturity Model for Software

The Capability Maturity Model (CMM) for software is a framework that helps organizations assess and improve how they develop and maintain software. It was developed by the Software Engineering Institute (SEI) at Carnegie Mellon University in the 1980s, and since then it has been widely adopted by the software industry.

The CMM consists of five levels of maturity, each representing a different level of organizational capability in software development and maintenance:

1. Initial: This level represents a chaotic and ad hoc approach to software development, with little or no process in place.
2. Repeatable: This level represents a more structured approach to software development, with processes in place for basic project management, requirements definition, and testing.
3. Defined: This level represents a fully illustrated and standardized approach to software development, with processes in place for all software life cycle stages.

4. Managed: This level focuses on continuous improvement and optimization of the software development process, with metrics in place to track and measure performance.
5. Optimizing: This level represents a focus on continuous innovation and improvement of the software development process, emphasizing maximizing value for the organization.

The CMM helps organizations identify areas for improvement in their software development process and guides how to make those improvements. Organizations worldwide use it as a benchmark for software development and maintenance practices [38].

Capture/Replay Tool

Capture/replay tools record the steps taken during a test scenario and then replay them to verify a system's functionality. These tools are handy for regression testing, where changes to a system may have unintended consequences on previously functioning features.

Test a capture/replay tool through the following steps:

1. Record a Test Scenario: This can be done by manually performing the steps required to test or by using a script to automate the process. The capture/replay tool should record all the actions taken during this process.
2. Replay the Recorded Scenario: Upon registering the procedure, a replay is possible to verify that the system functions correctly through manual or automation.
3. Compare the Results: A comparison of the replay's results to the expected results will reveal any discrepancies that should be investigated and resolved.
4. Repeat the Process: Repeat the process of recording and replaying the test scenario multiple times to ensure that the tool is working consistently and accurately.
5. Test the Tool's Features: Test the capture/replay tool also for its other features, such as the ability to edit recorded scenarios, skip specific steps, and so on.
6. Test the Tool's Performance: The tool's functioning should also be tested, including its speed and reliability.

Testing a capture/replay tool involves verifying its accuracy, reliability, and performance to ensure it is valuable for testing and debugging.

Cauchy Distribution

The Cauchy distribution is a continuous probability distribution with a symmetric bell-shaped curve, also known as a Lorentzian or Cauchy-Lorentz distribution (see Figure C.2). It is characterized by its heavy tails, meaning it has a higher probability of observing extreme values than other distributions, such as the normal distribution.

FIGURE C.2 An example of Cauchy distribution.

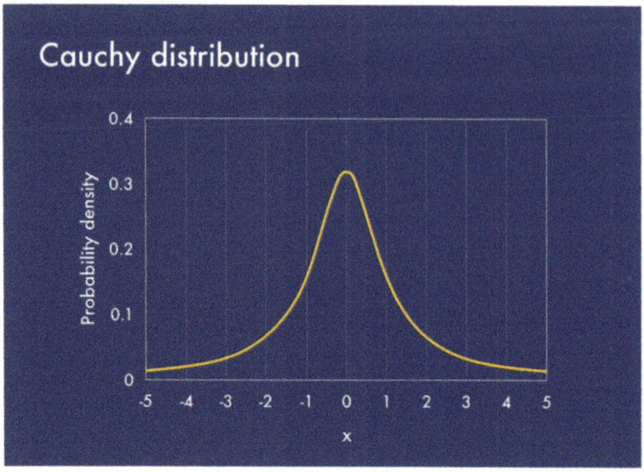

Dimitrios Karamitros/Shutterstock.com.

The probability density function of the Cauchy distribution is defined as follows:

$$f(x) = \frac{1}{\pi\left(1 + x^2\right)}$$

Where:

x is a random variable.

π is approximately equal to 3.14.

This equation describes the probability density function for a Cauchy distribution, a continuous probability distribution with a long, heavy tail. It is used to model the distribution of errors, performance, or residuals in statistical analysis.

The Cauchy distribution has no finite mean or variance, which makes it challenging to use in statistical analyses that rely on these measures. It is often used in modeling processes that exhibit extreme behavior or where the data do not conform to normal distribution [39].

Cause-Effect Analysis

The cause-effect analysis is used to identify the root causes of an observable symptom, problem, or issue and its effects on an organization or system found during testing. This analysis helps to understand the design issues and underlying causes of the problem rather than just address the symptoms. Identifying the root causes makes creating more effective and sustainable solutions possible.

Cause-effect analysis can be conducted using various methods, including the five whys, fishbone diagrams, and root cause analysis. These methods involve breaking down a problem into smaller pieces and identifying the causes and effects at each level. This exploration allows for a more thorough understanding of the problem and can help to identify multiple potential solutions.

Cause-Effect Diagram

A cause-effect diagram, also known as a fishbone diagram or Ishikawa diagram, represents the possible causes of a specific problem or issue. It is used to identify and analyze the root causes of problems to find solutions and prevent them from occurring in the future.

The diagram typically consists of a central line representing the problem or issue and several branches, representing the possible causes, extending from it (see Figure C.3). The branches can be divided into subcauses for a detailed problem analysis.

The use of a cause-effect diagram allows for a systematic approach to problem-solving. It helps to identify the underlying causes of issues rather than just address the symptoms. As a result, it is often used in quality improvement and process improvement efforts and in manufacturing, healthcare, and other industries [40].

FIGURE C.3 An example of a cause-effect (Ishikawa) diagram.

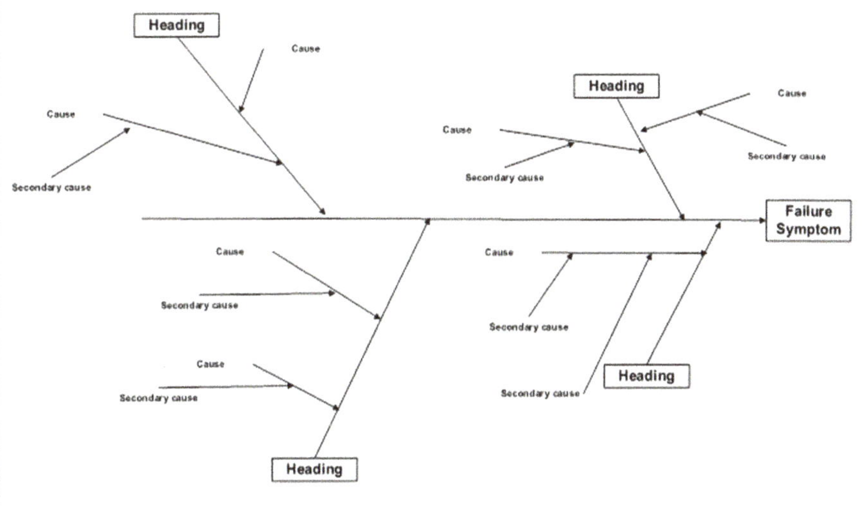

Central Tendency

Central tendency is the statistical measure representing a dataset's center or average. There are three standard measures of central tendency—mean, median, and mode (see Figure C.4):

1. Mean: The mean is the arithmetic average of a dataset, calculated by summing all the values and dividing by the total number of values.
2. Median: The median is the middle value in a dataset when the values are ordered from smallest to largest. If there is an even number of values, the median is the mean of the two middle values.
3. Mode: The mode is the value that occurs most frequently in a dataset. If more than one value occurs with the highest frequency, the dataset is said to have multiple modes.

FIGURE C.4 The three measures of central tendency, mode, mean, and median.

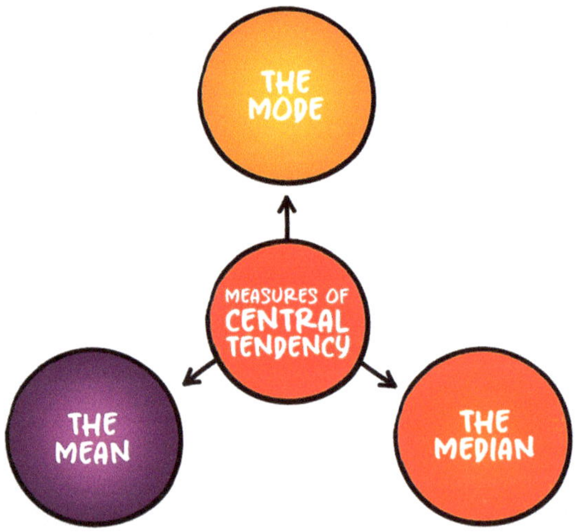

dizain/Shutterstock.com.

Central tendency is often used to summarize and describe a dataset's characteristics and can help identify trends or patterns within the data. No matter if they are continuous or discontinuous, all processes produce data. Therefore, there will be a pattern or distribution in all data. Normal, binomial, and Poisson distributions are a few examples. Three traits—central tendency, variance, and shape—describe or categorize the distributions.

Change Management (Product)

Product change management is a necessary process to ensure that changes made to a product are carefully planned, implemented, and tested before being released to customers. Testing plays a critical role in product change management, as it helps identify potential issues or unintended consequences that may arise from product changes. Additionally, change management articulates the product's contents, making effective and efficient testing possible. Without effective change management, the test group cannot develop test cases before the product delivery. Finally, change management applies to the product and tests artifacts and equipment; in this way, there is a close connection between change management and configuration management through release notes, hardware, and software.

In product change management, testing involves verifying that the changes made to a product have been appropriately implemented and do not introduce any new defects or issues. This comparison typically involves manual and automated testing and various testing techniques and tools.

By integrating testing into the product change management process, organizations can minimize the risk of introducing new defects or issues into the product and ensure that changes are correctly implemented and tested before being released to customers. This integration also helps testers understand the product's contents and make rational decisions about what is available for testing and what should be tested [41, 42].

Changeability

Changeability refers to the ease with which a product can be modified to adapt to meet changing requirements or conditions. Testing plays a vital role in ensuring changeability, as it can help identify areas of development that may be difficult or costly to modify.

By thoroughly testing a product, it is possible to identify areas of the product that are particularly complex or tightly coupled and may be difficult to modify without introducing unintended consequences. Testing can also help identify areas of development that may require extensive regression testing after modifications to ensure that the changes do not have unintended impacts on other parts of the product.

In addition, testing can help ensure that product changes do not introduce new defects or issues. Testing the original and modified products makes it possible to identify any new weaknesses introduced during the modification process.

Overall, testing is critical in ensuring that products are changeable and can be adapted to meet evolving requirements. In addition, by identifying potential issues and risks, testing can help to ensure that modifications are implemented smoothly and without unintended consequences and that the product remains stable and reliable throughout its life cycle.

Checklist-Based Review

Checklist-based reviews, also known as structured reviews, are a type of peer review process in which reviewers use a specific checklist to evaluate a document or artifact against a set of predetermined criteria. The checklist may include spelling and grammar, adherence to style guidelines, completeness, accuracy, and consistency.

Software development often uses checklist-based reviews to review code, design documents, and other artifacts. They can be performed by individuals or groups and conducted in person or remotely. The benefits of checklist-based reviews include increased efficiency, consistency, and objectivity in the review process. In addition, by using a predefined checklist, reviewers are less likely to overlook important issues or to be influenced by personal biases.

However, checklist-based reviews also have some limitations. For example, they may not be as effective at identifying more complex issues that require a deeper understanding of the product or system being reviewed. Additionally, missing important issues can occur if the checklist is not comprehensive or up-to-date.

Overall, checklist-based reviews can be a valuable tool for ensuring quality and consistency in the review process, coupled with other methods and careful attention to the specific needs and requirements of the product or system reviewed [12].

Chemical Exposure

Product chemical exposure testing evaluates the impact of potential chemical exposure, where such exposure can have serious product longevity or performance consequences.

Product chemical exposure testing aims to identify and quantify the presence of potentially hazardous chemicals in the product environment and to evaluate the likelihood and severity of any exposure. This identification and evaluation typically involve laboratory testing, modeling, and risk assessment.

The testing process may include the following steps:

1. Identify the Chemicals of Concern: This involves pinpointing the hazardous chemicals present in the vicinity of the product based on regulatory requirements and industry standards.
2. Test the Product: This involves conducting laboratory tests to determine the concentration and distribution of the hazardous chemicals in the product.
3. Model Exposure Scenarios: This involves developing models to estimate the potential exposure of users to hazardous chemicals based on various factors such as product use, frequency of exposure, and exposure pathways.
4. Evaluate the Risk: This involves conducting a risk assessment to evaluate the potential health impacts of exposure to hazardous chemicals based on the concentration and frequency of exposure.

5. Determine Regulatory Compliance: This involves comparing the test results and risk assessment to applicable regulatory requirements to determine whether the product complies with relevant regulations.

Product testing of chemical exposure results can inform product design, manufacturing processes, and marketing strategies, as well as help to ensure that products are safe for consumers. In addition, manufacturers can demonstrate their commitment to product safety and compliance with regulatory requirements by conducting thorough and rigorous product chemical exposure testing.

The table below is an example of the chemicals to which a vehicle may be subjected:

Engine Oils and Additives	Transmission Oil
Rear Axle Oil	Power Steering Fluid
Brake Fluid	Axle Grease
Window Washer Solvent	Gasoline
Diesel Fuel	Fuel Additives
Alcohol	Antifreeze-Water Mixture
Degreasers	Soap and Detergents
Steam	Battery Acid
Waxes	Kerosene
Freon	Spray Paint
Paint Strippers	Ether
Dust Control Agents (magnesium chloride)	Moisture Control Agents (calcium chloride)
Vinyl Plasticizers	Undercoating Material
Muriatic Acid	Ammonia
Diesel Exhaust Fluid (DEF)	

For more information, see **J1455** and **J1976** [20, 43].

Classification Tree

A classification tree is a machine learning algorithm to identify and sort items or classes based on certain features or characteristics. It works by building a treelike structure, where each internal node represents a decision based on the value of a particular feature, and each leaf node represents a class or category. The tree is built by starting at the root node and working through the branches, making decisions based on the values of the features at each node until a leaf node is reached.

Classification trees, also known as decision trees, can be used in software testing in several ways:

1. Test Case Design: Use classification trees to generate test cases by identifying combinations of input parameters likely to affect the system's behavior under test. By analyzing the decision rules in the classification tree, testers can identify test cases that cover all possible scenarios.
2. Test Prioritization: Upon identification of a set of test cases, classification trees prioritize test cases based on the likelihood of a particular scenario occurring. Testers can assign weights to each decision node in the classification tree based on the probability of reaching that node during testing. Prioritization helps testers focus on the most critical test cases first.
3. Fault Localization: If a test case fails, classification trees help localize the fault by identifying the traversed decision nodes during the test. Testers can then focus their debugging efforts on the code corresponding to those nodes.

Classification trees can be valuable for test case design, prioritization, and fault localization. However, they are most effective when combined with other testing techniques, such as exploratory testing and boundary value analysis.

Classification Tree Method

The process of applying the classification tree method (CTM) to automotive product testing involves several steps. Here's a step-by-step overview of the process:

1. Identify Test Objectives: Clearly define the objectives and goals of the testing process for the automotive product. Determine what aspects of the product need to be tested and what specific behaviors or functionalities should be covered.
2. Identify Input Parameters: Identify the key input parameters or variables that influence the behavior of the automotive product. These parameters include various features, settings, or conditions that are relevant to the product's functionality. For example, in testing a car's braking system, input parameters may include brake pedal pressure, vehicle speed, road conditions, and brake fluid temperature.
3. Define Parameter Values and Classification: Determine the possible values or ranges for each input parameter. Classify the values based on their characteristics or properties. For example, the parameter "brake pedal pressure" may be classified as "light," "moderate," or "heavy."

4. Construct the Classification Tree: Build a hierarchical tree structure that represents all possible combinations of the input parameters and their values. The tree should cover all meaningful combinations that need to be tested. The top-level branches represent the input parameters, and subsequent levels represent their possible values. The tree should be organized in a logical and structured manner.
5. Define Test Cases: Generate test cases based on the classification tree. Each test case represents a unique combination of input parameter values. Test cases should be designed to cover different combinations and scenarios, including positive and negative cases, edge cases, and boundary conditions. The objective is to ensure thorough coverage of the product's behavior.
6. Execute Test Cases: Execute the defined test cases on the automotive product. During the execution phase, inputs are provided based on the values specified in the test cases, and the corresponding behavior or output of the product is observed. Test cases should be executed in a systematic manner and the results should be recorded.
7. Analyze Test Results: Analyze the test results to identify any patterns, trends, or issues that emerge during testing. Compare the observed behavior or output with the expected behavior based on the test objectives. Identify any discrepancies, failures, or deviations from the expected results.
8. Iterate and Refine: Based on the analysis of the test results, refine the classification tree or test cases as needed. This iteration allows for continuous improvement of the testing process and enhances test coverage and effectiveness. Re-execute the updated test cases if necessary.
9. Documentation and Reporting: Document the testing process, including the classification tree, test cases, test results, and any issues encountered. Prepare a comprehensive report summarizing the testing activities, findings, and recommendations for improvement.

Client Server Applications

See Cloud-Based Testing.

Closed-Loop Testing

Closed-loop testing incorporates the use of feedback to improve the testing process. In closed-loop testing, the test results are fed back into the testing process and used to modify the test conditions or parameters, aiming to improve the test's accuracy and effectiveness.

The closed-loop testing process typically involves the following steps:

1. Define the Test Conditions and Parameters: Describe the test conditions and parameters, including the inputs and expected outputs of the test.
2. Execute the Test: Perform the test using the defined test preconditions, input data, and parameters.
3. Analyze the Results: Evaluate the test results to determine if performance matches expectations.
4. Modify the Test Conditions and Parameters: Based on the test results, adjust the test pre-conditions, test data, and parameters to improve the accuracy or effectiveness of the test.
5. Repeat the Test: Repeat the test using the modified test conditions and parameters.

The closed-loop testing process makes possible exploration of various testing scenarios, such as software testing, hardware testing, and system testing. In addition, it can help improve the quality and effectiveness of the testing process by identifying and correcting errors and issues in real time.

Closed-loop testing is a valuable tool for improving the testing process's accuracy and effectiveness and can help ensure that the products being tested meet required quality standards [44, 45].

Cloud-Based Testing

Modern vehicles communicate off vehicle via telemetry systems to the cloud, for example, with vehicle performance and error reports. Analysis of the component performance in the field is used to improve the product and better understand customer use of the product. In the case of commercial vehicles these systems are part of predictive maintenance.

Cloud testing of automotive components via a telematics system involves evaluating the performance, functionality, and reliability of automotive components that rely on cloud-based connectivity and telematics technology. Telematics systems in vehicles utilize wireless communication to transmit data to and from the cloud, enabling features such as remote diagnostics, software updates, infotainment services, and connected vehicle applications.

Here are key aspects of cloud testing for automotive components via a telematics system:

1. Connectivity and Communication Testing: Assessing the connectivity between the automotive components and the cloud infrastructure. This involves testing the reliability and stability of the wireless communication channels, verifying data transmission accuracy, and

evaluating the effectiveness of protocols and application programming interfaces (APIs) used for communication.

2. Functional Testing of Cloud Services: Evaluating the functionality and performance of cloud-based services accessed through the telematics system. This includes testing features such as remote diagnostics, over-the-air software updates, real-time data monitoring, location-based services, and vehicle-to-cloud interactions.
3. Security and Privacy Testing: Verifying the security measures implemented within the telematics system and cloud infrastructure. This involves conducting penetration testing, vulnerability assessments, and encryption checks to ensure data privacy, protection against cyber threats, and compliance with industry security standards.
4. Scalability and Load Testing: Assessing the scalability and load handling capabilities of the cloud infrastructure. This involves simulating high user traffic, data volume, and transaction loads to determine if the system can handle peak demands without performance degradation or service disruptions.
5. Resilience and Fault Tolerance Testing: Evaluating the system's resilience to failures and its ability to recover from disruptions. This includes testing failover mechanisms, backup and recovery processes, and disaster recovery plans to ensure continuous availability and fault tolerance of the cloud-based services.
6. Data Integrity and Synchronization Testing: Verifying the accuracy and consistency of data between the automotive components and the cloud infrastructure. This includes testing data synchronization, data integrity checks, and ensuring proper data handling and storage within the cloud environment.
7. Predictive Maintenance Testing: Throughout vehicle development estimates of component and vehicle life are made, and then tested with associated updates to the vehicle design. Even after their launch, the vehicles sold and used provide the original equipment manufacturer (OEM) with feedback on the quality of the vehicle, enabling quick responses to problems and continuous updates for product improvement.

Clustering Illusion

The clustering illusion is a cognitive bias by which people tend to see patterns or clusters in random or nonrandom data. This bias can lead people to draw false conclusions or overestimate the significance of a perceived pattern. The clustering illusion occurs when people try to find meaning in complex data, such as in financial markets, sports statistics, or medical research.

Mitigation of clustering illusion happens by taking a more systematic and objective approach to data acquisition and analysis, such as using statistical tests to determine if a pattern is significant or simply due to chance. Additionally, it is crucial to recognize that randomness and chance play an essential role in many complex systems and that patterns may not always have a meaningful explanation.

CMMI (Capability Maturity Model Integration)

CMMI is a framework for improving an organization's processes and practices. It provides guidelines and best practices for organizations to increase efficiency and effectiveness. This approach to operations is used in various industries and organizations, including software development, manufacturing, and government agencies. It aims to improve the quality and consistency of an organization's products and services and customer satisfaction. CMMI provides a benchmark for measuring an organization's performance and identifying areas for improvement.

CMMI 1.3 Product Development Knowledge Areas	Description
Requirements Development (RD)	Eliciting, analyzing, specifying, validating, and managing the requirements for the product or system
Technical Solution (TS)	Designing, developing, and implementing the technical solution that meets the requirements
Product Integration (PI)	Assembling the product or system from the components or subsystems and ensuring that it functions correctly
Verification (VER)	Evaluating the product or system to ensure that it meets the specified requirements
Validation (VAL)	Evaluating the product or system to ensure that it meets the customer's needs
Organizational Process Focus (OPF)	Establishing and maintaining a process improvement program that is aligned with the organization's business objectives
Organizational Process Definition (OPD)	Defining and maintaining the organization's standard processes and tailoring them for specific projects
Organizational Training (OT)	Providing training to the organization's personnel to ensure they have the skills and knowledge needed to perform their roles
Integrated Project Management (IPM)	Planning, monitoring, and controlling the project to ensure that it meets the customer's requirements
Risk Management (RSKM)	Identifying, assessing, and managing risks that could affect the project or the product

CMMI 1.3 Product Development Knowledge Areas	Description
Decision Analysis and Resolution (DAR)	Analyzing and resolving decision issues using a systematic approach
Measurement and Analysis (MA)	Collecting and analyzing data to identify trends and using the results to improve the organization's processes
Process and Product Quality Assurance (PPQA)	Evaluating the organization's processes and products to ensure that they comply with established policies and standards
Configuration Management (CM)	Managing the configuration of the organization's products and processes to ensure that they are consistent and traceable
Requirements Management (REQM)	Establishing and maintaining the requirements for the project or product and ensuring that they are appropriately managed throughout the development process

Capability Levels

Level	Description	Definition
0	Incomplete	Process is not performed or partially performed.
1	Performed	Process is a performed process that accomplished the needed work.
2	Managed	A performed process is planned, executed, monitored and controlled.
3	Defined	The process is tailored for each project according to the organizations tailoring guideline.

Maturity Level

Level	Description	Definition
1	Initial	Processes are ad hoc, unstable work process environment.
2	Managed	Processes are planned and executed to company policies.
3	Defined	Processes are clearly defined and understood described in standards, procedures, and tools.
4	Quantitative Managed	The project establishes quantitative objectives for quality and process performance.
5	Optimizing	The organization continuously improves the processes based on a quantitative understanding of the process.

Two CMMI process areas are associated with product testing, see Validation and Verification for those definitions [38].

Code Complete

Code complete refers to the point in software development when all necessary code has been written, and the software is ready for testing and debugging. It does not necessarily mean the software is entirely error-free or prepared for release, but all planned features and functionality have been implemented. Code completion marks the transition from the development phase to the testing phase. It is typically followed by debugging and fixing any remaining issues before the software is ready for release.

Code Coverage

Code coverage is used to identify areas of the software that have not been tested compared to the entire code. Typically calculated as a percentage, this metric can help developers identify the amount of the code tested and the remaining risk in the product.

$$Code\ Coverage = \frac{Tested_{statement}}{Total_{statements}}$$

In addition, this metric helps prioritize testing efforts to target the most critical or largest coverage area. Finally, code coverage tools help track changes in code coverage over time, which can help developers identify areas where testing has become less thorough and make adjustments to improve testing practices.

Code Inspection

Code inspection examines and reviews source code to identify and fix defects, improve code quality, and ensure that the code meets specific standards. This process can involve manual review by a team of developers or automated tools to analyze the code for issues. Code inspection is part of a software development process to catch defects early on and prevent them from becoming major problems later on. Code inspection can identify best context-specific practices and areas for improvement in the codebase [12].

Code Review

Code review evaluates code written by a developer or team to ensure it meets specific standards and requirements. These examinations include checking the code for correctness, efficiency, security, readability, and adherence to coding standards. A peer or a team leader often does code reviews, which can involve manual and automated tools to identify issues and suggest improvements. Code review aims to identify and fix problems early in the development process, improving the quality and reliability of the final product [12].

The steps for formal code review are as follows:

1. Set up a Code Review System: This initiation can be a simple process for requesting and conducting code reviews within your team.
2. Identify the Code to be Reviewed: This identification can be a new feature, bug fix, or any other code change that needs to be reviewed by your team.
3. Request a Code Review: This review can involve creating a pull request in a version control system, emailing the team, or simply asking a colleague to review your code.
4. Review the Code: This review involves carefully examining the code to ensure it is well-written, follows coding standards and best practices, and is free of errors and bugs.
5. Provide Feedback: If you find any issues with the code, provide constructive feedback to the author on how to improve it.
6. Make Any Necessary Changes: Based on the feedback received, the author should make any changes required to the code before it is accepted and merged into the main codebase.
7. Finalize the Review: Once the code is revised and all feedback addressed, the code review is finalized and merged into the main codebase [12, 47].

Code Walkthrough

A code walkthrough is a process in which a programmer or team of programmers reviews and analyzes a piece of code. This review identifies errors or issues with the code, understands how it works and fits into the overall system, and discusses potential improvements or optimizations [12, 47].

In a code walkthrough, one developer acts as the guide while the rest of the team members ask questions and look for deviations from the established development standards. The review meeting is often led by the author of the document under consideration and attended by other team members.

During a code walkthrough, the team will typically go through the code line by line, discussing and explaining each part of the code and its purpose. The team may also run the code and test it to see how it behaves in different scenarios, and they may make changes to the code as needed to fix any issues or improve its performance.

Overall, a code walkthrough is an essential part of the software development process, as it helps ensure that the code is of high quality and is working as intended. It also allows for collaboration and knowledge sharing among team members and helps identify areas for training or support.

The steps for conducting a code walkthrough are as follows:

1. Open the code file in your preferred code editor.
2. Begin by reading through the code, starting at the top and working your way down. This reading will give you a general understanding of the overall structure and purpose of the code.
3. As you read through the code, note any functions or variables that are defined or used.
4. Pay attention to any comments in the code, as these can provide valuable context and explanations for what the code is doing.
5. Look for any loops or conditional statements and try to understand their logic.
6. Take note of any external libraries or dependencies the code may use.
7. As you read through the code, try to understand the flow of execution. That is, try to understand the sequence exercises of different parts of the code.
8. If you encounter any unfamiliar code or concepts, consider researching to understand them better.
9. Continue reading through the code until you have a good understanding of how it functions as a whole.

Combinatorial Explosion

The term "combinatorial explosion" refers to the rapid increase in possible combinations or permutations as the number of elements or variables increases. It is common in computer science and mathematics, where algorithms and data structures handle many variables or factors.

For example, if we have a set of 5 elements, there are 5! (5 factorial) possible combinations or permutations. However, if we have 10 elements, there are 10! (10 factorial) possible combinations or permutations. As the number of elements increases, possible combinations or permutations increase exponentially [46].

A combinatorial explosion can make identifying defects or errors in a software system difficult, as testing every possible input combination is not feasible. However, some techniques can reduce the required tests but still ensure that the testing of the most critical combinations.

One such technique is pairwise testing, which involves testing all possible combinations of pairs of input variables. This approach can significantly reduce the required tests while providing good input space coverage. Other techniques include combinatorial testing tools and boundary value analysis to focus on the most critical input ranges [39].

Combinatorial Testing

A method of testing in which the input variables are systematically varied to cover all possible combinations. This technique identifies defects and vulnerabilities in software systems by subjecting the system to many simultaneous stimuli. Combinatorial testing is often used with other testing methods, such as boundary value testing and equivalence partitioning, to provide a comprehensive testing process.

Combined Environmental Testing

Combined environmental testing involves exposing a product or system to a range of environmental conditions, such as temperature, humidity, and vibration, to see how it performs and determine if it meets specified requirements. This type of testing is prevalent in the aerospace, automotive, and defense industries, where products must withstand extreme conditions. It is also used in other industries, such as consumer electronics, where products must operate reliably in various environments. Combined environmental testing ensures that a product or system can function properly and meet the intended performance standards under different environmental conditions.

Commercial Off the Shelf (COTS)

Commercial off-the-shelf (COTS) refers to products or services readily available for purchase from third-party vendors rather than being custom-built or developed in-house. COTS components are building blocks used to create larger systems or products; COTS products include hardware, software, and other components. In addition, COTS systems are standardized to be easily integrated into existing systems to ensure compatibility and interoperability across different platforms.

COTS products offer several advantages for organizations, including reduced development time and cost, increased reliability and quality, and greater flexibility and scalability. By using off-the-shelf components, organizations can focus their resources on developing the unique features and functionality that differentiate their products or systems from others in the market.

However, there are also some potential drawbacks to using COTS products. One challenge is ensuring compatibility and interoperability with other components and systems. In addition, COTS products may not always meet the specific needs or requirements of an organization and may require customization or modification to meet those needs entirely. Finally, the testing of these systems may not be congruent with the customer-specific application, requiring the customer to perform some testing to supplement.

In software development, testing is essential to ensure the integration of COTS components and that the overall system meets the required quality and reliability standards. This integration may involve testing individual components and conducting integration testing to ensure all parts work together as intended [47, 48].

Communication Testing

Automotive product testing of serial communication refers to the process of evaluating the performance, reliability, and functionality of communication protocols used in automotive electronic systems. Vehicles have numerous electronic control units (ECUs) that communicate to achieve various functions for the driver. Serial communication plays a crucial role in facilitating data exchange between various components and control units within a vehicle.

The following are some key aspects of automotive product testing for serial communication:

1. Protocol Compliance Testing: Ensuring that the communication protocol implemented in an automotive system adheres to the relevant industry standards and specifications. This involves verifying that the protocol operates correctly; handles data transmission, error detection and correction; and supports the required features and functionalities.
2. Interoperability Testing: Verifying that different components and control units within a vehicle can communicate seamlessly with each other using the serial communication protocol. This involves testing compatibility, message formatting, and proper interpretation of commands and responses between different devices.
3. Performance and Timing Analysis: Evaluating the timing and performance characteristics of the serial communication system. This includes assessing factors such as data transfer rates, latency, response times, and message transmission reliability. Performance testing helps identify any bottlenecks or issues that may affect overall system performance.
4. Error Handling and Resilience Testing: Assessing how the serial communication protocol handles errors and interruptions. This involves testing error detection, error recovery mechanisms, fault tolerance, and response to abnormal or unexpected conditions. The goal is to ensure that the communication system can handle and recover from errors effectively.

5. Noise and Interference Testing: Evaluating the resistance of the serial communication system to noise and interference commonly encountered in automotive environments. This includes testing for electromagnetic interference (EMI), radio frequency interference (RFI), voltage fluctuations, and other sources of signal degradation. The goal is to ensure reliable communication in real-world conditions.
6. Compliance with Automotive Network Architectures: Verifying that the serial communication protocol is compatible with the specific automotive network architecture, such as Controller Area Network (CAN), Local Interconnect Network (LIN), FlexRay, Ethernet, or other proprietary protocols. Testing ensures that the protocol can coexist and operate efficiently within the specified network.

Automotive product testing of serial communication is crucial to ensure reliable and efficient data exchange between components and control units in a vehicle. By conducting comprehensive testing, manufacturers can identify and resolve any issues related to protocol compliance, interoperability, performance, error handling, and noise immunity, thereby enhancing the overall functionality and reliability of the automotive system.

Comparison Testing

This type of testing involves comparing two or more products, services, or systems to determine which performs better or meets specific criteria more effectively. Comparison testing is used in various industries and contexts, including software development, marketing research, and product design.

The following are some standard methods of comparison testing:

1. A/B Testing: Testing two different product or service versions to see which one performs better. For example, a company might test two different website versions to see which one gets more traffic or leads.
2. Feature-by-Feature Testing: Comparing products or services based on specific features or functions. For example, a company might compare two software programs to see which one has more user-friendly features or better performance.
3. Multivariate Testing: Evaluating multiple variables at once to see how they affect the performance of a product or service. For example, a company might test different pricing combinations, marketing strategies, and product features to see which combination leads to the most sales.

Overall, comparison testing is a meaningful way to evaluate the effectiveness and efficiency of different products, services, or systems and identify improvement opportunities.

Compatibility

Compatibility refers to the ability of two or more systems, products, or components to work together without issues or conflicts. For example, for software development, compatibility testing ensures that an application or system can function correctly across different hardware platforms, operating systems, web browsers, and other components with which it may interact.

Compatibility Testing

Compatibility testing of software testing determines if a system, device, or application is compatible with other systems, apparatus, or applications, includes verifying that they can work together seamlessly and without issues. Compatibility testing is essential to ensure that a product or service can interface with other products or services when used in the real world. It is used to test the compatibility of the software or hardware with different hardware, operating systems, electronic control units, or other devices.

Compatibility testing typically involves testing a software application or system against various configurations, including hardware configurations, operating systems, web browsers, and other software components or subsystems. This testing may include applications or systems on multiple versions of the same operating system or different hardware platforms, such as desktop computers, laptops, tablets, and mobile devices.

Compatibility testing aims to identify any issues or conflicts arising when the software application or system is used in different environments or with various components. This compatibility testing includes incorrect formatting or information display, error messages, crashes, or other failures.

Compatibility ensures the software application or system works as intended for all users and environments. It can also help avoid potential issues or conflicts arising when users attempt to employ the software application or system in different configurations.

Compile

To compile means to take source code written in a programming language and transform it into a form that a computer can execute. This process involves translating the source code into machine code that the microprocessor or microcontroller can use to run the code. Then, the compiler is used to convert this source code to machine language. Compiling is an essential step in the software development process, as it allows developers to test and debug their code before it is deployed in a live environment.

Compiler

A compiler is a computer program that translates source code written in a programming language into a machine-readable form that a computer can execute. It converts the source code into a machine-readable format called object code, a sequence of instructions the computer can run. The compiler checks the source code for some logic and syntax errors and generates an error message. It also optimizes the code for better performance by removing unnecessary instructions and rearranging the code in a more efficient order. Compilers create executable files for applications, operating systems, and other software programs. Changing the compiler can change the generated software in such a way as to alter the machine code and, perhaps, the product performance.

Complexity

The complexity of a product refers to the number of components, features, and interactions that make up the product. As products become increasingly complex, they can be more challenging to design, develop, test, and maintain.

Several factors contribute to product complexity. One factor is the number of components or parts of the product. This includes hardware components, software components, mechanical components, and any other parts or systems required for the product to function.

Another factor is the number of features or functions that the product offers, including basic features like on/off switches or buttons and more advanced features like touchscreens, voice recognition, or artificial intelligence. Product complexity increases with the number of components, features, and interactions. For example, if one component fails or malfunctions, it can affect the performance of other components and lead to unexpected behaviors or errors.

Managing complexity in product design and development is vital to ensure that products are reliable, function properly, and meet users' needs. This may involve simplifying the design, reducing the number of features, or using modular design principles to make testing and maintaining individual components easier.

Compliance Testing

Compliance testing verifies that a product or system meets all relevant standards, regulations, and requirements. Compliance includes ensuring that the product or system meets legal or industry-specific needs. Compliance testing may be performed by a third-party testing laboratory or by the manufacturer or developer of the product or system. It is often a legal requirement or industry standard. Failure to pass compliance testing can result in fines, legal action, or the inability to sell or use the product or system [49].

Component Integration Testing

Component integration testing focuses on verifying the integration and communication between different components or modules in a system. It involves testing interfaces and interactions to ensure that they function correctly and meet the specified requirements.

Component integration testing typically happens after unit testing, which tests individual components in isolation, and before system testing. However, it is an essential step in the software development process. It helps identify any issues or defects at the minor level that may arise when different components are combined and interact with each other. Finding the source of a product anomaly while unencumbered by the system entirety is more straightforward than finding the defect when connected to other portions of the system.

Examples of component integration testing include evaluating the integration between a database and a user interface or between a web server and a database. Component integration testing is done via automated testing or manual testing techniques.

Component Specification

The component specification (also referred to as a spec) defines the requirements and characteristics of a specific component within a more extensive system or product. Specifications include determining the physical dimensions, electrical specifications, performance characteristics, and other relevant details necessary for the component to function correctly within the system [50, 51].

The component specification is an integral part of product design and development; it is the minor level of defining a system and ensuring that individual components meet the requirements of the overall system. It also helps to ensure that parts are compatible and easily integrated into the larger product. Specifications are written clearly and often in technical language. In addition, specifications will fall under control of change and configuration management because changes to specifications often lead to changes in the product.

The process of component specification typically begins with identifying the product's specific attributes based on the overall design and functionality requirements. Once the required components are identified, the specifications for each component can be defined in detail, often through collaboration between engineers, designers, businesspeople, and other stakeholders. The specification covers a range of information, such as physical dimensions, the weight of components, attachment points, materials used in its construction, and electrical characteristics, such as voltage, current, and frequency. They may also include information on the component's performance under different operating conditions, such as temperature, humidity, and vibration.

Component Testing

A component in software refers to a standalone piece of code or functionality that can be used within a more extensive software system or application. Components can be small, self-contained pieces of code that perform a specific task or more prominent, or they can be more complex modules with multiple functions and features. Components can be used to break down larger software projects into smaller, more manageable pieces, making it easier to develop and maintain the software. Components can also be reused in different software projects, saving time and effort in development. Examples of components in software include libraries, modules, and frameworks.

Compound Condition Testing

Compound testing tests the logical conjunction of two or more logical operators. Compound condition testing involves testing multiple conditions in a single test case. This type of testing ensures that a system functions correctly when various conditions are present.

For example, a heavy vehicle dump truck unloading on a site must meet specific requirements to be able to engage the dump mechanism:

- Vehicle speed below a certain level
- Vehicle on a level surface
- Engine running, transmission in neutral or park

If these conditions are not met, the system does not allow the dump to be engaged. Testing the amalgam of the states in a single test case makes assessing the specific function performance possible. Test other combinations to ensure the function becomes active only during the particular inputs.

Computational Fluid Dynamics (CFD)

Computational fluid dynamics (CFD) is a branch of fluid mechanics that uses numerical methods and algorithms to solve and analyze fluid flow problems. CFD uses mathematical models to simulate the behavior of fluids, such as gases and liquids, and to study the effects of various physical phenomena, such as turbulence, heat transfer, and chemical reactions, on fluid flow (see Figure C.5).

FIGURE C.5 An example of a CFD model.

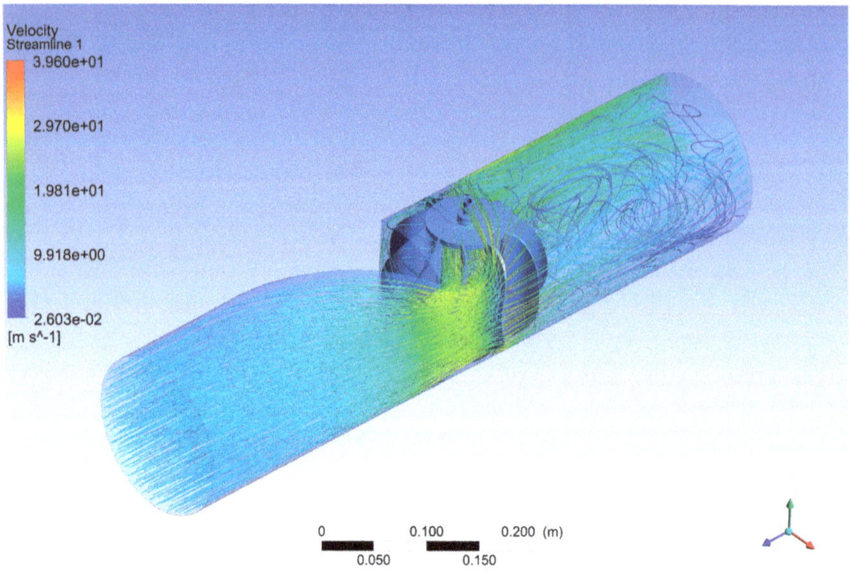

Shishir Gautam/Shutterstock.com.

CFD aims to provide a numerical solution to the equations that describe fluid flow, such as the Navier-Stokes equations, to predict fluid flow patterns, pressure and velocity fields, and other relevant properties.

Engineering and scientific applications, including aerodynamics, combustion, environmental science, and heat transfer, use CFD to explore products. It is a tool for optimizing and improving the performance of various systems and products, such as aircraft, power plants (engines), and automobiles. Engineers and scientists can analyze and understand fluid flow behavior before building and testing physical prototypes [52, 53].

Computer-Aided Software Testing

Computer-aided software testing (CAST) is the use of specialized software tools to assist in the testing process of a software application. These tools automate various aspects of testing, such as test case execution, test data generation, and test results analysis.

There are several benefits to using CAST tools:

1. Efficiency: CAST tools can automate repetitive and time-consuming tasks, allowing testers to focus on more complex, high-level jobs.
2. Accuracy: CAST tools can help ensure consistent and accurate tests, reducing human error risk.

3. Speed: CAST tools can execute tests much faster than human testers, allowing for more rapid testing and feedback.
4. Coverage: CAST tools can help testers ensure the execution of all relevant test cases, improving the scope of the testing process.

Many CAST tools are available, including test case management tools, test execution tools, and test data generation tools. Choosing the right CAST tools for your specific testing needs and goals is essential [54].

Computer Simulation

Computer simulation is a method of using computer software to replicate and analyze the behavior of a system or process. It allows researchers, engineers, and scientists to study complex systems and functions without physical experimentation.

Simulations are used in various fields, including physics, engineering, economics, and biology. They are often used to test new designs' performance and reliability, predict the outcome of different scenarios, and evaluate the efficiency of versions of products and processes.

Computer simulations can be highly detailed and accurate but have limitations. They rely on the accuracy of the data and assumptions used in the simulation, and they cannot replicate every possible real-world scenario. However, they are a valuable tool for understanding and predicting the behavior of complex systems [54, 55, 56].

Concurrency Testing

Concurrency testing is used to verify the performance and behavior of a software application when used by multiple users or processes simultaneously. This type of testing is essential for applications in high-concurrency environments, such as web servers and e-commerce platforms; for example, telematics systems in vehicles report vehicle performance data (predictive maintenance).

Concurrency testers create a test scenario that simulates multiple users or processes accessing the application simultaneously. For example, the test scenario may include logging in, adding items to a shopping cart, and purchasing. The tester then runs the test scenario multiple times, gradually increasing the number of users or processes involved in the test until the application starts to experience performance issues or errors.

Concurrency testing can help identify problems such as slow response times, resource contention, and data corruption. It can also help identify bottlenecks in the system and provide insights into the application's scalability.

Condition

Generally, a condition refers to the state (see State Diagram) or situation of something, often regarding a set of criteria or requirements. In addition, these conditions describe physical states, such as the condition of a piece of machinery or equipment, which represents the state of associated hardware and software.

Condition Coverage

Condition coverage is a test metric that evaluates the tested conditions juxtaposed with the number of total conditions in which the product or system can be found.

$$Condition\ Coverage = \left(\frac{Conditions_{tested}}{Conditions_{total}} \right)$$

The metric assesses the completeness of the testing activities.

Condition Determination Testing

Automotive condition determination testing refers to the process of assessing and evaluating the condition or state of various components and systems in vehicles. It involves conducting tests and inspections to determine if the components or systems meet the required standards, specifications, and performance criteria.

Condition determination testing in the automotive industry can cover a wide range of areas:

1. Mechanical Systems: Testing the condition of mechanical systems such as engines, transmissions, suspension systems, steering systems, and braking systems. This may involve performance testing, measurement of key parameters, and assessing the wear and tear of components.
2. Electrical and Electronics Systems: Assessing the condition of electrical and electronics systems, including wiring harnesses, sensors, actuators, control units, and communication networks. This may involve diagnostic testing, checking for proper functionality, and ensuring compatibility with various protocols and standards.
3. Safety Systems: Evaluating the condition of safety systems such as airbags, seat belts, antilock braking systems (ABS), electronic stability control (ESC), and collision avoidance systems. This may involve testing the responsiveness, accuracy, and reliability of these systems to ensure that they function as intended.

4. Emissions and Environmental Compliance: Testing the condition of emission control systems, exhaust systems, and other components related to environmental compliance. This includes evaluating emissions levels, checking for leaks or malfunctions, and verifying compliance with applicable regulations and standards.
5. Body and Interior: Assessing the condition of the vehicle's body structure, exterior components, interior features, and amenities. This may involve inspections for corrosion, damage, fit and finish, functionality of controls, and overall comfort and convenience.
6. Software and Firmware: Testing the condition and performance of software and firmware systems, including infotainment systems, navigation systems, vehicle control algorithms, and other embedded systems. This may involve functional testing, compatibility testing, and verification of software/firmware updates.

Condition determination testing aims to ensure that vehicles meet quality, safety, and performance standards. It helps identify any defects, malfunctions, or issues that may impact the vehicle's reliability, functionality, or regulatory compliance. By conducting thorough testing and inspection, automotive manufacturers and service providers can identify and address potential issues before vehicles reach customers or while in use, ensuring a high level of overall vehicle condition and performance.

Condition Testing

A method of designing (white-box) test cases to achieve predetermined outcomes, condition testing evaluates the performance or functionality of a system, product, or component under specific conditions. Condition testing develops test scenarios that explore the various states of a product or system, seeking to uncover defects, problems, or weaknesses and ensure that it meets desired performance standards. Condition testing may involve identifying and simulating different scenarios or environments to test the system's behavior and response to various conditions. It may also include monitoring the system's performance and collecting data to analyze and make recommendations for improvement. Condition testing is an essential step in developing and maintaining a system to ensure that it is reliable and meets the needs of its users.

Conditions often refer to specific circumstances or testing scenarios to ensure that a product or system meets its requirements and does not have errant extraneous performance. For example, a software testing team may need to test a product under different conditions, such as other operating systems, network configurations, or input datasets, to ensure that it works correctly under various scenarios. Part of the test planning will consider multiple scenarios and possible conditions of the product, component, or system—prioritization of the most common or likely riskiest states.

Conditioning

A logical statement that can be determined to be true or false, such as A > B. Also see Test Condition.

Confidence Interval

A confidence interval is a range of values within which a population parameter is estimated to fall at a certain confidence level. It is often used to estimate a population mean or proportion from a sample. The confidence level is typically set at 95%, meaning that if the same sample was taken multiple times, the population parameter would fall within the confidence interval 95% of the time.

$$CI = \overline{x} \pm z*\left(\frac{\sigma}{\sqrt{n}}\right)$$

Where:

CI is the confidence interval.
$\overline{x}$ is the sample mean
z^* is the z-score corresponding to the desired level of confidence (e.g., 1.96 for a 95% confidence interval).
σ is the population standard deviation.
n is the sample size.

If the population standard deviation is unknown and the sample size is large enough (typically $n \geq 30$), the formula becomes

$$CI = \overline{x} \pm z*\left(\frac{s}{\sqrt{n}}\right)$$

Where: s is the sample standard deviation.

If the sample size is small (typically $n < 30$) and the populations standard deviation is unknown, the formula becomes

$$CI = \overline{x} \pm z^{*}\left(\frac{s}{\sqrt{n}}\right)$$

If the sample size is small (typically $n < 30$) and the population standard deviation is unknown, the formula becomes

$$CI = \overline{x} \pm t^{*}\left(\frac{s}{\sqrt{n}}\right)$$

Where: t is the t-score corresponding to the desired level of confidence and degrees of freedom (n-1).

The size of the confidence interval depends on the sample size and the confidence level. A larger sample size and a higher confidence level will result in a smaller confidence interval, meaning the estimated population parameter is more precise.

Confidence intervals are often used in statistical analysis to give a range of possible values for a population parameter rather than a single-point estimate. This evaluation allows for a more realistic and nuanced understanding of the data rather than reliance on a single-point assessment which may not accurately reflect the true population parameter.

Configuration Accounting

Configuration accounting is a method of tracking and managing changes to the configuration of a system or product, commonly used in the automotive, aerospace, defense, and software development industries, where complex systems are constantly updated and modified. This updating and accounting occurs throughout the development process.

Configuration accounting involves creating a detailed record of all changes to a system or product's configuration, including the date of the change, the person responsible for the change, and the reason for the difference. This record is known as a configuration history.

Configuration accounting is essential because it allows testing groups to differentiate versions of and understand the changes made to a system or product, allowing effective testing. It also helps to ensure that the system or product complies with regulatory standards built according to the intended design. All of which is verified or refuted via testing.

Configuration accounting is typically managed by a dedicated team or individual responsible for maintaining the configuration history and ensuring that all changes are appropriately documented and tracked. This team/individual may verify that changes are correctly implemented and tested before being released to the end user [42, 62, 57, 58].

Configuration Audit

A configuration audit systematically reviews a system or product's configuration to ensure it complies with established standards and policies. Configuration auditing is used in the aerospace, defense, and software development industries, where complex systems are constantly updated and modified.

A configuration audit involves comparing a system or product's current configuration to a previously approved and documented baseline configuration. The audit may also include testing to verify that the system or product functions as intended and meets required specifications.

Configuration audits are essential because they help ensure that a system or product is built and maintained according to the intended design and is compliant with regulatory standards. They also help to identify and correct any issues or discrepancies that may have arisen during the development or maintenance process.

Experts with extensive knowledge of the system or product audited typically conduct configuration audits. The audit process may involve reviewing documentation, running tests, and interviewing team members to gather information about the system's configuration or product [42, 62, 57, 58].

Configuration Control

Configuration control is the process of managing changes to the configuration of a system or product. Automotive, aerospace, defense, and software development industries employ configuration control, constantly updating and modifying complex systems.

Configuration control involves establishing procedures and guidelines for making and reviewing system or product configuration changes. This control may include creating a configuration history to document all changes, reviewing and approving changes before implementation, and testing changes to ensure they are implemented correctly and do not introduce new issues.

Configuration control is essential because it helps ensure that a system or product is built and maintained according to the intended design and remains compliant with regulatory standards. As a result, configuration control improves the efficiency and effectiveness of the testing. In addition, this control and a record of each product iteration are available for the testers. As a result, it isn't easy to test what is not known. However, it also helps to prevent unintended consequences or issues that may arise from changing a system or product's configuration.

Configuration control is managed by a dedicated team or individual responsible for reviewing and approving system or product configuration changes. This team/individual may verify that changes are correctly implemented and tested before being released to the end user [42, 62, 57, 58].

Configuration Identification

Configuration identification documents a system or product's components, interfaces, and characteristics. It applies to many industries, including automotive, aerospace, defense, and software development, where complex systems are constantly being updated and modified.

Configuration identification involves creating a detailed record of all components and characteristics of a system or product, including the specifications, design, and configuration. This record is known as a configuration baseline.

Configuration identification is crucial because it helps ensure that a system or product is built and maintained according to the intended design and remains compliant with regulatory standards. It also helps to identify and track changes to the system or product's configuration over time.

Configuration identification is typically managed by a dedicated team or individual responsible for maintaining the configuration baseline and ensuring that all components and characteristics are appropriately documented and tracked. This team/individual may verify that the system or product is built according to the intended design and meets the required specifications.

An example of configuration identification in the development of a new software application might involve creating a detailed record of all software components, including the user interface, database structure, and algorithms used in the application. The configuration baseline would also include information about the software's design, such as the overall architecture and any specific requirements or constraints.

Once the configuration baseline is established, the software development team can use it as a reference for building and testing the application. Any changes to the software's configuration, such as adding new features or modifying existing ones, would be documented in the configuration history and reviewed and approved before being implemented.

By using configuration identification, the software development team can ensure that the software is built and maintained according to the intended design and remains compliant with regulatory standards. It also helps to prevent unintended consequences or issues arising from making changes to the software's configuration [42, 62, 57, 58].

Configuration Item

A configuration item is a specific component or characteristic of a system or product identified and tracked during the configuration management process. It is used in aerospace, defense, and software development industries, where complex systems are constantly updated and modified.

Configuration items include hardware components, software components, documents, or any other system or item. Each configuration item is identified, documented, and assigned a unique identifier. Changes to the configuration item are tracked and recorded in a configuration history.

Configuration items are essential because they help ensure that a system or product is built and maintained according to the intended design and remains compliant with regulatory standards. They also help to identify and track changes to the system or product's configuration over time.

Configuration items are typically managed by a dedicated team or individual responsible for maintaining the configuration baseline and ensuring that all configuration items are correctly identified and tracked. This team/individual may also verify that the system or product is built according to the intended design and meets the required specifications.

An example of a configuration item is a specific component of an aircraft, such as an engine or wing assembly. If the wing assembly of an aircraft was modified to include a new type of fuel tank, this configuration item would be updated in the configuration history to reflect the change. This helps ensure that the aircraft is built and maintained according to the intended design and remains compliant with regulatory standards.

Another example of a configuration item is a specific software component of a software application, such as a user interface module or database structure, that is identified and documented with a unique identifier, so that changes to the application are tracked and documented [42, 62, 57, 58].

Configuration Management

Configuration management is the process of identifying, controlling, and tracking changes to the configuration of a system or product. It is used in the aerospace, automotive, defense, and software development industries, where complex systems are constantly updated and modified.

Configuration management involves establishing procedures and guidelines for making and reviewing system or product configuration changes. It may also include creating a configuration baseline to document the specific components and characteristics of the system or product and a configuration history to track changes to the configuration over time.

Configuration management is essential because it helps ensure that a system or product is built and maintained according to the intended design and remains compliant with regulatory standards. It also helps to prevent unintended consequences or issues that may arise from changing a system or product's configuration.

Configuration management is typically overseen by a dedicated team or individual responsible for reviewing and approving changes to the system or product's configuration, maintaining the configuration baseline and history, and verifying that the system or product is built and maintained according to the intended design and meets the required specifications.

There are several steps involved in the configuration management process:

1. Establishing a Configuration Management Plan: This involves defining the policies, procedures, and guidelines used to manage the system's configuration or product.
2. Identifying Configuration Items: This involves identifying and documenting all components and characteristics of the system or product that are part of the configuration.
3. Establishing a Configuration Baseline: This involves creating a detailed record of the configuration of the system or product at a specific point in time.
4. Tracking and Controlling Changes: This involves reviewing and approving changes to the configuration before implementation and documenting the changes in a configuration history.
5. Verifying and Validating the Configuration: This involves testing the system or product to ensure that it is built and maintained according to the intended design and meets required specifications.
6. Maintaining the Configuration: This involves regularly reviewing and updating the configuration baseline and history to ensure they remain accurate and up-to-date.

Configuration Management Plan

A configuration management plan is a document that outlines the policies, procedures, and guidelines that will be used to manage the configuration of a system or product. It is commonly used in the aerospace, automotive, defense, and software development industries, where complex systems are constantly being updated and modified.

A configuration management plan typically includes information about the configuration management process, such as how changes to the system or product's configuration will be reviewed and approved, how the configuration will be tracked and documented, and how the configuration will be tested and verified. It may also include information about how the configuration baseline and history will be maintained.

Configuration management plans are important because they help ensure that a system or product is built and maintained according to the intended design and remains compliant with regulatory standards. They also help prevent unintended consequences or issues that may arise from changing a system or product's configuration.

Configuration management plans are typically developed by a dedicated team or individual responsible for configuration management. The plan is a reference for all team members building and maintaining the system or product.

A configuration management plan outline might include the following elements:

1. Introduction: This is a brief overview of the purpose and scope of the configuration management plan.
2. Configuration Management Process: This is a description of the steps involved in the configuration management process, including how changes to the system or product's configuration will be reviewed and approved, how the configuration will be tracked and documented, and how the configuration will be tested and verified.
3. Configuration Management Roles and Responsibilities: This is a description of the roles and responsibilities of the team members involved in the configuration management process, including the configuration manager, configuration review board, and configuration control board.
4. Configuration Management Tools and Techniques: This is a description of the tools and techniques that will be used to support the configuration management process, such as configuration management software, version control systems, and change management processes.
5. Configuration Management Deliverables: This is a list of the documents and artifacts that will be produced as part of the configuration management process, including the configuration baseline, configuration history, and test reports.
6. Configuration Management Processes: This is a description of the processes that will be used to maintain the configuration baseline and history, including how changes will be documented and how the configuration will be reviewed and updated.
7. Configuration Management Training and Support: This is a description of the training and support that will be provided to team members involved in the configuration management process, including guidance on how to use the configuration management tools and techniques.

The testing plans are coupled to the configuration management preparing for iterations of the product going to test, making it possible to develop test fixtures, test plans, and test cases in advance of the version of the product available for testing.

Confirmation Bias

Confirmation bias can be especially problematic in product testing, where objective and accurate results are crucial for ensuring safety, reliability, and performance. Product testing typically involves evaluating products based on standardized criteria and specifications to ensure that the product meets minimum safety and quality requirements.

Confirmation bias can influence technical product testing in several ways. For example, a tester familiar with a particular brand or product type may be likelier to overlook or downplay technical issues or to attribute problems to external factors rather than the product itself. Similarly, a tester who has a vested interest in the success of a particular product or company may be more likely to interpret results positively, even if the product does not meet objective standards.

Minimize the effects of confirmation bias in product testing through test protocol standardization and development of a common lexicon for the test team. Additionally, where possible, use clear pass/fail criteria (metric-based). In addition, testers should be encouraged to remain neutral and objective throughout the testing process and to report positive and negative results transparently and consistently.

Product testing should also be conducted by independent, perhaps third-party organizations with no stake in the product's success or failure. This level of independence can help ensure that the testing process is free from bias and that results are objective and reliable (see independent test).

Confirmation Testing

Confirmation testing, or retesting, is performed to verify the resolution of a previously identified defect or issue. In addition, confirmation testing evaluates developer changes to a product or system to address the problems found in previous testing. Confirmation testing aims to ensure that an earlier problem has been resolved and that the product or system functions as expected. This testing may involve running the same test cases that initially identified the issue or running additional tests to ensure the changes have not affected related functionality.

Confirmation testing is essential to the software testing process, as it helps ensure the resolution of defects and issues before the product release. By performing thorough confirmation testing, testing teams can help ensure that products are reliable and functional and they meet users' needs. Confirmation testing should not be confused with regression testing.

Conformance Testing

Conformance testing evaluates a system or product to determine whether it meets documented standards or specifications. In addition, conformance testing evaluates project delivery and is part of contract closure. Conformance testing is prevalent in the automotive, aerospace, defense, and software development industries.

Conformance testing involves creating a set of test cases to evaluate documented aspects of the system or product's performance. These test cases may include functional tests to verify that the system or product is operating as intended and nonfunctional tests to evaluate reliability, usability, and security factors.

Conformance testing is vital because it helps ensure that a system or product is built and maintained according to the intended design and meets required specifications and regulations. It also helps to identify any issues or discrepancies that may have arisen during the development or maintenance process.

The testing process may involve reviewing documentation, conducting tests, and analyzing the results to determine whether the system or product complies with established standards and specifications.

Confounding

Confounding in technical product testing refers to extraneous variables affecting the ability to test or the test results. Confounding makes it difficult to determine the genuine cause-and-effect relationship between the product and the measured outcome. Confounding variables can arise from various sources, such as environmental factors, individual differences in test subjects, and testing equipment or protocols.

Test engineers can use various techniques to mitigate confounding effects in technical product testing, including randomization, blinding, and control groups. Randomization involves assigning test subjects to different groups to minimize the impact of individual differences. Blinding withholds information about the product being tested from the test subjects or evaluators to reduce bias. Finally, control groups of test subjects do not receive the product tested so as to provide a baseline for comparison.

Technical product testing is valuable for ensuring product quality and identifying potential issues. Still, it is essential to carefully design and conduct tests to minimize the effects of confounding variables and obtain accurate results.

Confusion Matrix

A confusion matrix is a table used to evaluate the performance of a machine learning model on a test dataset. The matrix shows the number of correct and incorrect predictions made by the model compared to the actual outcomes in the test dataset. The matrix is presented in a tabular format, with rows representing the true class labels and columns representing the predicted class labels.

Figure C.6 is an example confusion matrix for binary classification (i.e., two classes).

The four entries in the matrix represent the following:

1. True Positive (TP): This field contains the number of samples that belong to the positive class, correctly predicted by the model as positive.
2. False Positive (FP): This field contains the number of samples that belong to the negative class, incorrectly predicted by the model as positive.
3. True Negative (TN): This field contains the number of samples that belong to the negative class, correctly predicted by the model as negative.
4. False Negative (FN): This field contains the number of samples that belong to the positive class, incorrectly predicted by the model as negative.

FIGURE C.6 An example of a confusion matrix.

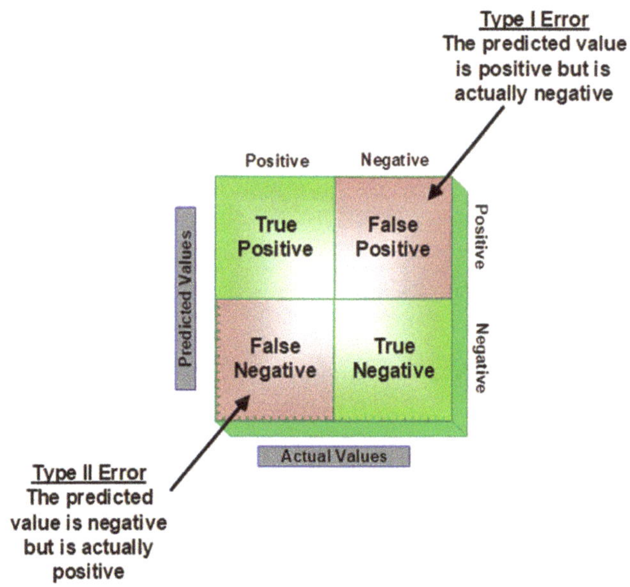

The model's performance was evaluated using various metrics calculated from the confusion matrix, such as accuracy, precision, recall, and F1 score (see F1 Score).

Testing is the process of evaluating the performance of a machine learning model on a previously unseen dataset. The confusion matrix assesses the performance of a model on a test dataset. By comparing the predicted outcomes of the model with the actual results in the test dataset, we can calculate the various metrics to assess the model's performance.

A confusion matrix is a valuable tool for evaluating the performance of a machine learning model on a test dataset, and it provides insights into the model's strengths and weaknesses.

Conjunctive Fallacy

The conjunctive fallacy is a cognitive bias in which people assume that the conjunction of two events is more probable than one of the events alone. This fallacy often arises when people are presented with a confluence of events that seem to fit together well but may not be likely to occur together. Testing comes with an evaluation of combined events and stimuli. This testing process typically involves a range of tests designed to assess a product's performance under various conditions.

Cognitive biases can quickly impact testers and testing. For example, a person conducting a product test may be prone to the conjunctive fallacy when evaluating the product's performance under certain conditions, assuming that because the product performed well under one condition, it is more likely to perform well under another state.

To avoid the conjunctive fallacy in product testing, it is essential to approach each test condition as a separate event rather than to assume that the product's performance under one condition will necessarily predict its performance under another. Conducted tests should be evaluated independently, and the results should be analyzed as a whole to determine the overall quality and reliability of the product. This evaluation includes test cases that may have a combination of stimuli as part of the test.

Context-Driven Testing

Context-driven testing prioritizes adapting testing practices to the project, product, team, and environment context. It recognizes that each software project and product application is unique and that no testing methodology fits all situations.

This approach is based on the Context-Driven School of Testing, a group of software testing professionals who advocate for a more flexible and adaptive approach to testing. According to this school of thought, testing is not a set of prescribed activities but rather a way of learning about the software and providing feedback on its quality.

In context-driven testing, the testers rely on their knowledge, experience, and expertise to design and execute testing activities appropriate for the project's specific context. The testing process is continuously adapted and refined based on feedback from testing results, user needs, and project constraints.

Context-driven testing emphasizes collaboration and communication between testers, developers, and stakeholders. This approach recognizes that testing is a team effort involving all stakeholders. Context-driven testing is a more flexible and adaptable approach to software testing that can help improve the software's quality and reduce the risk of defects and failures.

Continuing Conformance Testing

Continuing conformance is a periodic sampling and testing approach to product testing that involves ongoing monitoring and testing of products from a manufacturing line to ensure that they continue to meet the desired quality standards and conform to the relevant specifications, standards, and regulations.

This approach recognizes that products and manufacturing are not static but dynamic, constantly evolving and changing. As a result, it is essential to continuously monitor and test products to ensure that they remain functional, reliable, and secure.

Ongoing conformance testing involves product sampling, followed by automated testing tools, scripts, and procedures designed to test the software's functionality, performance, security, and other quality attributes. These tests are executed regularly, semi-annually, or annually to evaluate material and manufacturing impacts on the final product quality.

Continuing conformance is performed less frequently than in the past, as the automated end-of-line testing of manufactured parts supplants.

Contractual Testing

Contractual testing ensures that a product or system meets the requirements and specifications of a contract or agreement between two parties. These are contractual obligations that a project will use as part of the contract closure activities.

The goal of contractual testing is to ensure that the product or system meets the requirements agreed upon in the contract. Contractual testing may involve running particular test cases or scenarios outlined or defined in the contract. These test cases directly trace to the requirements and confirm or refute meeting requirements. In addition, the testing may include specific features or functionality to meet the client's needs.

Contractual testing is a business and development process for many products and systems, particularly in contracting for the automotive, defense, and aerospace industries, where strict adherence to identified requirements and specifications is crucial. By performing thorough contractual testing, vendors can ensure that their products or systems meet their clients' needs and that they are delivered on time and within budget. It can also help reduce disputes or legal issues if the product or system does not meet the agreed-upon specifications.

Contradiction Testing

Contradiction testing is used to evaluate the logical consistency of a statement or argument. It involves comparing a given requirement with other known conditions or facts to determine whether they are consistent with or contradict each other.

In natural language processing and artificial intelligence, contradiction testing determines whether two given statements are logically consistent. This is typically done by comparing the meaning and structure of the two statements and looking for any contradictions or inconsistencies.

One common approach to contradiction testing is using a formal logic system, such as propositional or predicate logic, to represent the statements and evaluate their logical consistency. Another method uses machine learning algorithms, such as neural networks, to detect contradictions and inconsistencies in natural language statements.

Control Flow Testing

Control flow testing is a white-box method in which the tester knows the structure of the code. Control flow testing is software testing that focuses on the order in which certain events or actions occur within a program or system. It involves verifying that the program follows the correct sequence of steps and branches according to the control flow diagram or flowchart.

Control flow testing ensures that the program executes the correct logic, makes the right decisions based on input, and takes the proper actions at the correct times. In addition, it helps to identify errors in the program's control flow, such as loops that don't terminate, conditional statements that are always true or false, or incorrect branching.

There are several methods of control flow testing:

1. Statement Coverage: This method ensures that each statement in the program is executed at least once during testing.
2. Branch Coverage: This method ensures that each possible branch of a conditional statement is executed at least once during testing.

3. Condition Coverage: This method ensures that each possible combination of conditions within a conditional statement is executed at least once during testing.
4. Path Coverage: This method ensures that each possible execution path through the program is executed at least once during testing.

Control flow testing is typically done manually by a tester, which follows the control flow diagram and executes the program with different input values to verify that the program behaves as expected. Alternatively, automated testing tools can generate test cases and run the program.

Control flow testing is an integral part of the software testing process, as it helps ensure that the program functions correctly and meets the users' requirements.

Control Plan

A control plan is a document that outlines the strategies and methods used to monitor and maintain the quality of a product or process, usually manufacturing. It is typically used in manufacturing and production environments to ensure that an end product meets all required specifications and standards.

The control plan typically includes the following:

1. The specific quality characteristics or requirements to monitor
2. The measurement methods or tools used to assess these characteristics
3. The frequency at which measurements are taken
4. The acceptance criteria for each characteristic
5. The actions to be taken if the characteristics do not meet the acceptance criteria (e.g., adjusting the process, reworking the product, or rejecting the product)

The control plan is an integral part of the quality management system, as it helps to ensure that the end product meets the required standards and the customer's expectations. It also helps to identify and address potential issues in the production process before they result in defective products.

Corner Cases

Corner cases are unpredictable or unknown events that seldom happen, reflect a serious problem, and are rarely included in test cases. Also known as "corners," they are specific conditions or scenarios used to stress a product and evaluate its performance. These conditions may be extreme or unusual, including testing the product under high or low temperatures, high humidity or dry conditions, or heavy load or stress. By testing a product in various corners, manufacturers and engineers can ensure that the product performs consistently and reliably in multiple conditions to which the product may be subjected. This testing helps identify potential issues or weaknesses in the product and can also help improve the product's overall quality and performance.

Corrective Actions

Corrective actions are steps to address and fix a problem or issue within a system or process. These actions follow identifying an issue and actions taken to remediate and prevent the problem from recurring. Corrective actions include making changes to procedures or processes, implementing new controls or safeguards, and providing additional training or education to employees. It is essential to carefully plan and execute corrective actions to ensure the resolution of the issue and minimize the risk of recurrence.

Corrosion Testing

Corrosion testing is a process used to evaluate the corrosion resistance of materials and coatings in various environments. Corrosion can lead to material degradation, affecting the performance and safety of products and structures. Therefore, corrosion testing is essential in many industries, including automotive, aerospace, marine, and construction [20, 59].

Several types of corrosion testing methods can be used to evaluate materials and coatings:

1. Salt Spray Testing: This involves exposing test specimens to a saltwater mist in a controlled environment to evaluate their resistance to corrosion.
2. Humidity Testing: This typically involves exposing the materials or coatings to a high-humidity environment for a fixed period while monitoring the effects of humidity on the materials. The exposure time and humidity levels used in testing depend on the specific application and standards.
3. Condensation Testing: This involves exposing the materials to a high-humidity environment and monitoring the formation of condensation on the material's surface, which can lead to corrosion.
4. Immersion Testing: This involves placing test specimens in a corrosive solution (often an iced saltwater bath) and monitoring their performance post testing for saltwater intrusion.
5. Electrochemical Testing: This involves applying an electrical potential to a test specimen and monitoring the resulting electrochemical reactions, which can provide information about the material's resistance to corrosion.

6. Cyclic Testing: This involves exposing test specimens to various corrosive environments, including temperature and humidity variations, to simulate real-world conditions.
7. Accelerated Testing: This involves artificially accelerating the corrosion process to evaluate the material's resistance to corrosion in a shorter time frame.

Corrosion testing evaluates various materials, including metals, alloys, plastics, and coatings. The results of corrosion testing can be used to inform material selection, design decisions, and maintenance strategies to ensure the durability and longevity of products and structures. Additionally, products and structures often require corrosion testing to ensure that meet safety and performance requirements.

Co-Simulation

Co-simulation enables analysts to separate a problem into subsystems and utilize specialized tools to deal with each sub system's unique dynamics. On the other hand, these subsystems must be connected through the exchange of coupling variables at specific times in time, known as communication points.

Co-simulation in testing, specifically in the automotive industry, refers to the collaborative simulation of different subsystems or components of a vehicle or automotive system. It involves integrating and running multiple simulation models simultaneously to assess their combined behavior and interactions accurately.

In the automotive industry, vehicles are complex systems with numerous interconnected subsystems, such as powertrain, chassis, electrical systems, and control units. Co-simulation enables the evaluation of how these subsystems interact with each other and how they collectively impact the overall performance and behavior of the vehicle.

Key aspects of co-simulation in the automotive industry include the following:

1. Integration of Multiple Simulation Models: Co-simulation involves combining and integrating simulation models representing different subsystems or components of a vehicle. These models may be developed by different teams or suppliers and employ various simulation techniques and software tools.
2. Time and Data Exchange: Co-simulation requires exchanging information and data between the simulation models to ensure synchronization and accurate representation of interactions. This includes exchanging inputs, outputs, and states of the respective simulation models at each time step.
3. Real-Time Simulation: Co-simulation often aims to simulate the behavior of the interconnected subsystems in real time or near real

time. This allows for the assessment of system-level performance and the evaluation of dynamic responses, such as vehicle dynamics, power distribution, or control system interactions.

4. Validation and Verification: Co-simulation is used to validate and verify the performance and behavior of the integrated system. It helps identify potential issues, conflicts, or suboptimal interactions between subsystems early in the development process, allowing for corrective measures to be taken.
5. Functional Testing and Optimization: Co-simulation enables functional testing of the integrated system under various scenarios, such as different driving conditions, environmental factors, or control strategies. It aids in optimizing the design, performance, and efficiency of the vehicle or system.
6. Virtual Prototyping: Co-simulation serves as a virtual prototyping tool, enabling engineers to assess and refine the performance of the vehicle or system before physical prototypes are built. It reduces development costs, time, and risks associated with physical testing and enables iterative design improvements.

Co-simulation plays a significant role in the development and testing of complex automotive systems, allowing for comprehensive evaluations of system interactions and performance. It facilitates early detection and resolution of integration issues, leading to more efficient and reliable vehicle development processes.

Cost-Benefit Analysis

Cost-benefit analysis (CBA) is used to evaluate the potential costs and benefits of a proposed project or investment, including the costs and benefits of testing. CBA is a tool for decision-making, as it allows stakeholders to compare the potential costs and benefits of different options and make informed decisions.

CBA can be used to evaluate the potential costs and benefits of testing a product or system. This includes the costs of testing, such as personnel and equipment costs, as well as the potential benefits, such as increased product reliability and reduced liability costs. In addition, CBA helps evaluate the limits of testing effort and expenditure.

Some of the potential benefits of testing include the following:

1. Improved Product Quality: Testing can help to identify and address defects and weaknesses in a product, leading to higher quality and reliability.

2. Reduced Costs: By identifying and addressing issues early in the development process, testing can help to reduce the costs associated with fixing defects and addressing issues later on.
3. Increased Customer Satisfaction: Testing can help ensure that products meet customer needs and expectations, increasing customer satisfaction and loyalty.

However, there are also costs associated with testing, including the cost of personnel, equipment, and time. To determine whether testing is cost-effective, stakeholders can use CBA to evaluate the potential costs and benefits of testing and compare these with the costs and benefits of other options, including risk acceptance.

CBA is a valuable tool for evaluating the potential costs and benefits of testing to make informed decisions about product development and investment. By considering the potential costs and benefits of testing, stakeholders can make informed decisions using data, likely leading to the best outcomes.

Cost of Quality

The cost of quality refers to the expenses incurred to ensure that products or services meet the desired level of quality (see Figure C.7). In addition, the cost of quality includes costs associated with prevention, inspection, and failure. Prevention costs refer to the expenses incurred to prevent defects from occurring in the first place, such as training and education for employees, quality control processes, and implementing quality management systems.

FIGURE C.7 The cost of the quality curve, cost of good and bad quality.

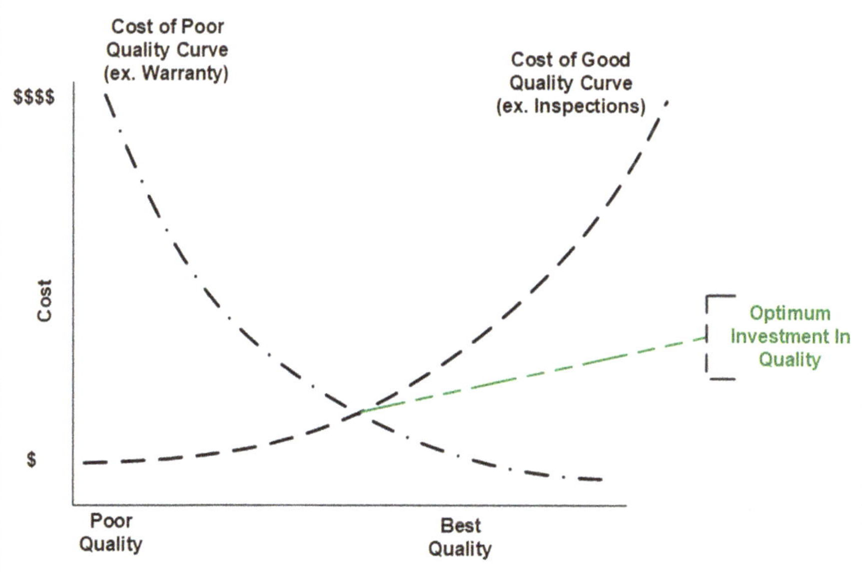

Inspection costs refer to the expenses incurred to identify defects during production, such as testing and inspection of products or services.

Failure costs refer to the expenses incurred due to defects, such as repairs (rework), warranties, and customer complaints.

The cost of quality in product testing can significantly impact a company's profitability and reputation. Reducing the cost of quality can increase profits, while increasing the cost of quality can decrease profits.

Costs of Detection

The time and effort of detecting a particular issue or problem can vary widely depending on the specific circumstances and the available resources. In general, the more spent on quality improvement efforts early or throughout development, the better the quality. Some of the factors that may affect the cost of detection include the following:

1. The Complexity of the Problem or System: More complex issues may require more time and resources to detect and diagnose, resulting in higher costs.
2. The Resources Required for Detection: The tools and equipment needed for detection and the personnel to operate them can add to the overall detection cost.
3. The Level of Expertise Needed: Specialized expertise or training may be required to detect certain problems, which can add to the cost.
4. The Availability of Resources: If the resources needed for detection are scarce or in high demand, this can drive up the cost.
5. The Urgency of the Situation: In some cases, the cost of detection may be higher if the problem needs to be detected and addressed quickly.

Detection costs can vary significantly depending on the specific circumstances and the available resources. Therefore, it is essential to consider all these factors carefully when planning for and budgeting for detection efforts (see Figure C.8).

FIGURE C.8 Cost of defect detection and prevention quality curve.

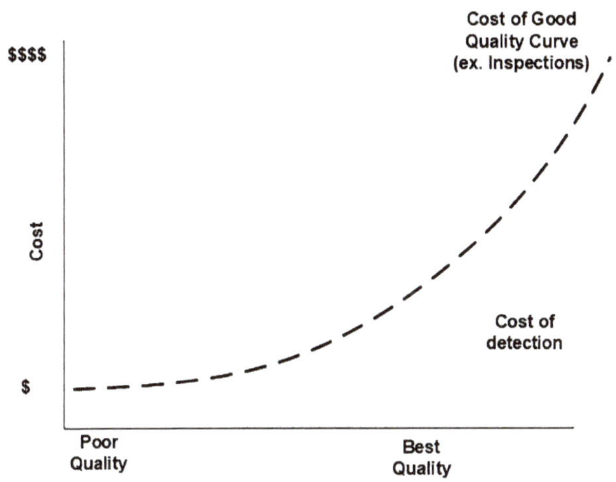

Costs of External Failure

Costs incurred to fix issues customers find after purchasing a product or using a service. Processing client returns, warranty claims, and product recalls are a few examples. Also, a part of the cost of poor quality includes external and internal quality failures (see Figure C.9).

FIGURE C.9 Quality costs in the field; as quality improves, the costs of poor quality decrease.

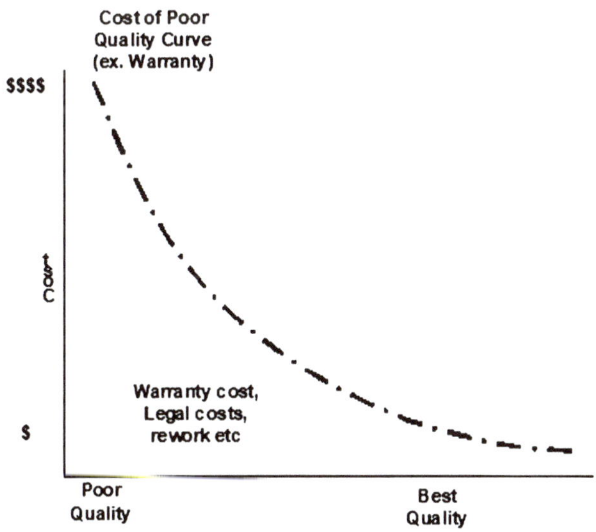

Costs of Internal Failure

The cost of internal failure refers to the financial and nonfinancial costs associated with defects or problems identified within a company's processes or products before they reach the customer. These costs include the following:

1. Rework Costs: These represent the cost of fixing defects or errors in products or processes before they are released to the customer.
2. Scrap Costs: These are the cost of discarding or reworking defective products or materials.
3. Lost Productivity: This is the cost of time and resources spent on fixing problems instead of focusing on more productive activities.
4. Delay Costs: These include delays caused by rework or repairs, including missed deadlines or lost sales opportunities.
5. Lower Morale: Recurring product failures impact employees through decreased employee motivation and engagement due to frustration or disappointment caused by internal failures.
6. Damage to Reputation: This is the cost of negative publicity or loss of customer trust due to product defects or process failures.

The cost of internal failure can be high and harm a company's bottom line, reputation, and customer relationships. To minimize these costs, companies should focus on improving the quality of their products and processes, implementing effective quality control measures, and continuously monitoring and improving their internal systems and procedures.

Costs of Prevention

There are several costs associated with the prevention of product quality issues:

1. Quality Control Testing: This includes the costs of hiring quality control personnel, purchasing testing equipment and supplies, and conducting quality control tests.
2. Raw Material Inspection: This includes the costs of inspecting raw materials for defects and ensuring they meet the required standards.
3. Process Improvements: This includes the costs of implementing process improvements or upgrades to prevent quality issues.
4. Training: This includes the costs of training personnel on quality control measures and best practices.

5. Quality Management Systems: This includes the costs of implementing and maintaining quality management systems, such as ISO 9001 certification.
6. Warranty Repairs: This includes the costs of repairing or replacing products with quality issues covered under warranty.

Prevention of product quality issues can be a significant investment, but businesses must prioritize this to maintain customer satisfaction and avoid costly recalls or legal matters.

Courtesy Bias

Courtesy bias occurs when people give positive or socially desirable responses to avoid offending or displeasing others. For example, in the testing context, courtesy bias can manifest in testers providing positive feedback or avoiding reporting issues or defects that they perceive may reflect poorly on the development team or product.

Courtesy bias can negatively impact testing, leading to issues and defects being overlooked or ignored, ultimately resulting in a lower-quality product or system. To mitigate the effects of courtesy bias, testers must remain objective and honest in their assessments and report issues and defects promptly and transparently. In addition, the organization's culture should prioritize clear articulation of circumstances or situations.

Testing teams can also implement strategies to reduce the impact of courtesy bias, such as providing anonymous reporting channels, establishing clear testing criteria and standards, and encouraging open and honest communication between testers and developers. By addressing courtesy bias and other cognitive biases that may impact testing, teams can help ensure that their testing efforts are effective and that their products or systems meet the needs and expectations of their users.

Coverage Tool

A coverage tool is software used to measure the amount of code executed during testing. This is known as code coverage (see Code Coverage), the percentage of code the tests have executed. Code coverage tools identify areas of the code that have not been tested or tested inadequately. By measuring the code coverage, testers can determine the effectiveness of their tests and identify areas of the code that may require additional testing.

Coverage tools typically provide a report showing the percentage of code the tests have executed. The resulting report identifies areas of the code that need additional testing and evaluates the effectiveness of the testing effort.

Coverage tools can be integrated into the testing process in various ways. For example, some testing frameworks provide built-in coverage tools, while others require third-party tools. Additionally, some development environments may provide built-in code coverage analysis tools.

Cross-Continuity Test

A wire harness cross-continuity test checks for unintended electrical continuity between different wire components in a harness. This testing ensures that the wires are correctly connected and that the circuit has no connection to other undesired links [60, 61, 62, 63].

To perform a wire harness cross-continuity test, you will need a multimeter and a wiring diagram or schematic for the harness you are testing and the harness. Then, use the probe of the voltmeter to test one wire end cycling the other meter probe through the other wire harness ends.

Curse of Knowledge

The curse of knowledge refers to the cognitive bias whereby people who are knowledgeable about a subject have difficulty imagining what it is like for someone who does not have the same knowledge or expertise. This cognitive bias can lead to communication breakdowns, misunderstandings, and other problems when conveying information or ideas to others.

Regarding testing, the curse of knowledge can be a significant obstacle to effective communication and collaboration between testers and other stakeholders, such as developers, product owners, and end users. Testers may assume that others understand the technical details and intricacies of the system or product being tested, leading to miscommunication, missed defects, and other issues.

To overcome the curse of knowledge in testing, it is essential for testers to do the following:

1. Put themselves in the shoes of other stakeholders and think about how they might perceive or understand the system or product tested.
2. Use clear and straightforward language when communicating with others, avoiding technical jargon or complex terminology that may be confusing or overwhelming.
3. Use visual aids and other tools to help explain complex concepts or processes.
4. Seek feedback from others and actively listen to their input and concerns.
5. Collaborate with other stakeholders throughout the testing process rather than work in isolation.

By recognizing and overcoming the curse of knowledge, testers can improve communication, collaboration, and the overall effectiveness of the testing process, leading to better-quality products and systems.

Customer Evaluation

Customer evaluation provides customer critique through the use of the product. This evaluation happens late in product development, quality assurance, and product qualifying. Customer evaluation is a precursor to customer acceptance but allows developers to respond to errors and deviations from expected or desired performance. This evaluation enables businesses to gather feedback from actual users and customers, identify issues and opportunities for improvement, and ensure that their products and services meet the needs and expectations of their target audience.

There are several ways in which customer evaluation is conducted:

1. Usability Testing: This involves observing how customers use a product or service and gathering feedback on its ease of use, functionality, and overall user experience. Usability is performed through in-person or remote testing sessions and involves various tools and techniques, such as surveys, interviews, and task-based testing.
2. Beta Testing: This involves releasing a product or service to a select group of customers before its official launch and then gathering feedback on its performance, features, and usability. Beta testing can help businesses identify issues and opportunities for improvement before the product is released to a broader audience.
3. Customer Surveys: This involves gathering customer feedback through surveys or questionnaires, which can be distributed via email, social media, or other channels. Surveys can help businesses collect data on customer preferences, satisfaction, and pain points, which can inform product development and improvement efforts.
4. Focus Groups: This involves gathering a small group of customers or potential customers to discuss their opinions, experiences, and preferences about a product or service. Focus groups can provide businesses with valuable qualitative information about customer needs and preferences.

Customer evaluation and testing are critical components of product development and quality assurance, enabling businesses to gather feedback, identify issues, and improve their products and services to meet their target audience's needs and expectations.

Cyclomatic Complexity

Cyclomatic complexity is a measure of the complexity of a program based on the number of independent paths through the code. A higher cyclomatic complexity indicates more potential for errors and a higher maintenance cost. It can be calculated by counting the number of decision points (such as if statements) and adding 1. A cyclomatic complexity of 1 indicates a simple, straightforward program with no decision points, while a higher number means a more complex program with multiple decision points. Cyclomatic complexity can be used as a tool to identify areas of a program that may require refactoring or simplification.

Cyclomatic complexity measures the number of independent paths through a program's source code. It is calculated using the following equation:

$$M = E - N + 2P$$

Where:
E is the number of edges in the control flow graph (CFG).
N is the number of nodes in the CFG.
P is the number of connected components (i.e., independent paths) in the CFG.

The higher the cyclomatic complexity, the more complex and challenging to understand the code is likely to be.

In other words, the cyclomatic complexity of a program is equal to the number of edges in its control flow graph minus the number of nodes plus two times the number of connected components. The control flow graph visually represents the program's control flow, showing the different paths the program can take based on different conditions and inputs.

The cyclomatic complexity metric helps evaluate the maintainability and testability of a software program. A higher cyclomatic complexity indicates a more significant number of independent paths through the program, which can make the program more difficult to understand, test, and maintain. Generally, a cyclomatic complexity of 10 or less is considered good. In contrast, a complexity of 20 or more is considered high and may indicate that the program should be refactored to reduce complexity.

D

"If you want to make decisions, then eliminate all the alternatives with the power of factual data. If you do not want to make decisions, then do us all a favor by staying out of the way."

—John Mott, president, AMR Travel Services

Daily Build

The act of finishing a software build of the most recent version of a program each day is known as a daily build or nightly build. The software build is compiled and tested (daily) to detect an introduction of errors and ensure all necessary dependencies are present. The testing of the daily build may be constrained to the contents of that daily build. Clean compilation should not be confused with daily build testing. The daily build may be made public, giving everyone access to the newest improvements and allowing for input. In this sense, a build happens after compiling and linking every file that makes up a program. When numerous programmers work on a single piece of software in large organizations, following rigorous processes like daily builds is especially important. Daily builds make it possible for developers to work with the knowledge that any new issues that arise result from their work completed the previous day.

Data-Driven Testing

Data-driven testing is an approach whereby test cases are designed and executed based on data inputs and expected outputs. The test data is stored in a database or spreadsheet, and the test cases are automated to retrieve the data and run the tests.

This approach is helpful in testing applications with many input combinations or scenarios, as it allows for consistently and efficiently testing multiple datasets. It also provides easier maintenance and updates to the test cases.

Changing the test data without requiring changes to the test code is easy. Test data provides a consistent base for the testing.

Data-driven testing is often used with other testing methods, such as exploratory or automated testing, to ensure the application is effectively explored and potential scenarios are covered.

Data Flow

Data flow refers to the movement of data within a system or process. It involves transferring data from one location or component to another, typically through input/output devices or networks. Data flow is important because it helps to ensure that data is appropriately processed and stored within a system or process. It also helps identify potential bottlenecks or issues impacting the data flow.

Data flow can be visualized using a diagram or model showing the flow between different components or locations. This visualization can help reveal the relationships between various components and the data dependencies within the system or process.

Data flow can be managed and optimized using data modeling, data integration, and data management techniques. These techniques help ensure that data is properly structured, accurate, organized, and available when needed.

Data Flow Testing

Data flow testing involves examining the input, processing, and output of data flowing through a software system to ensure it is correct and consistent. This type of testing focuses on how the data is transformed and moved within the system. For example, consider data that begins as sensor input, analog fuel level. This analog input may become a digital representation over a data bus and into other ECUs that will use the data.

Data flow testing typically creates test cases involving different input and output scenarios to see if the system processes the data correctly and produces the expected output. Data flow testing is vital to software testing because it helps identify system data processing logic errors. It can also help to identify any data inconsistencies that may occur due to incorrect data flow.

Debugger

A debugger is a software tool used to identify and fix errors or bugs in computer programs. It allows a programmer to pause the execution of a program, examine the state of variables and memory, and step through the program line by line to identify the cause of an error. Some debuggers provide additional features such as breakpoints, which allow a programmer to pause the program at a specific point, and watchpoints, which enable the programmer to monitor the value of a particular variable. Debuggers are an essential tool for software development, as they help identify and fix problems that can prevent a program from functioning correctly.

Debugging Tools

Programmers use a variety of debugging tools to study program states, reproduce errors, and identify related flaws. Programmers can use debuggers to run programs step-by-step, stop them at any program statement, and set and check program variables.

1. Debuggers: These tools allow developers to step through code execution, set breakpoints, and inspect variables to identify and fix issues. Some examples include GDB, LLDB, and Visual Studio Debugger.
2. Debug Logs: These are logs generated by the application or system that contain detailed information about the execution of the code and the data used to identify and troubleshoot defects.
3. Profilers: These tools allow developers to analyze the performance of their code and identify areas that may be causing issues or bottlenecks. Some examples include Xdebug, Blackfire, and AQtime.
4. Network Analyzers: These tools allow developers to monitor and analyze network traffic to identify issues with communication between systems or devices. Some examples include Wireshark, Fiddler, and Charles.
5. Error Reporting Tools: These tools allow developers to capture and report errors that occur during code execution automatically. These reports help developers quickly identify and fix issues. Some examples include Sentry, Rollbar, and BugSnag.

Decision Condition Testing

Decision condition testing involves evaluating the conditions and decision points in a software program to ensure that they are functioning as intended. This evaluation includes testing the logic of the program, verifying that the appropriate outcomes are produced based on different inputs, and ensuring that the program can handle different scenarios and edge cases.

We start with the decision tree outline (Figure D.1 shows a two decision tree) from that it is possible to determine specific decisions and test cases that would exercise those decisions. Decision testing ensures that each possible decision within the software system is exercised at least once. In addition, this technique helps identify potential flaws or bugs that may occur within the decision.

FIGURE D.1 An example of a decision tree.

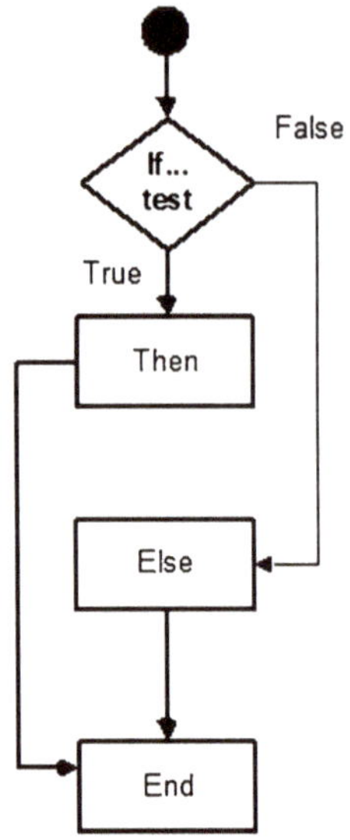

Decision condition testing is an essential part of the software development process, as it helps to ensure that programs are reliable and perform as intended under different conditions. In addition, by thoroughly testing the decision points in a program, developers and testers can identify and fix issues that could cause problems or errors in the final product.

Decision Coverage

Decision coverage is a software testing metric that assesses the testing coverage. It involves recording the decisions tested against the total decisions possible in the code expressed as a percent decision coverage. This expression makes it possible to evaluate the level of risk that may remain in the product due to missed decision tests.

$$Decision_{coverage} = \frac{Decisions_{tested}}{Decisions_{total}}$$

Decision Table

A decision table is a tool used to represent a set of conditions and their corresponding actions in a logical and organized manner. It is commonly used in decision analysis to help evaluate options and provide a structure for test scope decisions via prioritization of test cases.

A decision table consists of a matrix with rows representing different scenarios or conditions and columns representing other actions or outcomes. Each cell in the matrix represents the action or outcome that should be taken based on the corresponding scenario or condition.
Decision tables are useful because they clearly and concisely represent the relationships between different conditions and actions. They also allow for easy evaluation of other options or courses of action based on a given scenario's specific requirements and constraints.

Criteria	Option 1	Option 2	Option 3
Engine	2.0 L, 4-cylinder	1.8 L, 4-cylinder	2.5 L, 6-cylinder
Transmission	6-speed manual	8-speed automatic	CVT
Drive Type	Front-wheel drive (FWD)	All-wheel drive (AWD)	Rear-wheel drive (RWD)
Safety Features	Standard: ABS, airbags	Optional: backup camera, blind-spot monitoring	Optional: adaptive cruise control, lane departure warning
Entertainment Features	Standard: AM/FM radio, USB ports	Optional: touchscreen infotainment system, satellite radio	Optional: premium sound system, rear-seat entertainment system
Exterior Features	Standard: LED headlights, alloy wheels	Optional: sunroof, heated mirrors	Optional: spoiler, sport-tuned suspension

Decision tables are commonly used in a variety of fields, including business, finance, and engineering, to help make decisions based on complex data and scenarios. In addition, they can be used in conjunction with other decision-making tools, such as decision trees or decision matrices, to provide a more comprehensive analysis of different options or courses of action.

Decision Testing

White-box testing focuses on the internal structure, design, and implementation of a software system. Decision testing, also known as decision coverage or branch coverage testing, is a specific type of white-box testing technique. It aims to ensure that all possible decision outcomes or branches within the code are tested at least once.

In decision testing, the objective is to exercise both the true and false outcomes of every decision point (such as if statements, switch statements, or loops) in the code. The goal is to achieve 100% decision coverage, meaning that every possible decision outcome has been tested.

Deembrittlement Verification Test

Deembrittlement verification tests are commonly conducted on materials to assess their resistance to hydrogen embrittlement, a phenomenon whereby materials become brittle and susceptible to failure under the influence of hydrogen. This standard outlines test methods and practices which can detect embrittlement of steel parts. It is a process control or referee verification test. The risk of embrittlement of steel is minimized by using best practices in the finishing/coating process. One such practice is described in **SAE/USCAR-5, Avoidance of Hydrogen Embrittlement of Steel** [64].

For example, a torque/tension test is used for threaded parts and a tensile test is used for nonthreaded parts. The test consists of three steps:

1. Determine the ultimate torque or tensile stress to failure for threaded and nonthreaded parts, respectively.
2. Load the parts to some percentage of the ultimate torque or tensile stress.
3. Maintain the torque or tensile stress for some determined length of time. Whenever possible, the test fixture should simulate the intended application [74].

One of the commonly used standards for hydrogen embrittlement testing is SAE's "Hydrogen Embrittlement Testing of Ultra High Strength Steels and Stamping." This standard describes a test method for evaluating the susceptibility of uncoated cold rolled and hot rolled ultra high-strength steels (UHSS) to hydrogen embrittlement. The thickness range of materials that can be evaluated is limited by the ability to bend and strain the material to the specified stress level in this specification. For more information, see **J3215**.

Hydrogen embrittlement can occur with any steel with a tensile strength greater than or equal to 980 MPa. Some steel microstructures, especially those with retained austenite, may be susceptible at lower tensile strengths under certain conditions. The presence of available hydrogen, combined with high stress levels in a part manufactured from high-strength steel are necessary precursors for hydrogen embrittlement.

Due to the specific conditions that need to be present for hydrogen embrittlement to occur, cracking in this test does not indicate that parts made from that material would crack in an automotive environment (see Figures D.2 and D.3). Results from this test should be considered in conjunction with the strain state of the material and the operating environment of the part when selecting any UHSS. Since this test method is comparative, the most information can be gained if a control sample of known performance is evaluated along with the material being studied [65].

FIGURE D.2 Test specimen layout in test container before acid immersion [65].

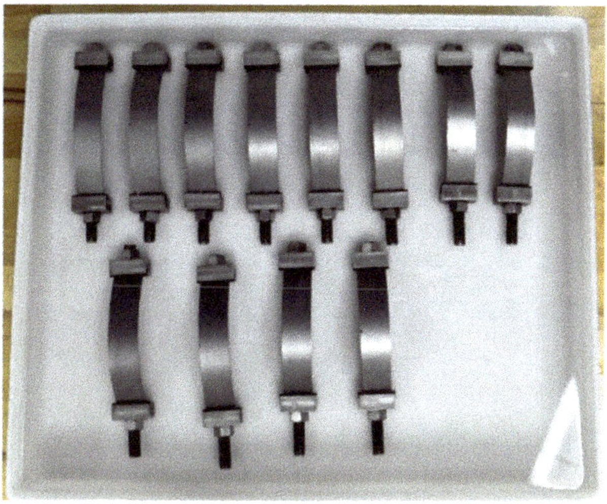

FIGURE D.3 Posttest example material [65].

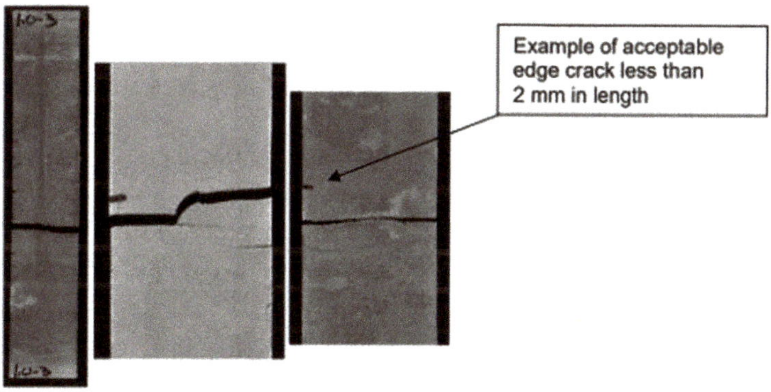

Defect

In software and hardware development and testing, a defect refers to a flaw or fault in a product application that causes it to behave unexpectedly or unintentionally (see Figure D.4). A defect can be a mistake or error in the code, design, or requirements or a problem with the software environment or configuration.

FIGURE D.4 An example of a hardware defect due to manufacturing.

Defects can manifest in many ways, such as a crash or freeze, incorrect or intermittent output or behavior, performance issues, interface issues, security vulnerabilities, or other issues that prevent the software from meeting its intended functionality or quality goals. In addition, the severity of defects can range widely, from minor problems that have little impact on the user experience to critical issues that can cause data loss or system failure, or even compromise system security.

To manage defects, software development teams typically use defect tracking and management tools that allow them to report, track, prioritize, and resolve defects throughout the development life cycle. Defect management aims to identify, isolate, and eliminate defects as early in the development process as possible to minimize their impact on project timelines and quality.

Defect Arrival Rate

The defect arrival rate is a test metric of the frequency at which defects or errors often appear in a product or system during testing. This rate is often used as a measure of quality and is often used to determine the effectiveness of quality control processes. For example, a high defect arrival rate may indicate the need for improvement in the manufacturing or design process. In contrast, a low defect arrival rate suggests that the product or system is high quality. This metric can help assess the remaining defects in the product before testing is complete. For example, say that testing is 50% complete and the defect arrival rate has been three defects per day. It is almost certain there are more defects in the remaining testing. The defect arrival rate is usually measured over a specific period, such as per unit produced or per operation hour (see Figure D.5).

FIGURE D.5 An example of defect arrival rate, including severity category.

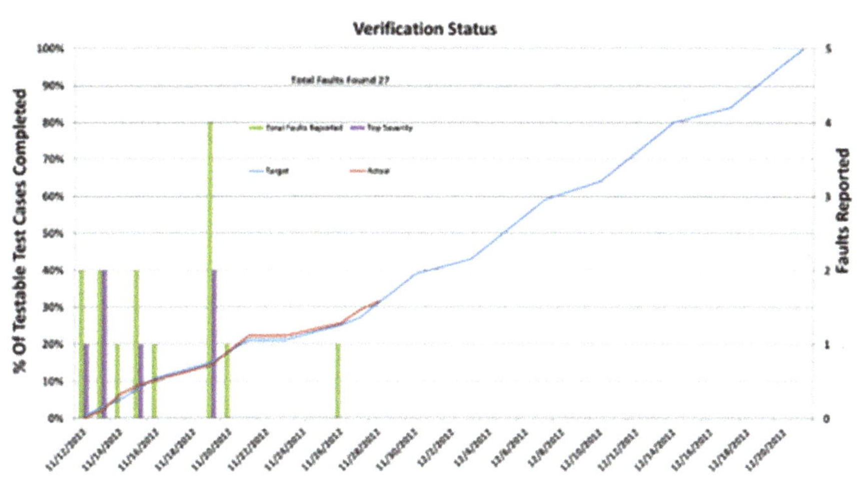

Defect Closing Rate

The defect closing rate is a measure of the efficiency of a software development team in fixing defects (also known as bugs) in their code. It is calculated by dividing the number of defects closed in a given period by the total number open at the beginning.

$$Defect\ Closing\ Rate = \left(\frac{defects}{output_{tested}}\right) \times 100$$

For example, if a team starts a sprint with 100 open defects and closes 75 by the sprint's end, the sprint's defect closing rate would be 75%. The defect-closing metric can be used to track a team's progress in addressing defects and identify any issues that may impede their ability to fix defects efficiently.

$$Defect\ Closing\ Rate = \left(\frac{75}{100}\right) \times 100 = 75\%$$

It is important to note that the defect closing rate is only one metric used to evaluate the performance of a software development team. Other metrics, such as the number of new defects introduced, the time it takes to fix defects, and the codebase quality, may also be essential to consider when evaluating the team's effectiveness.

Defect Containment

Defect containment refers to identifying and addressing defects or errors in a product or system to prevent further issues or damage. Defect containment may involve isolating the defective item, implementing corrective actions, and testing to resolve the issue. The goal of defect containment is to minimize the impact of defects on dependent systems and individuals or customers. The post-containment process may involve implementing quality control measures, such as inspections and testing, to identify and fix defects before they cause problems.

Defect Density

Defect density measures software code quality and is expressed as the number of defects or bugs found per code unit, such as per a thousand lines of code.

To calculate defect density, follow these steps:

1. Determine the Number of Defects: This involves identifying all the defects or issues found during testing, such as coding errors, design flaws, or functionality issues.
2. Determine the Code's Size: The code's size is measured in terms of lines of code or function points.
3. Calculate the Defect Density: To calculate the defect density, divide the number of defects by the size of the code and multiply the result by a constant factor, such as 1,000 (for defects per 1,000 lines of code). If there are 50 defects found in 10,000 lines of code (loc), the defect density would be calculated as follows:

$$Defect_{density} = \left(\frac{Defect_{found}}{Tested_{LinesOfCode}}\right) \times 1{,}000$$

$$Defect_{density} = \left(\frac{50}{10,000}\right) \times 1,000 = 5$$

Defect density is a valuable metric for measuring software quality and can help identify areas of the code that require improvement. However, it is essential to note that defect density alone does not provide a complete picture of the quality of the code, and other metrics, such as code complexity and code coverage, should also be considered.

High defect densities can indicate poor quality or manufacturing issues, while low defects may indicate higher reliability or durability. Defect density assesses a product's or material's quality and identifies manufacturing process trends or improvement areas.

Defect Detection Percentage

The defect detection percentage measures the effectiveness of a quality control system in identifying defects in a product or process. It is calculated by dividing the number of defects found by the total number of units inspected and multiplying by 100.

$$Defect\ Detection\ Percentage = \left(\frac{Defects_{found}}{Number\ Units_{inspected}}\right) \times 100$$

For example, if a quality control team inspected 100 units and found 10 defects, the defect detection percentage would be 10%. This means that the quality control system identified 10% of the defects in the inspected units.

A higher defect detection percentage is seen as a sign of a more effective quality control system. It indicates that the system can identify a more significant number of defects. On the other hand, a lower defect detection percentage suggests that the quality control system is not performing as well as it should be and may need to be improved or refined.

Defect Management

Defect management identifies, tracks, and resolves defects or errors in a product or system. It is an essential part of the software development process, as it helps ensure that the final product is of high quality and meets the requirements and specifications of the user. Defect management includes reporting defects, assigning them to developers for resolution, and tracking their progress until fixed (see Figure D.6). It also involves creating and maintaining a defect database and implementing and following standard processes and procedures for identifying, reporting, and resolving defects.

1. Define Defect Management Process: Establish a clear set of procedures for identifying, tracking, and resolving defects in your product or service.
2. Identify Defects: Use various testing methods and tools to identify defects in your product or service. This identification process may include manual testing, automated testing, or user feedback. Is this a valid defect, or is it an interpretation of specification?

FIGURE D.6 The flow of determination between defect and specification error.

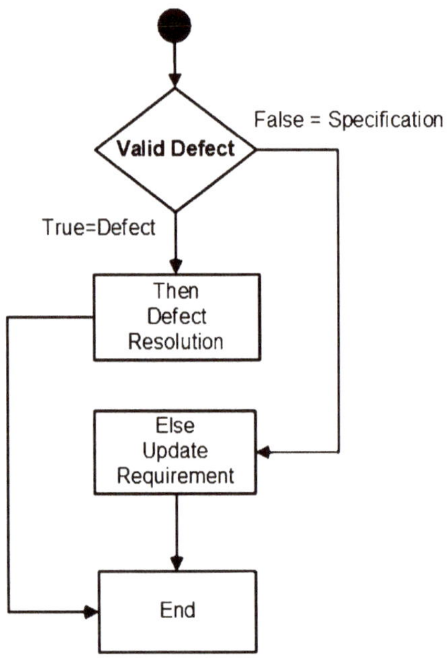

3. Document Defects: Record all faults in a central database or tracking system, including information about the defect, its impact, and any relevant details.
4. Assign Defects: Appropriate team members or individuals receive defect information for resolution.
5. Prioritize Defects: Determine the priority of each defect based on its impact on the product or service and the resources required to fix it.
6. Fix Defects: Work to resolve defects as quickly as possible, using the appropriate resources and techniques.
7. Test Fixes: Test any fixes to ensure they are effective and do not cause additional defects.
8. Close Defects: Once a defect has been resolved, close it in the tracking system and update any relevant documentation.

9. Review Defects: Regularly review the defects that have been identified and resolved to identify any patterns or trends that may indicate a larger issue or systemic problem.
10. Continuous Improvement: Use the insights gained from reviewing defects to improve your defect management process and prevent similar defects from occurring.

Defect Masking

Defect masking is a process in which defects or flaws in a product or material are hidden or covered up to present a more appealing appearance or meet specific quality standards. This can be done through various methods, such as painting, coating, or adding a material layer to cover the defect. However, it is essential to note that defect masking does not actually fix the problem, and the underlying defect may still exist and potentially cause issues in the future.

Defect Taxonomy

Defect taxonomy refers to categorizing defects or issues in a product or system. It is a way to organize and classify defects based on specific characteristics, such as type, severity, impact, or cause. This taxonomy helps identify patterns and root causes and prioritize defects for resolution. Some specific categories in a defect taxonomy include the following:

1. Functional Defects: These are related to a product's intended functionality, such as incorrect output or errors in the user interface.
2. Performance Defects: These impact a product's performance, such as slow response times or high resource usage.
3. Security Defects: These defects compromise a product's security, such as vulnerabilities to hacking or data leakage.
4. Compatibility Defects: These impact a product's compatibility with other systems or devices, such as issues with integration or interoperability.
5. Usability Defects: These make a product complex or confusing, such as unclear instructions or navigation.
6. Quality Defects: These relate to a product's overall quality, such as defects in materials or manufacturing.

Dependency Testing

Dependency testing focuses on verifying the relationships between different components or modules of a software system (applies to software but also hardware). This type of testing ensures that changes made to one component do not negatively impact the functionality of other components. Dependency testing evaluates the system's functions providing proper system performance with different parts used.

Dependency testing is used to identify and fix issues related to data flow, communication, and integration between other constituent parts of the system. It is typically performed during the integration testing phase of the software development process.

Dependent Failures

Dependent failures refer to situations in which the failure of one system or component leads to the failure of another system or component. For example, an open circuit on a wire harness power line might cause an electronic control unit, such as the antilock brake system (ABS), which is responsible for controlling a vehicle's braking and traction control process, to also fail. Dependent failures can cascade, creating a chain reaction and leading to multiple other failures. Various factors, including poor design, inadequate maintenance, and unexpected environmental conditions, can cause dependent failures. They can be difficult to anticipate and prevent and, moreover, have significant consequences for the reliability and functionality of a system.

Design-Based Testing

Design-based testing that involves creating test cases based on a software system's design. This approach focuses on testing the system's design rather than the actual implementation. Design-based testing ensures the system functions as intended and meets the specified requirements. This test can involve testing the system's inputs, outputs, and interactions, as well as the overall structure and flow of the design. Design-based testing is often used with other testing techniques, such as functional and regression testing, to assess the system's performance comprehensively.

Design of Experiments (DoE)

Design of experiments (DOE) is a methodical approach to identifying the relationship between the inputs and outputs of a process, usually associated with hardware and manufacturing exploration. The design of an experiment refers to the plan and strategy used to experiment, including the specific variables tested and the methods and procedures used to collect and analyze data. The approach includes determining the sample size, appropriate controls and treatments, and proper statistical analysis. It is essential to carefully plan and design an experiment to ensure it is well-controlled and provides reliable and accurate results:

1. Define the Problem or Research Question: Clearly define the problem or research question that will be addressed through the experiment.
2. Identify the Variables: Identify the variables that you want to test or measure in your experiment, including the independent variable (the factor that you are manipulating), the dependent variable (the factor that you are measuring), and any confounding variables (factors that could potentially affect the outcome of the experiment).

3. Develop a Hypothesis: Formulate a hypothesis about the relationship between the variables and how you expect them to affect each other.
4. Select the Experimental Design: Choose the appropriate experimental design for your study, such as a randomized controlled trial, a case-control study, or a cohort study.
5. Determine the Sample Size: Determine the appropriate sample size for your experiment, considering factors such as the size of the population, the desired level of precision, and the potential impact of confounding variables.
6. Randomize the Sample: Randomly assign subjects to different treatment groups or control groups to ensure any differences between the groups are due to the experimental treatment rather than preexisting differences.
7. Administer the Treatment: Administer the experimental treatment or intervention to the appropriate treatment group(s) and observe the effect on the dependent variable.
8. Collect and Analyze Data: Collect data on the dependent variable and analyze it using appropriate statistical techniques to determine the relationship between the independent and dependent variables.
9. Draw Conclusions: Draw conclusions based on the experiment's results and compare them to the original hypothesis. If the results support the hypothesis, the experiment is considered successful; if the results do not, the experiment is considered unsuccessful. Either way, the team has learned something.

Destructive Testing

Destructive testing involves intentionally damaging a material, component, or system to determine its strength, durability, or performance. This type of testing verifies a product's integrity or identifies any weaknesses or defects. It is typically conducted on a small number of samples, and the results are used to improve the design or manufacturing process of the product. Some standard destructive testing methods include bending, tensile, impact, short circuit, vibration, and fatigue testing.

Development Models

Product development models are frameworks that guide the process of developing a new product or improving an existing one. Common product development models include the following:

1. Waterfall Model: This is a linear and sequential approach to product development whereby each stage of the process must be completed before moving on to the next (see Figure D.7). The stages include planning, design, implementation, testing, and deployment.

FIGURE D.7 An illustration of the waterfall model.

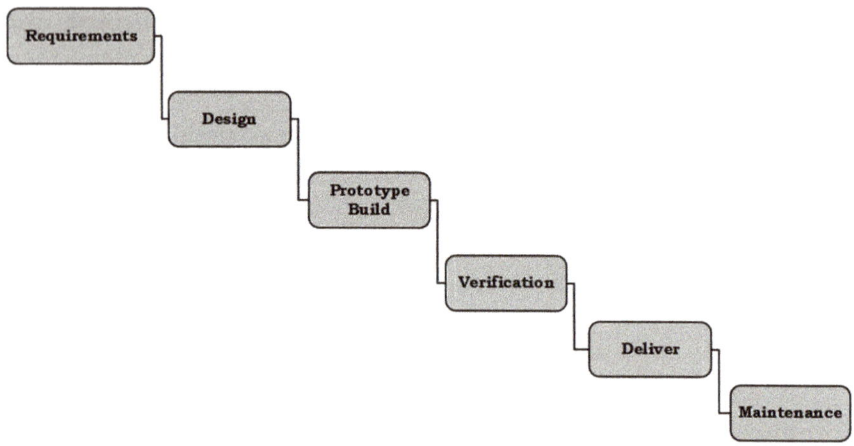

2. V-Model: The V development model, also known as the verification and validation model, emphasizes the relationship between testing and development. The V-model is an extension of the waterfall model and provides a more detailed and structured approach to the testing phase of development (see Figure D.8).

FIGURE D.8 An illustration of the V-model of product development.

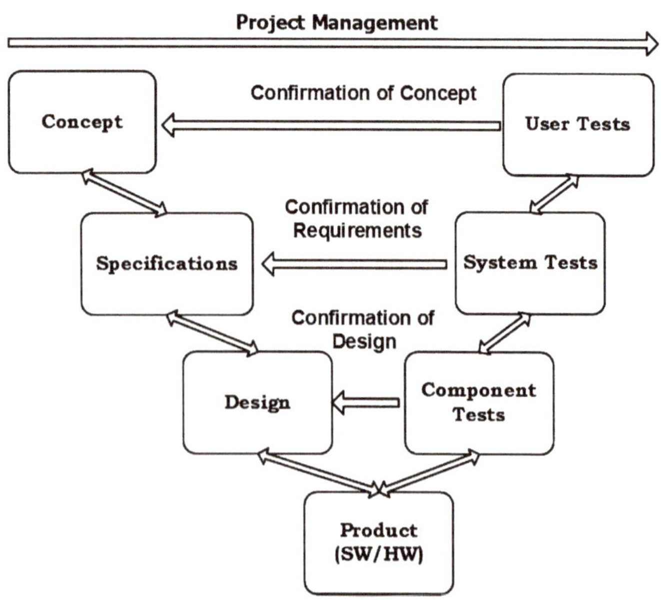

3. Agile Model: This is an iterative and incremental approach to product development. The development process is broken down into small, manageable parts that can be completed in short cycles called sprints (see Figure D.9). This model emphasizes collaboration and flexibility.

FIGURE D.9 An illustration of the Agile model of product development.

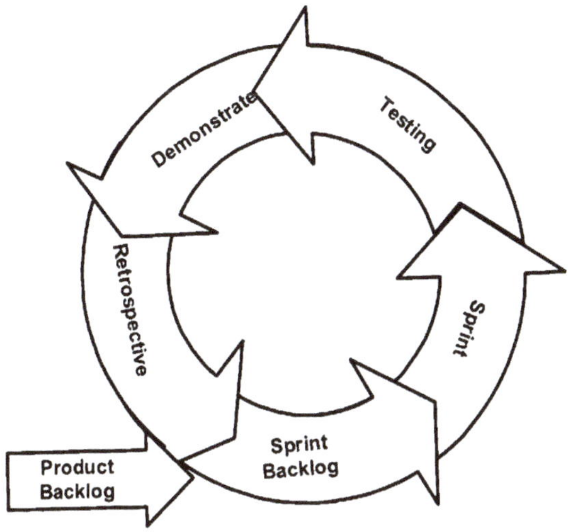

4. Stage-Gate Model: This model involves breaking the product development process into distinct stages or gates, each with its own set of deliverables and criteria for success (see Figure D.10). A gate review is conducted at the end of each stage to determine whether the project should proceed to the next stage.

FIGURE D.10 An illustration of the stage-gate model.

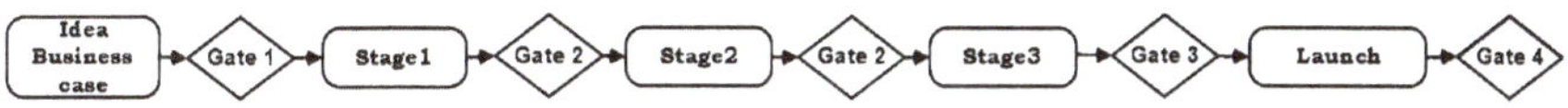

5. Lean Startup Model: This customer-focused approach to product development emphasizes rapid experimentation, continuous feedback, and iteration (see Figure D.11). This model is used for developing new products or services in uncertain or rapidly changing markets.

FIGURE D.11 An illustration of the lean startup model of development.

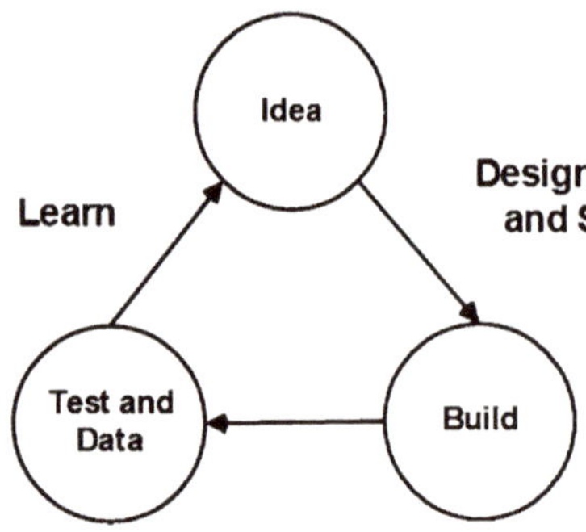

6. Design Thinking Model: This human-centered approach to product development emphasizes empathy, creativity, and experimentation (see Figure D.12). This model involves a deep understanding of the user's needs and preferences and focuses on prototyping and testing early and often.

FIGURE D.12 An illustration of the design thinking model of development.

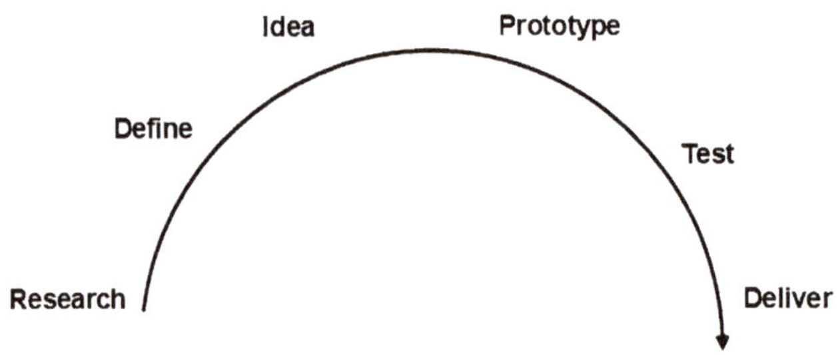

7. Spiral Model: This risk-driven model involves a series of iterations, each building on the previous one (see Figure D.13). The model begins with a small set of requirements and progresses through several iterations, each expanding on the previous one until the product is completed.

FIGURE D.13 An example of the spiral model of development.

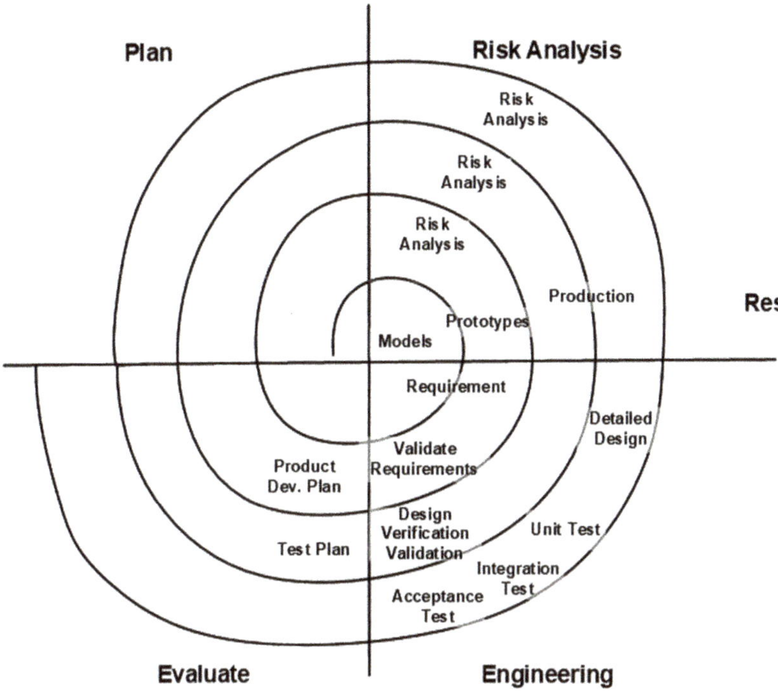

Each model has its strengths and weaknesses. The choice of which model to use will depend on the product's nature, the development team's size, the available talent and resources, and the project timeline.

Deviations

There are two definitions for the term "deviation": the first is product and configuration management centric, the other is measurement focused.

The first definition of deviation refers to the degree to which a value or set of values differs from the average or mean value in a set of data. It can be used as a measure of variability or dispersion in a set of data and is often used in statistical analysis to describe the distribution of data. Deviation can be calculated using the formula: deviation = (value – mean)/standard deviation. Testing uses measurements to compare the desired performance against actual performance. This applies from product design through manufacturing and past manufacturing.

The second definition of deviation is associated with configuration management. In configuration management, a deviation refers to a departure or divergence from the approved or established configuration baselines or standards. It occurs when there is a change or modification made to the configuration items that deviates from the predefined configuration.

Deviation in configuration management can happen due to various reasons:

1. Unauthorized Changes: Changes made to the configuration items without proper authorization or approval. This could include modifications made by individuals who are not authorized to make such changes.
2. Errors or Mistakes: Unintended or accidental modifications to the configuration that do not align with the approved configuration baselines. These deviations can result from human errors during the configuration management process.
3. Emergency Changes: Changes made to the configuration in response to urgent situations or critical incidents without going through the standard change management procedures. This deviation may be allowed in certain circumstances to address immediate concerns, but it should be properly documented and reviewed afterward.
4. Workarounds or Temporary Modifications: Temporary changes made to the configuration to address specific operational needs, troubleshoot issues, or accommodate special requirements. These deviations should be monitored and properly documented to ensure they are either reverted or incorporated into the formal configuration if necessary.
5. Vendor Updates or Patches: Changes introduced by software or hardware vendors through updates, patches, or upgrades. While these changes are external to the organization, they may still require evaluation and documentation to determine their impact on the overall configuration.

It is essential to manage and track deviations in configuration management to ensure that changes are properly authorized, evaluated, and controlled. Deviations should be recorded via the version release notes, analyzed, and either reverted or appropriately incorporated into the configuration baselines through change management processes. This process will either update the specifications and drawings, when the deviation is desired, or update the product when the deviation is not. Proper documentation and communication are crucial to maintain transparency and accountability in configuration management. It is difficult and inefficient to test a product when the content of the product is not reliably known. The result is testing features that are not present in the product.

Device under Test

A device under test (DUT) is a product or part of a product put to the test to see how well it performs (see Figure D.14). For example, a DUT might be a part of a larger module or unit tested (UUT). An examination (test) of a DUT uncovers flaws to ensure functionality and performance. The device under test will be the subject matter for our testing and verification efforts.

FIGURE D.14 An example of a device undergoing a voltage test.

Matveev Aleksandr/Shutterstock.com.

DFMEA

DFMEA stands for design failure mode and effects analysis. It is a process used in engineering and manufacturing to identify potential failure modes in a product or system and evaluate the possible consequences of those failures. DFMEA aims to identify and address potential issues before they occur, leading to a higher-quality product and reducing the risk of product failures. The process involves identifying the potential failure modes, evaluating the effects of those failures, and determining the likelihood and severity of those failures. The results of the DFMEA are used to refine the testing scope and implement corrective actions to prevent or mitigate the identified failure modes.

The DFMEA steps (see Figure D.15) are as follows:

FIGURE D.15 An overview of the steps in the DFMEA.

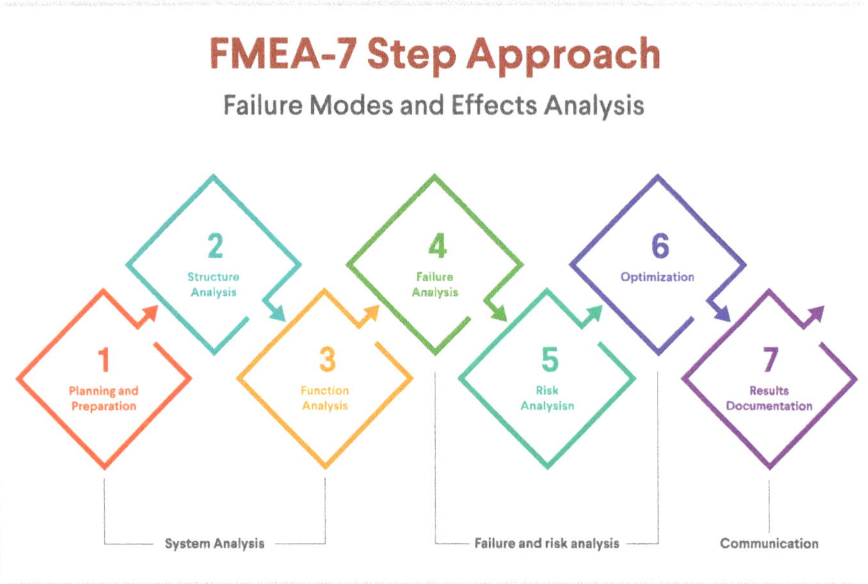

Master_shifu/Shutterstock.com.

1. Identify the Purpose and Scope of the DFMEA: Determine the product or process being analyzed and the potential failure modes that must be evaluated.
2. Identify the Potential Failure Modes and Their Causes: This may involve brainstorming sessions with team members or reviewing past failure data.
3. Evaluate the Severity, Occurrence, and Detection of Each Failure Mode: Use a standardized rating system to determine the potential impact of each failure mode on the product or process.
4. Identify Potential Countermeasures for Each Failure Mode: These include design changes, process improvements, or testing methods.
5. Prioritize the Failure Modes Based on Severity, Occurrence, and Detection Ratings: This prioritized effort addresses the most critical failure modes first.

6. Develop a Plan for Implementing the Countermeasures: This may involve updating design documents, training team members, or implementing new processes or testing methods.
7. Review and Update the DFMEA Regularly as Necessary: As the product or process evolves, it may be necessary to revisit the failure modes and countermeasures identified in the DFMEA to ensure they are still effective.

Digital Model

Digital models are an essential tool for testing, as they enable businesses and organizations to simulate and test their designs, products, and systems in a controlled and efficient manner. By creating a digital model of a physical object or system, companies can test and analyze its performance under different conditions, identify potential issues and weaknesses, and optimize its design and functionality.

Some ways in which digital models to use for testing include the following:

1. Prototyping and Testing: Digital models can create virtual prototypes of products or systems, which can be explored via augmented reality (AI) and refined before building physical prototypes (see Figure D.16). Virtual exploration and testing can save time and money and enable more efficient and effective testing and optimization.
2. Simulation and Analysis: Digital models can simulate and analyze the performance of products or systems under different conditions, such as varying loads, temperatures, or environmental factors. Simulation can enable businesses to identify potential issues and optimize their designs for better performance and efficiency.
3. User Testing: Digital models can create virtual testing environments in which users can interact with and provide feedback on products or systems in a controlled and realistic setting. User testing enables businesses to identify usability issues and optimize their designs for better user experience.

Digital models are a powerful tool for testing and optimization, enabling businesses and organizations to more effectively and efficiently design, test, and refine their products and systems for better performance and optimal user experience.

FIGURE D.16 An example of a virtual version of a product.

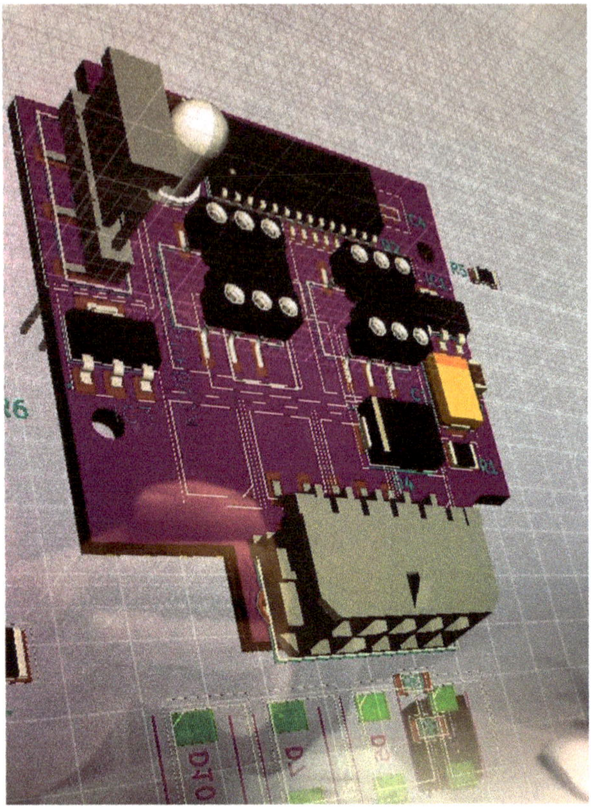

Digital Twin

A digital twin is a virtual representation of a physical object, system, or process used for simulation, analysis, and testing (see Figure D.17). It is used in the manufacturing, aerospace, and defense industries, where complex systems are constantly updated and modified.

FIGURE D.17 The digital version of the product reflects that of the physical product.

mechichi/Shutterstock.com.

Digital twins are used in testing to evaluate different scenarios and configurations without physical testing. They allow for the simulation of other conditions and inputs and can provide valuable insights into the performance and behavior of the physical object, system, or process.

Digital twins are created by collecting data from sensors and other sources and using this data to build a virtual model of the physical object, system, or process. The virtual model then simulates and analyzes different scenarios and configurations, identifying potential issues or improvements.

Digital twins are helpful for testing because they allow for evaluating different scenarios and configurations in a controlled and cost-effective manner. They also provide valuable insights into the performance and behavior of the physical object, system, or process and can help identify potential issues or improvements.

Documentation Testing

Documentation testing is a type of software testing that evaluates the quality and completeness of a product's documentation, including user manuals, technical guides, and other written materials that provide information about how to use and maintain the product.

Documentation testing is essential in software development because it helps developers and users quickly understand and use a product. It also helps to identify any discrepancies or omissions in the documentation, which can be corrected before the product is released to the public.

There are several approaches to documentation testing:

1. Manual Review: This involves a team of testers manually reviewing the documentation to ensure it is clear, concise, and accurate.
2. User Testing: This involves testing the documentation with actual users to see if they can understand and follow the instructions.
3. Automated Testing: This involves using automated tools to check for formatting and spelling errors and to confirm that the documentation is up-to-date and consistent with the product.

Overall, documentation testing helps to ensure that users have access to accurate and useful information about the product, which can help improve the user experience and increase customer satisfaction.

Domain Testing

Domain testing (software/hardware) focuses on verifying the functionality of a system within a specific domain or area of expertise. It involves testing the system's behavior and functionality within the context of the business or technical domain in which it operates. Domain testing includes testing the system's performance, reliability, security, and compliance with industry standards and regulations.

Domain experts or testers typically have a deep understanding of the business or technical domain in which the system operates. They use their knowledge and expertise to design and execute tests that verify the system's functionality and performance within the domain.

Examples of domain testing include testing the functionality of a telemetry system from vehicle to back office, testing the performance of the vehicle systems throughout the range of domains, and vehicle security.

Domain testing is an essential part of the software testing process. It helps ensure that the system functions correctly within the specific domain in which it operates and meets the needs and expectations of the users within that domain.

Driver-in-Loop Simulator

When developing and verifying a vehicle's chassis, drivetrain, and ADAS systems, the simulator offers an immersive driving exploration of the system in which the actions of the test vehicle and the driver become intertwined. This exploration enables crucial and reliable driver feedback allowing refinement in requirements, product and test strategies, and artifacts [66].

Drop Test

Drop testing is a type of environmental testing that evaluates the performance of a product or system when subjected to sudden impact, usually caused by a drop or a fall. Drop testing aims to determine a product's ability to withstand the effects of a drop/fall and ensure that it will function properly and safely under real-world conditions [20].

Drop testing is performed on electronics, medical devices, packaging, and consumer goods. The test involves dropping the product from a specified height onto a hard, flat surface and observing its behavior during and after the impact. The test results are a basis for evaluating the product's structural integrity, functional performance, and overall reliability.

The height (e.g., 1 meter), orientation (product axis), and the number of drops for a drop test are specified based on the product's intended use and the requirements of the relevant industry standards (see Figure D.18). Drop testing can be performed using specialized equipment and can be done in a laboratory or in real-world conditions.

FIGURE D.18 Illustrates the three orientations for product drop.

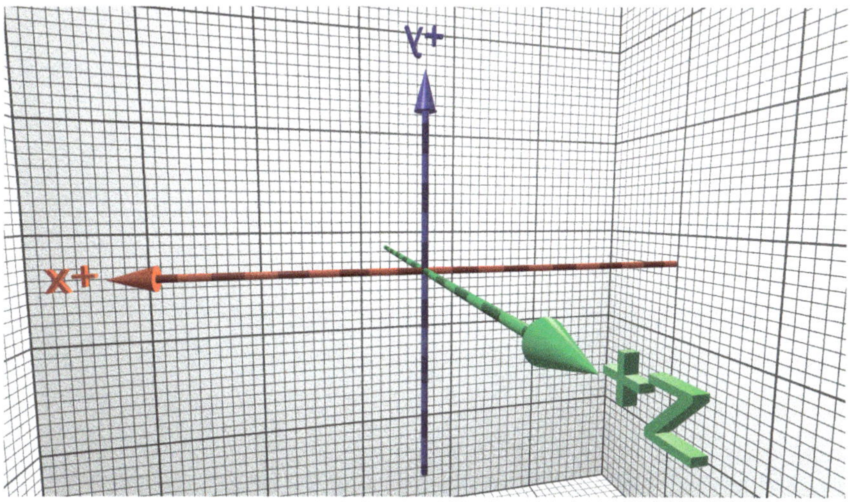

sbellott/Shutterstock.com.

Drop testing is an important aspect of product design and development, as it helps to identify potential weaknesses and improve the overall quality and reliability of the product. Drop testing provides valuable information about the product's performance and durability by simulating real-world conditions and subjecting the product to controlled impact [20].

Dunning-Kruger Effect

The Dunning-Kruger effect refers to the phenomenon whereby people with low ability in a particular area tend to overestimate their competence and skills. This bias can lead to a lack of awareness of one's own limitations and overconfidence in one's abilities.

In the context of product testing, the Dunning-Kruger effect can manifest in testers who are not skilled or experienced enough in a particular area but still believe they can accurately assess the product. These experience limits can lead to inaccurate or incomplete testing and may result in missed bugs or issues impacting the product's overall quality.

To mitigate the Dunning-Kruger effect in product testing, ensure that testers have the necessary skills, knowledge, and experience to assess the product accurately. Mitigation may involve training or hiring experienced testers familiar with the particular domain or technology tested.

Additionally, having a diverse team of testers with different backgrounds and perspectives may help to ensure a comprehensive and well-rounded testing process. Diversity and reviews can help identify potential issues that may be missed by testers who are overconfident in their abilities.

It is essential to recognize the potential impact of the Dunning-Kruger effect in product testing and take steps to mitigate it to ensure that the product is thoroughly tested and meets the desired quality standards.

DVP&R (Design Verification Plan and Report)

The DVP& outlines the processes and methods that will be used to verify that a product or system meets its design specifications and requirements. The DVP&R is part of the automotive product development effort and is used to ensure that the design has been thoroughly tested and is ready for production. The DVP&R includes a list of tests to perform, the number of units, pass criteria for the product, and the results of the tests. The DVP&R is a critical quality assurance tool that helps ensure a product's or system's reliability and performance.

Dynamic Analysis Tool

Dynamic software analysis tools are used to evaluate the behavior and performance of software during runtime. They are commonly used to identify issues or vulnerabilities in software systems, such as bugs, security vulnerabilities, or performance bottlenecks.

Dynamic software analysis tools analyze the software as it is executed and collect data about its behavior and performance. This data identifies any issues or vulnerabilities in the software and provides recommendations for improving the software's performance or security.

There are several types of dynamic software analysis tools:

1. Code Profilers: These tools measure code performance and identify bottlenecks or issues.
2. Debuggers: These tools identify and fix bugs in software.
3. Security Scanners: These tools identify security vulnerabilities in software.
4. Load Testers: These tools simulate high levels of traffic or usage to test the performance of software under load.

Dynamic software analysis tools help identify issues and improve the performance and security of software systems. Software developers and quality assurance teams commonly use them to ensure that software is of high quality and that it meets required specifications.

Dynamic Testing

Dynamic testing involves evaluating the behavior and performance of a system or product during runtime. It is used to identify issues or vulnerabilities in the system or product, such as bugs, security vulnerabilities, or performance bottlenecks.

Dynamic testing involves executing the system or product and collecting data about its behavior and performance. This data is then used to identify any issues or vulnerabilities in the system or product, and to provide recommendations for improving its performance or security.

There are several types of dynamic testing techniques:

1. Load Testing: This involves simulating high levels of traffic or usage to test the performance of the system or product under load.
2. Stress Testing: This involves examining the system or product under extreme conditions to identify its limits and capacity.
3. Performance Testing: This involves evaluating the performance of the system or product under various conditions to identify any bottlenecks or issues.
4. Security Testing: This involves evaluating the security of the system or product to identify any vulnerabilities or weaknesses.

Dynamic testing is essential in developing and maintaining any system or product, as it helps identify and address issues or vulnerabilities before they become significant problems. In addition, software developers and quality assurance teams commonly use it to ensure that systems and products are high quality and meet required specifications.

Dynamometer Testing

Dynamometer testing is a standardized method for evaluating the performance of electric and hybrid vehicle powertrains. The Society of Automotive Engineers (SAE) has outlined the test procedures and requirements for conducting dynamometer testing, including the following:

1. High-Speed Performance Testing: This test evaluates the power and torque output of the vehicle at high speeds.
2. Acceleration Testing: This test evaluates the vehicle's acceleration performance, including 0–60 mph (0–97 km/h) and quarter-mile times.
3. Regenerative Braking Testing: This test evaluates the regenerative braking performance of the vehicle, including the amount of energy recovered during braking.
4. Range Testing: This test evaluates the vehicle's range under various driving conditions.
5. Thermal Performance Testing: This test evaluates the thermal performance of the vehicle's battery and other components under different operating conditions.

According to the SAE, dynamometer testing is a crucial tool for evaluating the performance and efficiency of internal combustion and electric and hybrid vehicles, as well as for benchmarking different models against each other. In addition, it can help automakers and researchers identify improvement areas and optimize the design and performance of electric and hybrid vehicles. For more information, see **J3152**.

E

"If an organization is to work effectively, the communication should be through the most effective channel regardless of the organization chart."

—David Packard, founder, Hewlett-Packard

Edges

An edge case is a particular kind of software bug that won't be a common problem in software development. Instead, it impacts only a few users and devices or can only be replicated in specific situations. However, that does not automatically imply that the bug is challenging to reproduce. See Corner Case.

Efficiency Testing

Efficiency testing is used to evaluate the performance and efficiency of a system or product. It helps to identify any bottlenecks or issues that may be impacting the performance of the system or product and to provide recommendations for improving its efficiency.

Efficiency testing involves collecting data about the performance and resource utilization of the system or product under various conditions. This data is used to evaluate the system or product's efficiency and identify any areas for improvement.

There are several types of efficiency testing techniques:

1. Load Testing: This involves simulating high levels of traffic or usage to test the performance of the system or product under load.
2. Stress Testing: This involves examining the system or product under extreme conditions to identify its limits and capacity.

3. Performance Testing: This involves evaluating the performance of the system or product under various conditions to identify any bottlenecks or issues.
4. Resource Utilization Testing: This involves evaluating the use of resources, such as CPU, memory, and storage, to identify any areas of inefficiency.

Efficiency testing is essential in developing and maintaining any system or product, as it helps to address any issues or bottlenecks impacting its performance. In addition, software developers and quality assurance teams commonly use it to ensure that systems and products are efficient and meet required specifications.

Electromagnetic Compatibility (EMC)

EMC testing evaluates how well an electronic device or system can function in its intended environment without causing electromagnetic interference (EMI) or being susceptible to EMI from other sources.

EMC testing is typically required for electronic products to comply with regulatory standards and requirements, such as those set by the Federal Communications Commission (FCC) in the United States or the European Union's EMC Directive. The testing may involve measuring the device's emissions of electromagnetic radiation and its susceptibility to external radiation sources. There are four methods:

1. Capacitive coupling clamp (CCC) method
2. Direct capacitive coupling (DCC) method
3. Inductive coupling clamp (ICC) method
4. Capacitive/inductive coupling (CIC) method

There are various types of EMC testing, including radiated emissions testing, emissions testing, radiated immunity testing, and conducted immunity testing. Radiated emissions testing measures the amount of electromagnetic radiation that a device emits. In contrast, emissions testing measures radiation through the device's power or signal cables. Radiated immunity testing measures how well a device can operate in the presence of external electromagnetic radiation, and conducted immunity testing measures how well a device can work in the presence of conducted interference.

EMC testing is essential for ensuring that electronic products can function properly in their intended environments without causing or being affected by EMI, which can cause problems such as signal interference, data loss, or device malfunction. For more information, see **J1113**, **J1812**, and **J551** [67, 68, 69, 70, 71].

Electromagnetic Immunity

Electromagnetic immunity refers to the ability of a device or system to withstand electromagnetic interference (EMI) and continue functioning properly. In the automotive industry, electromagnetic immunity is of utmost importance due to various electronic systems and components in vehicles that are susceptible to interference from external electromagnetic sources.

FIGURE E.1 Example of a parallel plate field generator setup from SAE.

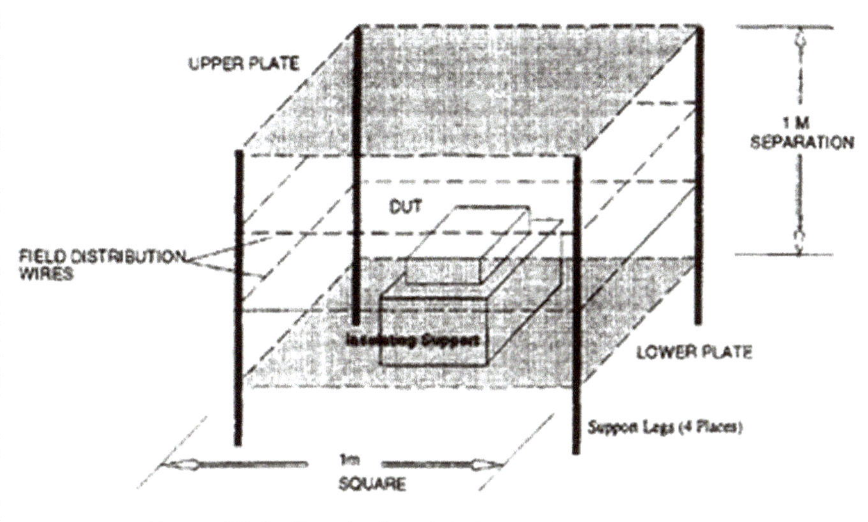

Automotive standards play a crucial role in ensuring electromagnetic immunity in vehicles. These standards define the requirements and test methods for electromagnetic compatibility (EMC) in automotive systems. Here are some notable automotive standards related to electromagnetic immunity:

1. SAE specifies automotive test methods and levels of the electrical stimulus to which the component or product is to be subjected and must withstand (see Figure E.1). For more information, see **J113**.
2. ISO 7637 specifies test methods and levels for electrical disturbances from conduction and coupling in automotive vehicles. It covers various transient and voltage disturbances that can occur in automotive electrical systems, including those caused by load dump, alternator field decay, and electrostatic discharge.

3. ISO 11452 is a series of standards that outlines test methods and procedures for measuring the electromagnetic immunity of electronic components and vehicle systems. It includes radiated emissions, conducted disturbances, and immunity to electrical transients.
4. CISPR 25 was developed by the International Special Committee on Radio Interference (CISPR) to set the limits and measurement methods for radiated and conducted electromagnetic emissions from vehicles and their components. It ensures that automotive electronics do not emit excessive EMI that can affect other electronic systems.
5. AEC-Q100, though not explicitly focused on electromagnetic immunity, this Automotive Electronics Council (AEC) standard sets the reliability qualification requirements for automotive integrated circuits (ICs). It includes stress tests that help assess the robustness of ICs against environmental factors, including EMI.
6. Various automotive manufacturers also have specific standards and requirements for electromagnetic immunity in their vehicles. These OEM-specific standards may incorporate elements from international standards and further refine them based on their particular needs and technologies.

Electrostatic Discharge (ESD)

Electrostatic discharge (ESD) occurs when two objects with different electrostatic potentials come into contact or proximity. The resulting discharge can damage electronic components, such as integrated circuits, transistors, and printed circuit boards. ESD testing determines how much electrostatic charge a product can withstand before being damaged.

ESD testing is performed on electronic products during the product and manufacturing development process and after the product is assembled to ensure that it can withstand the electrostatic discharges it may encounter during use. The testing may involve exposing the product to an electrostatic discharge, which is simulated using an ESD generator, and measuring the amount of charge on the product before and after the discharge (see Figure E.2).

FIGURE E.2 Test setup for powered mode component test [72].

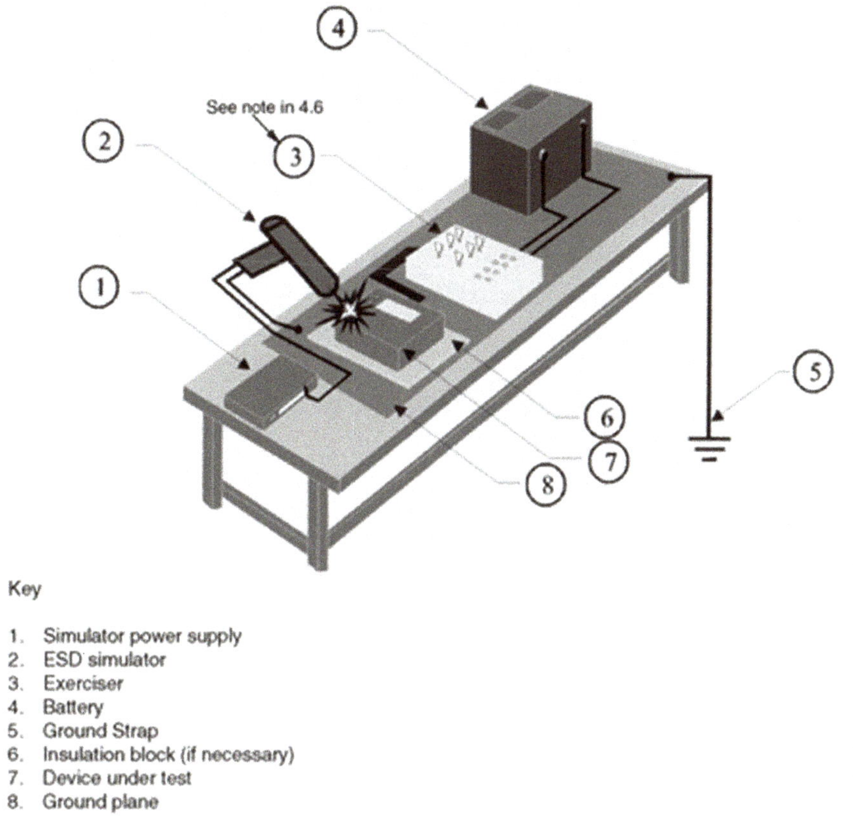

There are various industry standards and test methods for ESD testing, including the International Electrotechnical Commission (IEC) 61000-4-2 standard, which outlines test procedures and requirements for ESD testing. ESD testing is important for ensuring the reliability and longevity of electronic products and reducing the risk of premature failure due to electrostatic discharges. For more information, see **J551/15** and **J1113/13** [73, 74, 75].

Elementary Comparison Testing (ECT)

Software development uses the white-box, control-flow, test design process known as elementary comparison testing (ECT). The goal of ECT is to make it possible to test complicated and crucial software thoroughly. Testing is done on software code or pseudocode to determine how well it handles all decision-related outcomes. Modified condition/decision coverage (MC/DC) covers all independent and isolated conditions like multiple-condition coverage and basis path testing. Formal test cases are created by connecting isolated circumstances into connected scenarios. A condition's independence can be demonstrated by altering the condition value alone. Test cases include coverage for all pertinent condition values.

ECT is a statistical method used to compare two sets of data to determine if they are statistically different or if the observed differences between them are due to random chance. This method is also known as a two-sample t-test.

The basic idea behind ECT is to calculate the difference between the means of the two datasets and then compare that difference to the variability within each dataset. If the observed difference between the means is more extensive than expected due to random chance, we can conclude that the two datasets are statistically different.

To perform an elementary comparison test, we need to follow a few steps:

1. Define the Null Hypothesis: The null hypothesis states no significant difference between the two datasets.
2. Collect Data: We need to collect data from both sets we want to compare.
3. Calculate the Means and Standard Deviations: We need to calculate each dataset's mean and standard deviation.
4. Calculate the Test Statistic: We need to calculate the test statistic, which is the difference between the means divided by the standard error.
5. Determine the Critical Value: We need to determine the critical value, which depends on the level of significance and the degrees of freedom.
6. Make a Decision: If the test statistic exceeds the critical value, we reject the null hypothesis and conclude that the two datasets are statistically different. Conversely, suppose the test statistic is less than the critical value. In that case, we fail to reject the null hypothesis and complete that there is no significant difference between the two datasets.

$$t = (x1 - x2) / \left(s^{2\,pooled} \times \left(\frac{1}{n1} + \frac{1}{n2} \right) \right)^{0.5}$$

Where:
t is the test statistic.
$x1$ and $x2$ are the sample means of the two sets of data being compared.
$s^{2pooled}$ is the pooled estimate of the variance, calculated as $\frac{\left[(n1-1)s1^2 + (n2-1)s2^2\right]}{n1+n2-2}$.
$n1$ and $n2$ are the sample sizes of the two sets of data compared.
s1 and s2 are the sample standard deviations of the two sets of data compared.

Once the test statistic is calculated, we can compare it to a critical value from a t-distribution with (n1 + n2 – 2) degrees of freedom to determine if the observed difference between the two datasets is statistically significant. Elementary comparison testing is a valuable tool for comparing two datasets and contrasting if there is a substantial difference between them.

Elephant Foot Testing

Elephant foot testing is a method of evaluating the stability and durability of a product by simulating the weight and pressure of an elephant standing on it. This type of testing is often used in the construction and engineering industries to ensure that structures, such as bridges and buildings, can withstand heavy loads and remain stable under extreme conditions. The test involves placing a simulated elephant foot (usually made of concrete or steel) on the product and measuring its response to the applied pressure and weight. As a result, elephant foot testing ensures the safety and reliability of structures and other products designed to withstand heavy loads.

Embedded Software

Embedded software is designed to run on a specific device or system, such as a vehicle, aircraft, or industrial control system. It is typically used to control the functions and features of the device or system and is often a critical component of its operation.

Embedded software is written in a low-level programming language, such as C or C++. It is designed to be highly optimized for the device or system's specific hardware and operating environment. It may also be designed to operate in real time, meaning it must promptly respond to inputs and events.

Embedded software is often used in aerospace, defense, automotive, and manufacturing industries, where complex systems must be highly reliable and precise. It is also commonly used in consumer electronics, such as smartphones and smart home devices.

The development and testing of embedded software are typically more complex than traditional software development due to the device or system's specific hardware and operating environment. It often requires specialized tools and techniques to ensure that the software is reliable and efficient and meets required specifications.

Embedded Systems

Embedded systems in the automotive industry are computer systems designed to control specific functions or features of a vehicle. They are typically integrated into the vehicle's electrical and mechanical systems and are often a critical component of its operation (see Figure E.3).

FIGURE E.3 An example of an embedded product printed circuit board.

Mentor57/Shutterstock.com.

Examples of embedded systems in the automotive industry include the following:

1. Engine Control Systems: These systems control the engine's fuel injection, ignition, and emission control systems.
2. Brake Control Systems: These systems control the vehicle's brake system, including the brake pedal, brake fluid, and brake pads.
3. Transmission Control Systems: These systems control the vehicle's transmission, including the gears, clutches, and torque converter.
4. Navigation Systems: These systems provide navigation and route guidance to the driver.
5. Infotainment Systems: These systems provide entertainment and information to the driver and passengers, such as music, movies, and traffic updates.

Embedded systems in the automotive industry are designed to operate with high reliability and precision and are typically subjected to rigorous testing to ensure that they meet required specifications. As a result, they are an essential component of modern vehicles and are constantly evolving to meet the increasing demand for advanced features and technologies.

Emulation

Emulation is the process of simulating the behavior and functionality of one system or device using another system or device. It is commonly used in software development and testing to evaluate the performance and compatibility of software on different platforms or environments.

Emulation involves creating a virtual environment that replicates the behavior and functionality of the target system or device. This virtual environment executes the software and evaluates its performance and compatibility.

There are several types of emulation techniques:

1. Hardware Emulation: This involves simulating the hardware of a system or device using software or another hardware device.
2. Software Emulation: This involves simulating the software of a system or device using another software application.
3. Operating System Emulation: This involves simulating the operating system of a system or device using another operating system or software application.

Emulation is helpful for testing and evaluating software on different platforms or environments, as it allows for the simulation of other conditions and scenarios without the need for physical hardware. Software developers and quality assurance teams commonly use it to ensure that software is compatible with and performs as intended on different platforms and devices.

Emulator

Also called an in-circuit emulator (ICE), an emulator is a device used to debug and test electronic circuits and systems (see Figure E.4). It is used in developing and testing microcontroller-based systems, such as those found in automotive, industrial, and aerospace applications.

An ICE works by connecting to the target circuit or system and simulating the operation of the microcontroller. The developer can test and debug the circuit or system in real time and evaluate the microcontroller's performance and behavior.

These devices are typically used with a software development environment, such as a debugger or simulator, to provide a more comprehensive analysis of the circuit or system. They are an essential tool for developers working on microcontroller-based systems, as they allow for the testing and debugging of circuits and systems in a controlled and cost-effective manner.

FIGURE E.4 An example of an in-circuit emulator (ICE).

End-of-Line Test

The purpose of end-of-line testing is to ensure that a product is functioning correctly and meets all required specifications and standards (see Figure E.5). End-of-line testing encompasses a variety of tests, including functional and environmental testing to verify that the product operates as expected. This test ensures that the product can withstand various conditions (e.g., temperature, humidity, vibration), and safety testing ensures that the product does not pose any hazards to users.

FIGURE E.5 End-of-line testing samples points on the board to confirm the build.

The specific tests performed during end-of-line testing will depend on the type of product manufactured and the industry standards and regulations that apply. However, the results of end-of-line testing identify issues to correct the product, which helps to ensure customer satisfaction and minimize the need for costly returns and repairs.

End-to-End Testing

End-to-end testing examines an entire software application from start to finish to ensure that all components work together as expected and that the application meets all requirements. In addition, this type of testing helps identify defects and issues that can arise when different application parts that can affect the entire system's functionality are integrated.

End-to-end testing aims to validate the application's behavior and performance in a realistic environment that simulates the user experience. This type of testing involves testing the application as a whole, including all of its components, such as the user interface, database, network communication, and all other integrated systems.

End-to-end testing typically involves the following steps:

1. Define Test Scenarios: Define the end-to-end test scenarios to be executed, including the various inputs, expected outputs, and the steps required to run the tests.
2. Prepare Test Data: Prepare the necessary test data to simulate different user scenarios and test cases.
3. Execute Tests: Execute the defined test scenarios, including the integration of different components of the application.
4. Verify Results: Verify the results of the executed tests and identify any defects or issues that require correction.
5. Report Defects: Report any defects or issues found during the testing process to the development team for fixing.
6. Retest: Once the defects are fixed, retest the application to ensure that the fixes are effective and have not caused any new issues.

End-to-end testing is an essential part of the software development process, as it helps to ensure that the application is working as intended and meets the requirements of end users. End-to-end testing is performed toward the end of the development cycle once all the system's components have been tested and integrated. An automotive example is telemetry or telematics systems that perform a vehicle's functions and connect to external devices such as back-office data collection and presentation.

Endurance Testing

Endurance testing in the automotive industry evaluates a vehicle's or its components' durability and reliability over an extended period. It identifies any issues or weaknesses that may arise during the vehicle's life and ensures that it meets all required specifications and standards.

Endurance testing in the automotive industry typically involves subjecting the vehicle or its components to tests and simulations designed to replicate actual conditions and stresses that the vehicle is likely to experience over its lifetime. These tests include the following:

1. Road Load Simulations: These simulations replicate the loads and stress the vehicle will likely experience during normal driving conditions.
2. Climate Testing: This involves exposing the vehicle or its components to extreme temperature and humidity conditions to evaluate their performance and durability.
3. Vibration Testing: This involves exposing the vehicle or its components to vibration and shock loads to evaluate their performance and durability.
4. Corrosion Testing: This involves exposing the vehicle or its components to corrosive environments to evaluate their performance and durability.

Endurance testing is an essential step in developing and maintaining any vehicle, as it helps ensure that the vehicle is reliable and meets required specifications and standards. In addition, automotive manufacturers and suppliers commonly use it to ensure that their products are of high quality and meet the customer expectations.

Environmental Simulation Tools

Environmental simulation tools are software programs that allow engineers and scientists to simulate and analyze the effects of various environmental factors on proposed products and systems, thus creating a form of virtual testing of the product. These tools are used in multiple industries, including aerospace, automotive, electronics, and consumer goods, to test and optimize the performance of products and systems under different environmental conditions.

Examples of environmental simulation tools include the following:

1. Finite Element Analysis (FEA) Software: This software is used to simulate and analyze the structural behavior of products and systems under different environmental loads, such as temperature, vibration, and shock. FEA can help engineers optimize designs and materials to withstand extreme conditions and reduce the risk of failure.

2. Computational Fluid Dynamics (CFD) Software: This software is used to simulate and analyze the fluid flow and heat transfer of products and systems under different environmental conditions, such as airflow, pressure, and temperature. CFD can help engineers optimize the design of cooling systems, HVAC systems, and other fluid-based components.
3. Environmental Test Chambers: These test chambers are physical systems used to simulate and test the effects of various environmental factors on products and systems, such as temperature, humidity, and altitude. Environmental test chambers can be used to validate the results of simulation tools and to test the performance of products and systems under extreme conditions.
4. Electromagnetic Simulation Software: This software is used to simulate and analyze the behavior of electromagnetic fields in products and systems under different environmental conditions, such as radiation, interference, and electromagnetic compatibility (EMC). This software can help engineers optimize the design of electronic components, antennas, and other electromagnetic devices.

Environmental simulation tools are essential to product development and testing, as they allow engineers and scientists to predict and optimize the behavior of products and systems in real-world environments. These tools can help reduce costs, improve performance, and ensure the safety and reliability of products and systems.

Environmental Testing

Environmental testing evaluates how a product or system will perform and endure in its intended operating environment. This type of testing simulates the environmental conditions a product or system may encounter in real-world use, including factors such as temperature, humidity, pressure, vibration, mechanical shock, UV exposure, and other physical and mechanical stressors.

Environmental testing is typically used in aerospace, automotive, defense, electronics, and medical devices to ensure that products and systems perform as expected and meet established performance and reliability standards. In addition, environmental testing aims to identify potential problems and ensure that products and systems are robust enough to withstand the challenges of their operating environment.

Environmental testing includes temperature, humidity, vibration, shock, and altitude testing (see Figure E.6). These tests use specialized equipment in a controlled laboratory environment or real-world conditions using prototypes or final products. For more information, see **J1455** [20].

FIGURE E.6 A thermal chamber exposes the product to extended temperature exposure.

Audrius Merfeldas/Shutterstock.com.

Equivalence Partition Coverage

Equivalence partition coverage is a testing technique in which test cases are created based on equivalence partitions or groups of inputs expected to behave similarly. This technique aims to generate test cases that cover a wide range of possible inputs while minimizing the number of test cases needed. This reduction is achieved by creating test cases that cover each partition rather than making a separate test case for every possible input. Through this technique, testers can ensure that their tests are thorough and efficient and that the software tested is robust and reliable.

$$Partition_{coverage} = \frac{Partitions_{tested}}{Partitions_{total}}$$

Equivalence Partitioning

Equivalence partitioning is a testing technique that involves dividing input data into distinct groups, called equivalence classes, to test the functionality of a system or application. Each equivalence class represents an input dataset expected to produce the same output or result. This technique helps identify and test the most critical cases while reducing the number of test cases needed. In addition, by focusing on the equivalence classes, testers can ensure that the system works correctly for all possible inputs and is free of defects.

Error Guessing

Error guessing is a technique in which a tester attempts to anticipate and identify potential errors or defects in a system or software using their knowledge and experience. This method involves the tester making educated guesses about the likely locations or causes of errors and testing those areas to see if the guess was correct. Error guessing is used with other testing techniques, such as exploratory or black-box testing, to identify as many defects as possible before the software is released.

Error Seeding

Error seeding is a problem that occurs when there is an issue with the initial data or starting point used to initiate a process or system. This can lead to problems with the accuracy and reliability of the process or system, as the initial errors can propagate and cause issues throughout the system. To fix this issue, it is important to identify and correct the source of the error and then reseed the system with accurate and reliable data.

Executable Statement

An executable statement is a line of code in a programming language that performs a specific action or task. It is a statement that the computer can execute to produce a result. Examples of executable statements in Python include a variable assignment, function call, conditional statement, or loop. For instance, consider the Python code in Figure E.7.

FIGURE E.7 An example of executable statements in software.

The first line assigns the value 5 to the variable x in this code. The second line is a conditional statement that checks if x is greater than 3. If the condition is true, the third line executes the print statement, which outputs the "x is greater than 3" string to the console. These are all examples of executable statements because they perform actions that the computer can execute.

Exercised

In programming, exercising or testing refers to running code to ensure it works as intended and meets its requirements. There are different types of testing, including unit testing, integration testing, and acceptance testing, which involve testing various aspects of the code, such as its functions, logic, and user interface.

Unit testing involves testing individual units or components of the code to ensure that they work correctly and meet their specifications. Developers usually perform it as part of the development process, and it can be automated using testing frameworks such as PyTest or JUnit.

Integration testing involves testing how different code components work together and can be used to identify and resolve issues arising when the components interact. Testers typically perform this type of testing, which can be manual or automated.

Acceptance testing involves testing the code against user requirements to ensure that the code meets the needs of its intended users. This type of testing can involve both manual and automated testing performed by users or user representatives. Exercising and testing are critical aspects of software and hardware work products from the development process. They help ensure that products work as intended, are reliable, and meet specified requirements.

Exhaustive Testing

Exhaustive testing is an approach in which every possible input combination and scenario is tested to ensure that software performs as intended under all possible conditions. In other words, it involves trying every possible variety of inputs and circumstances the software may encounter during its operation.

However, exhaustive testing is often impractical or not feasible, especially for complex systems, because the number of possible test cases can be large or even infinite. For example, consider a software application that takes user input for a text field. If the text field can contain any character, the number of possible input combinations can be infinite.

Instead, testers prioritize what is tested and use techniques such as equivalence partitioning, boundary value analysis, and random testing to select a representative set of test cases that provide high coverage and identify potential issues.

These techniques aim to identify and test the most critical or likely failure scenarios rather than exhaustively testing all possible scenarios.

Exit Criteria

Exit criteria refer to the conditions that must be met before a project or task can be considered complete and ready for final review or approval. Some common exit criteria include the following:

1. All project deliverables are complete and reviewed by the relevant stakeholders.
2. All identified project risks have been considered and, where appropriate, mitigated.
3. The project has met all performance and quality requirements.
4. User acceptance testing was completed successfully.
5. All necessary documentation was completed and approved.
6. All project resources have been released or reassigned.
7. All project debts are paid or settled.
8. The project team has completed a thorough post-project review and identified any lessons learned.

Testing uses exit criteria to plan when to cease testing and to report against.

Expectation Bias

Expectation bias occurs when an individual's expectations or beliefs influence their perceptions, judgments, and actions. In the testing context, expectation bias can lead to testers overlooking defects or assuming that certain aspects of the software work as intended without thorough testing.

For example, suppose a tester strongly believes a specific software feature works correctly; they may unconsciously overlook issues or defects in that feature during testing. Similarly, imagine a tester expects certain errors or failures to occur in a specific software part. In that case, they may focus their testing efforts on that area, neglecting other potential issues.

Mitigating the effects of expectation bias in testing can be accomplished with a variety of techniques, such as exploratory testing, which encourages testers to approach testing with an open mind and without preconceived notions of how the software should behave. Testers can also collaborate with other team members to share perspectives and identify potential biases.

In addition, using automated testing tools, such as unit tests and integration tests, can help reduce the effects of expectation bias by providing a consistent and objective way to test software functionality. Awareness of expectation bias and appropriate testing techniques and tools can help testers identify defects and ensure that software functions as intended.

Expected Result

Expected results are the anticipated or desired outcomes of a particular action or function in software. They are typically defined in the software requirements or specifications and represent the behavior the software is intended to exhibit under specific conditions or inputs.

During testing, expected results are a benchmark against which the actual results of a test are compared. Testers will execute a test case, typically providing inputs or performing actions within the software and verifying that the resulting output or behavior matches the expected results. If the actual results match the expected results, the software is considered to have passed the test case. On the other hand, suppose the actual results do not check the desired results. In that case, the software has failed the test case, and the tester will typically investigate the cause of the failure and report it to the development team for resolution.

Expected results are essential in software development because they provide a clear target for developers and testers to aim for. As a result, they help ensure the software meets all requirements and specifications and functions correctly. In addition, expected results can also be used to identify potential defects or issues in the software, as discrepancies between the desired results and the actual results may indicate a problem that needs addressing. Expected results are a crucial component of software testing and development.

Experienced-Based Testing

Experienced-based testing relies on the testers' knowledge, expertise, and intuition to identify potential defects and issues in the software. This approach is based on the assumption that experienced testers have accumulated knowledge and skills over time that enable them to identify problems that may not be apparent to less experienced testers or automated testing tools.

Experienced-based testing can take many forms: exploratory, scenario-based, and error guessing. Exploratory testing involves testers trying out different scenarios and inputs and using their intuition and experience to identify software issues. Scenario-based testing consists in creating test scenarios based on real-world scenarios or uses to identify potential defects. Finally, error guessing involves testers deliberately trying to break the software by thinking about where possible errors or issues may occur.

While experienced-based testing can effectively identify defects, it is subjective and can be influenced by individual biases and assumptions. Therefore, experienced-based testing is often supplemented with other approaches, such as automated testing and peer reviews, to provide a more comprehensive testing strategy.

Experienced-based testing can complement other testing methods to provide a more effective approach. It is beneficial when software is complex or has limited documentation or product specifications and available time to perform testing.

Exploratory Testing

Exploratory testing is an approach that emphasizes creativity, learning, and adaptability. It involves testers trying different scenarios and inputs and using their intuition and experience to identify software issues. Rather than following a predefined set of test cases, exploratory testing is more fluid and allows testers to adapt their testing approach based on what they discover during testing.

Exploratory testing is beneficial when testing complex or poorly documented software, where traditional testing approaches may not be as practical. In addition, exploratory testing allows testers to use their knowledge and experience to identify defects not considered in a predefined test plan.

Testers may use ad hoc, session-based, and heuristics-based testing techniques during exploratory testing. Ad hoc testing involves testers exploring the software and trying different inputs and scenarios without a specific test plan. Session-based testing involves organizing exploratory testing into defined periods or sessions, where testers work collaboratively to explore the software and identify issues. Finally, heuristics-based testing involves using heuristics, or rules of thumb, to guide testing and identify potential problems.

Exploratory testing is a practical approach to testing that emphasizes flexibility, creativity, and adaptability. It can complement other testing approaches and help identify defects and issues that more traditional testing methods may miss.

Extensibility

Extensibility refers to the ability of software to be easily extended or modified to meet changing requirements or to integrate with other software systems. Extensibility is a crucial consideration in software development, as it can improve the usability and longevity of software systems.

Testing plays a crucial role in ensuring the extensibility of software systems. As new features are added, or modifications are made, it is essential to ensure that the software continues functioning as intended and that new features do not introduce defects or impact existing functionality.

Testers may use regression, integration, and performance testing techniques to test extensibility. Regression testing involves retesting existing functionality to ensure new changes do not introduce defects or impact existing functionality.

Integration testing involves reviewing how the software integrates with other systems or components. Finally, performance testing involves evaluating how the software performs under different loads or conditions to ensure it can handle increased usage or user demands.

In addition, testers may use exploratory testing to identify potential areas where extensibility may be impacted and to ensure that the software can handle new features or modifications. Testers may also collaborate with developers to identify areas where software architecture or design changes may be necessary to improve extensibility.

Testing plays a crucial role in ensuring the extensibility of software systems. Testers can identify potential issues through appropriate testing techniques and ensure that the software can be easily extended or modified to meet changing requirements or integrate with other systems.

F

"My great concern is not whether you have failed, but whether you are content with your failure."

—Abraham Lincoln

F-Number

In software testing, F-number (also known as F-score, F1 score) measures the accuracy of a binary classification model. It is used to evaluate the performance of a machine learning algorithm in predicting positive and negative cases. It is a metric for evaluating the performance of machine learning models in technical testing. Technical testing examines the functionality and performance of software or hardware systems, and machine learning models are often used to automate or assist in technical testing tasks.

In testing, the F-number score is used to assess the performance of machine learning models for tasks such as defect prediction, fault detection, and anomaly detection. For example, in defect prediction, a machine learning model is trained to predict whether a code change will likely introduce defects into the system. The F-number can be used to evaluate the model's performance by comparing its predictions to the actual defects found in the system.

The F-number is calculated based on two primary measures: precision and recall. Precision refers to the proportion of true positive cases out of all optimistic predictions, while recall refers to the proportion of true positive cases out of all actual positive cases.

The F-number is calculated by taking the harmonic mean of precision and recall and provides a single measure of the algorithm's performance that balances precision and recall.

F-number is useful in software testing for evaluating the performance of software components or systems that involve binary classification, such as spam filters or fraud detection systems. It can also help identify areas where the system may be producing false positives or negatives and guide improvements to its accuracy and performance.

F-number is a helpful metric in software testing for evaluating the accuracy and performance of binary classification systems and can help guide improvements to the system's design and implementation.

Facilitator

In software development, a facilitator is essential in expediting code or document reviews. Code or document reviews are a process in which a team of developers, testers, or subject matter experts review and evaluate a software component or documentation to identify defects, potential issues, and areas for improvement.

The role of the facilitator in a code or document review is to help guide the process, ensure that all relevant team members are involved, and provide guidance and support to the team throughout the review process. This includes developing review plans and checklists, facilitating discussions and feedback sessions, and helping to ensure that all review comments and feedback are appropriately documented and addressed.

Facilitators can help ensure that the review process is organized and efficient and that all team members are actively engaged and contributing to the review process. They can also help identify and address any issues or concerns during the review and work with the team to develop strategies for improving the quality and accuracy of the software component or documentation reviewed [12].

Failure

In the testing context, a failure refers to a deviation from a component or system's expected behavior or resulting performance. This can occur when the product does not behave as intended, exhibits unexpected behavior, or produces incorrect output.

Failures can occur for various reasons, including defects in the software code, hardware, prototype or manufactured part quality; incorrect input or configuration settings; issues with the hardware or network environment; or errors in the software or hardware design or requirements.

When a failure occurs during testing, it is typically logged and reported to the development team, who will investigate the cause of the failure and work to resolve it. This root cause may involve debugging the software code, identifying and correcting errors in the design or requirements, or making changes to the configuration or environment.

Identifying and resolving failures is an integral part of the software development and testing process, as it helps ensure that the software is of high quality and functions correctly. By identifying and correcting failures, developers can improve the reliability and usability of the software and minimize the risk of defects or issues impacting users.

Failures are an essential part of product development and testing, as they help identify areas where products can be improved and help ensure that products are of high quality and function correctly.

Failure Mechanism

A failure mechanism refers to the specific process or sequence of events that lead to the failure or breakdown of a system or component. It can be physical, such as wear and tear or corrosion, or related to design or manufacturing defects. Understanding the failure mechanism is vital for identifying the root cause of a failure and preventing future failures.

Failure Mode

In engineering and quality assurance, a failure mode is when a system, component, or process fails to meet its intended design or performance specifications. Various factors, such as design flaws, material defects, manufacturing errors, or external factors, such as environmental conditions or user error, can cause failure modes.

Understanding failure modes is an integral part of quality assurance and risk management. It helps identify potential points of failure in a system or component and develop strategies for mitigating these risks. Failure modes can be analyzed using techniques such as failure modes and effects analysis (FMEA), which involves systematically identifying and analyzing potential failure modes and their associated risks and developing strategies for preventing or mitigating them.

In product (software or hardware) engineering and testing, failure modes can refer to how software components or systems may fail to meet their intended design or performance specifications. This includes errors or bugs in the software code, issues with compatibility or integration with other systems, or problems with usability or user experience.

Understanding failure modes in software is an integral part of software testing and quality assurance, as it helps identify potential issues and risks that may impact the software's performance, usability, and reliability. By identifying and addressing these failure modes, software developers and testers can help ensure that software systems and components meet their intended design and performance specifications and deliver value to end users [76].

Failure Mode Effects Analysis

Failure mode and effects analysis (FMEA) is a systematic method used to identify and evaluate potential failure modes and their associated risks in a system, process, or product. FMEA is used in various manufacturing, engineering, and software development industries. The failure mode effects analysis comes in a few incarnations:

1. Design FMEA: Products
2. Process FEMA: Manufacturing processes
3. Systems FEMA: Systems design and configurations
4. Machinery FEMA: Tooling and equipment [77]

See DFMEA and PFMEA. For more information, see **J1739** [76, 78].

Fatigue Failure

Fatigue failure is a type of mechanical failure that occurs when a material is subjected to repeated loading and unloading cycles over time. As a result, the material becomes weaker and more prone to failure as cycles increase. Factors that can contribute to fatigue failure include the intensity and duration of the loading, the frequency of the cycles, and the type of material used. It is a mode of failure in metals and other materials that are subjected to stress in everyday use, such as in aircraft, bridges, and automobiles. For more information, see **J2562**, **J2409** and **J2649** [79, 80, 81].

Fault Injection Testing

Fault injection testing is a technique used to evaluate a system's or application's resilience to various faults or errors. This type of testing aims to identify potential failure modes and weaknesses in the system and assess how the system responds to these faults.

There are several types of fault injection testing techniques:

1. Hardware Fault Injection: This technique involves introducing faults or errors into the hardware components of the system, such as the CPU, memory, or input/output devices.
2. Software Fault Injection: This technique involves introducing faults or errors into the software components of the system, such as the code, data, or configuration settings.
3. Environment Fault Injection: This technique involves introducing faults or errors into the system's environment, such as network or power failures, to evaluate the system's resilience to these events.
4. Timing Fault Injection: This technique involves introducing faults or errors related to timing, such as delayed responses or timing violations, to evaluate the system's performance and resilience under different timing scenarios.

Fault injection testing can be used to evaluate various types of systems, including hardware, software, and networked systems. It is beneficial for identifying failure modes and weaknesses in complex systems and assessing their resilience to unexpected events.

The fault injection testing results can inform system design decisions, improve system reliability, and develop contingency plans for potential failure scenarios. Additionally, regulations and standards often require fault injection testing to ensure the safety and reliability of critical systems and applications.

Fault Insertion

Fault insertion is a testing method that involves intentionally introducing faults or defects into a system to test its tolerance and resilience. It evaluates the system's ability to detect and respond to failures or malfunctions and identify potential weaknesses or vulnerabilities in its design or implementation.

Fault insertion can be performed in various ways, depending on the nature of the system tested and the type of faults introduced. For example, it may involve simulating failures or defects in software or hardware components or introducing physical failures or defects into the system.

Fault insertion is a valuable tool for identifying and addressing potential issues in a system before they occur. It can help to ensure that the system can handle failures or malfunctions in a controlled and predictable manner and can help improve its overall reliability and performance. However, it can be resource-intensive and require careful planning and coordination to ensure it is conducted safely and effectively.

Fault Report

A fault report is a document or message that describes a problem or malfunction in a system, equipment, or service. It informs the responsible parties of the issue and requests assistance in fixing it. Fault reports include details such as the location of the fault, a description of the problem, any symptoms or effects of the fault, and relevant information about the system or equipment involved. Fault reports may also include recommendations for corrective action and may be accompanied by supporting documentation or other relevant information [18, 82].

The fault report should contain the following information:

1. Date and time of the fault occurrence
2. Hardware and software version part numbers as well as specific parameter configuration items
3. Description of the fault, including any error messages or warning signs displayed
4. The location and environment when the fault occurred, including relevant details such as temperature, humidity, or equipment used
5. The steps taken to identify and troubleshoot the fault

6. Any actions taken to resolve the fault, including any repairs or maintenance performed
7. The results of the fault resolution include whether the fault has been fully resolved or is still present
8. Any recommendations or actions to prevent similar faults in the future
9. Contact information for the person submitting the report and any relevant technical support or maintenance personnel who assisted in resolving the fault

Through three processes, the maintenance concept of fault reporting lowers operating costs while increasing operational availability.

1. Reduce time-consuming diagnostic testing
2. Reduce downtime from diagnostic tests
3. Notify the management of a deteriorated operation.

Fault Seeding

Fault seeding is a method for assessing a testing procedure's effectiveness. Without alerting the testers, one or more flaws are intentionally inserted into a code base. Seeded flaws gauge how well the test procedure works during testing. For example, if ten faults are planted and the test group finds eight of those induced faults, this suggests that there are roughly 20% latent defects in the product.

Fault Severity

Fault severity refers to the impact or seriousness of a problem or error within a system or process. It can be measured as the consequences of the fault, such as loss of revenue, decreased efficiency, or harm to individuals. The higher the fault severity, the more significant the impact of the fault is on the system or process. To prioritize the corrections, fault severity determines the priority of repairs or fixes [7, 18, 19, 92].

Feature-Driven Development

Feature-driven development (FDD) is a software development process focused on delivering small, incremental changes to a system or product in a highly iterative and gradual manner. It is commonly used in the automotive industry to develop and maintain complex software systems, such as those found in vehicles or support systems that enable vehicle development, testing, and operation.

In FDD, software development is organized around delivering small, discrete features or functionality. These features are developed in short, focused development cycles called "features" or "features iterations." Each feature iteration involves developing and delivering a specific set of features, which is followed by testing and feedback to ensure that those features meet required specifications and standards.

FDD is useful for the automotive industry because it allows for the rapid development and delivery of small, incremental changes to a system or product. It also provides for the integration of feedback and changes from users or stakeholders, which can help to ensure that the final product meets the needs and expectations of its users.

FDD is often used with Agile development methodologies, such as Scrum or Kanban, to provide a more flexible and iterative approach to software development. Automotive manufacturers and suppliers commonly use it to develop and maintain complex software systems that support vehicle development, testing, and operation.

Feature Interaction Testing

Feature interaction testing in the automotive industry evaluates the interactions and dependencies between different features or functions of a vehicle or its support systems. It is used to identify any issues or conflicts between various features or functions and to ensure that they work together as intended.

Feature interaction testing in the automotive industry typically involves testing the interactions and dependencies between different features or functions in a controlled and simulated environment. This exploration includes test vehicles, simulators, and specialized testing equipment and tools.

Feature interaction testing is essential in developing and maintaining any vehicle and its support systems. It helps ensure that the different features or functions work together as intended and meet all required specifications and standards. In addition, automotive manufacturers and suppliers commonly use feature interaction testing to identify and address any issues or conflicts between different features or functions and to ensure that the final product meets the needs and expectations of its users.

Feature Tuning/Calibration

Feature tuning or calibration in the automotive industry involves adjusting and optimizing the performance and behavior of a vehicle or its components to meet all required specifications and standards. It is used to fine-tune a vehicle's or its components' performance, fuel efficiency, and emissions.

Feature tuning or calibration in the automotive industry typically involves adjusting various parameters and settings of the vehicle or its components to optimize performance and behavior. This tuning includes the use of specialized tools and equipment, as well as software and diagnostic systems.

Feature tuning or calibration is essential in developing and maintaining any vehicle or its components. Automotive manufacturers and suppliers commonly use it to optimize their products' performance and ensure that they meet the expectations of their customers.

Fiberboard Testing

Fiberboard is a commonly used material in the automotive industry, particularly for interior components such as headliners, door panels, and trim. Fiberboard is a composite material made from wood fibers and other materials that are compressed and formed into sheets.

FIGURE F.1 An example of a test fixture for fiberboard testing [83].

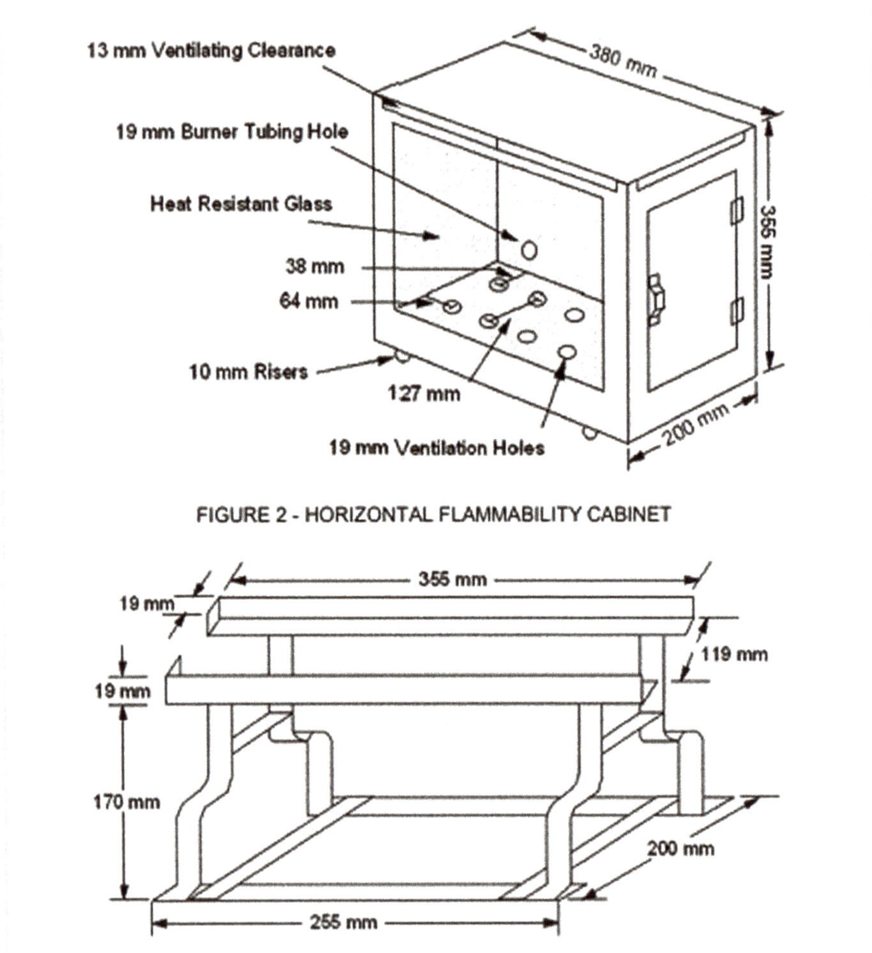

Testing fiberboard materials used in automotive applications is critical to ensure they meet the required performance standards for durability, strength, flammability, and other properties (see Figure F.1). SAE International has a published standard related to fiberboard testing in the automotive industry. This standard covers laminated fiberboard used for motor vehicle headliners and specifies requirements for their materials, construction, and performance, including tests for dimensional stability, strength, and flammability. For more information, see **J369** [83].

Fleet Testing

Fleet testing in the automotive industry involves examining and evaluating vehicles in real-world driving conditions. It is a critical step in developing and validating new vehicle models and technologies and is used to assess a vehicle's performance, durability, reliability, and safety (see Figure F.2).

FIGURE F.2 Fleet data monitoring and analytics main component [84].

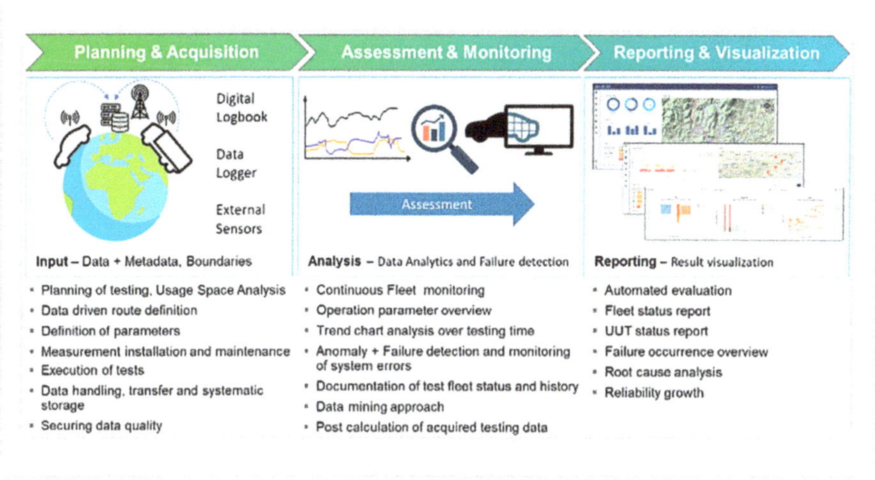

Fleet testing typically involves deploying a fleet of vehicles, either preproduction or production models, on public roads or test tracks for a predetermined period. The vehicles are equipped with data logging and monitoring systems to capture a wide range of performance data, including fuel economy, emissions, acceleration, braking, handling, and ride comfort.

During fleet testing, the vehicles are subjected to various driving conditions, such as different road surfaces, temperatures, altitudes, and driving styles, to evaluate their performance in various real-world scenarios. Fleet testing can also identify potential issues or failures that may not be detectable in laboratory testing or simulations; that is, real-world stimuli while operating the vehicle can shake out other defects.

Fleet testing is crucial in developing and validating new automotive technologies like electric and autonomous vehicles. It allows engineers to collect and analyze real-world data to improve performance, efficiency, and safety. It also evaluates the durability and reliability of components and systems, such as engines, transmissions, and suspension systems.

The data collected during fleet testing is analyzed to identify areas for improvement and validate the vehicle's performance and safety. This information is used to refine the vehicle design and engineering and ensure that the final product meets regulatory and industry standards [84].

Focusing Effect

The focusing effect, also known as the anchoring effect, is a cognitive bias in which people rely too heavily on the first piece of information they receive and use it as an anchor to make subsequent judgments or decisions. This bias can affect testing in several ways.

First, the focusing effect can lead testers to place too much emphasis on their initial assumptions or hypotheses. Focusing can cause them to overlook or discount other potential explanations or possibilities that may arise during testing, and result in a compromise of testing validity and reliability.

Second, the focusing effect can impact the interpretation of testing results. For example, when testers are focused on a particular hypothesis or expectation, they may interpret the results in a way that supports their preconceived notions rather than objectively analyze the data.

It is crucial to approach the testing process with an open and flexible mindset to mitigate the focusing effect. Testers should be willing to consider multiple hypotheses and interpretations of the results and avoid fixating on a particular assumption or expectation. Additionally, it can be helpful to involve multiple testers or experts in the testing process to provide diverse perspectives and reduce the impact of individual biases.

Framing Effect

The framing effect is a cognitive bias in which people are influenced by how information is presented. This bias can impact testing in several ways. First, the framing effect can influence how testers design and conduct tests. For example, by framing the test in a certain way, testers might develop test cases and equipment that focus on specific aspects of the system and/or overlook other important factors. Errant framing of the test can lead to biased or incomplete testing results.

Second, the framing effect can impact how test results are interpreted. For example, if the results are framed in a particular way, testers may be more likely to draw certain conclusions or make certain decisions based on those results, even if they are not the most accurate or appropriate conclusions.

To mitigate the framing effect in testing, clearly articulate the goal of the specific tests. This framing will trace back to the requirements. Therefore, it is crucial to be aware of the manner of presentation of the information and consciously try to avoid being influenced by it. Testers should strive to design tests that are as objective and comprehensive as possible and to interpret the results in a way that is not biased by how they are presented.

Additionally, involving multiple testers or users in the testing process can help to reduce the impact of individual perspectives and biases. Open discussion of the particular test objectives increases the likelihood that test results will be objective. Finally, by being aware of the framing effect and mitigating its impact, testers can improve the quality and reliability of their testing results.

Frozen Test Basis

A frozen test basis refers to a set of requirements or specifications used for software testing that is considered stable and unchanging; it is related to configuration management, change management, and release notes. Once a test basis is deemed frozen, it should not be modified unless necessary, as any changes to the test basis could invalidate the test results and make it difficult to compare different test runs.

A frozen test basis ensures that the testing process is consistent and repeatable and that the testing results can be compared over time. However, if the test basis constantly changes, it can be challenging to determine whether observed defects result from software changes or the test basis.

In practice, a frozen test basis typically consists of requirements or specifications that have been reviewed and approved by all stakeholders and are not expected to change significantly during testing. The frozen test basis is used to develop test cases, test scripts, and other testing artifacts.

Once the testing process is completed, the results are compared to the frozen test basis to determine whether the software meets the specified requirements. Identified defects are recorded and tracked until they are resolved. Suppose the test basis needs to be modified at any point during the testing process. In that case, the changes should be carefully documented and communicated to all stakeholders to ensure that everyone is aware of the potential impact on the testing process and the validity of the test results.

Function Coverage

Function coverage is a testing metric used to measure the degree to which the functions or subroutines of a software system are executed during testing. Function coverage is a type of structural coverage that focuses on a software system's code implementation rather than its external behavior.

$$Function\,Coverage = \left(\frac{Functions_{tested}}{Functions_{total}}\right) \times 100$$

Function coverage is a percentage of the system's total number of functions or subroutines executed during testing. For example, if a software system has

100 functions or subroutines and 90 of them are executed during testing, the function coverage would be 90%.

The purpose of function coverage is to evaluate the percentage of functions or subroutines in a software system tested and to identify any functions or subroutines that may have been overlooked or neglected in the testing process.

Function coverage can be measured using various testing techniques, such as unit testing, integration testing, and system testing. These techniques involve executing the functions or subroutines of the system under controlled conditions to ensure that they behave as expected and that all prioritized scenarios and inputs are considered.

It should be noted that achieving 100% function coverage does not guarantee that a software system is entirely error-free or will behave correctly in all possible scenarios. Therefore, other types of coverage, such as branch and path coverage, should also be considered in conjunction with function coverage to ensure a comprehensive and effective testing process.

Function Generator

Typically, a function generator is a piece of electronic test equipment that produces various electrical waveforms across multiple frequencies (see Figure F.3).

FIGURE F.3 An example of a function generator that can provide signals for system inputs in the lab.

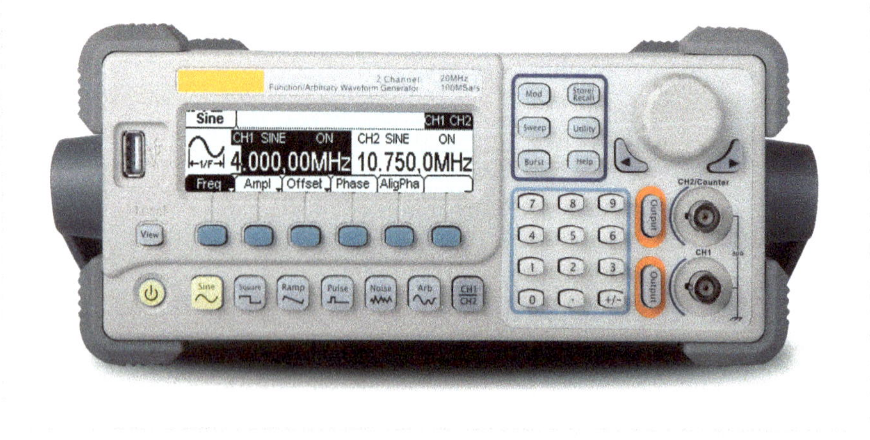

Khotenko Vladymyr/Shutterstock.com.

Function Point

Function points are a metric used in software engineering to measure the functional size of a software application or system. The concept of function points was developed in the 1970s by Allan Albrecht at IBM. Function points measure a software system's functionality, regardless of the programming language or technology used to develop it. This is achieved by identifying and counting the inputs, outputs, inquiries, files, and external interfaces required to support the software's functionality.

Function points are used in software testing to measure the size and complexity of a software application or system, which can in turn inform the test planning and estimation process. The number of function points in a system are used to estimate the amount of testing effort required to test the system thoroughly.

In addition, function points allow for identifying the most critical and complex parts of a software system, which can inform the prioritization of testing efforts. For example, if a particular function or feature of a software system has many function points, it may require more testing and scrutiny than other system parts.

Function points are used to track the progress of testing efforts and measure the testing process's effectiveness. By comparing the actual number of defects found during testing to the expected number based on the number of function points in the system, it is possible to identify areas where the testing process may need to be improved.

However, it should be noted that function points are just one metric used in software testing. They should be used with other metrics and techniques to ensure a comprehensive and effective testing process. It is also essential to recognize that function points only measure the size and complexity of a system and do not provide any information about the quality or correctness of the system's functionality. Therefore, other testing techniques, such as functional and nonfunctional testing, should also be employed to ensure the overall quality and reliability of the software system.

Functional Decomposition

Functional decomposition is a technique used in systems engineering and software development to break down a complex system into smaller, more manageable parts or functions. Functional decomposition aims to identify the different functions or tasks a system must perform to achieve its overall goals and break these into smaller, more manageable subfunctions.

The process of functional decomposition typically involves the following steps:

1. Identify the System Goals: First, identify the overall goals or objectives of the system. These goals will provide the basis for identifying the different functions that the system must perform.
2. Identify the Primary Functions: Next, identify the principal roles the system must perform to achieve these goals. These significant functions are often referred to as "top-level functions."
3. Break Down the Top-Level Functions: Next, break down the top-level functions into smaller, more manageable subfunctions. This process is repeated until the subfunctions are simple enough to be implemented and tested.
4. Define the Interfaces: Each subfunction must be defined, including its inputs and outputs and its interface with other functions. These interfaces are necessary to ensure that the subfunctions can be integrated to form the overall system.
5. Verify and Validate the Decomposition: The functional decomposition must be verified and validated to accurately reflect the system requirements and goals.

Functional decomposition is a technique for breaking down a complex system into smaller, more manageable parts. It helps identify the different functions a system must perform and provides a structured approach for developing and integrating these functions into a cohesive system.

Functional Fixedness

Functional fixedness is a cognitive bias that refers to the tendency to think of an object or problem only in terms of its typical use or function and not consider other possible benefits or solutions. Functional fixedness occurs in software testing when testers approach a software component or system with a preconceived notion of how it should function or behave and fail to consider other possible scenarios or use cases.

Functional fixedness can lead to testing that is narrow in scope and misses potential issues or risks. For example, if a tester only focuses on testing a software system's most common or expected use cases and fails to consider less common scenarios or edge cases that could lead to problems or failures, the test may miss defects or issues and compromise the software system's quality and reliability.

Testers can overcome functional fixedness through techniques such as exploratory testing or boundary testing, which involve deliberately seeking out unusual or edge cases to test. Testers can also try to approach the software system or component from different perspectives, such as a user with limited technical knowledge or a user with a disability, to identify potential issues or usability problems.

By recognizing and addressing functional fixedness in testing, testers can help ensure that they are testing software components and systems thoroughly and effectively and identifying and mitigating potential issues and risks before they impact end users.

Functional Requirement

A functional requirement is a specification for what a system or software component should do or achieve regarding its behavior or functionality. The applicable requirements describe the manner the product will meet the needs of its users or stakeholders. Functional requirements define a system or software component's specific features or capabilities to perform its intended tasks or functions.

Functional requirements are specified in various forms, such as use cases, user stories, or available specifications. However, they typically describe the expected behavior or output of a system or software component under different input or usage scenarios. For example, an applicable requirement for an e-commerce website might be to allow users to search for and purchase products or provide recommendations based on browsing history or purchase behavior.

Functional requirements are a critical part of the product development process, as they clearly define the scope and functionality of the system or component developed. In addition, functional requirements establish product and quality assurance, as they clearly define what a system or component should do. In addition, they help ensure that the product meets the needs of its intended users and stakeholders and provide a basis for testing and validation to ensure that the system or component performs as intended.

Functional Safety

Functional safety refers to the ability of a system or equipment to operate safely and reliably without causing harm to people or damaging the environment. This concept is fundamental in industries with risks of severe accidents or other hazardous events, such as the automotive, aerospace, and manufacturing sectors.

Functional safety is achieved through hardware and software measures, including redundancy, fault tolerance, fail-safe mechanisms, and safety-critical software development processes. These measures ensure that the system or equipment can detect and respond appropriately to any faults or failures and that it can shut down or enter a safe state if necessary.

The international standard for functional safety is IEC 61508, which provides a framework for developing safety-related systems and equipment. Other industry-specific standards, such as ISO 26262 for automotive functional safety and DO-178C for avionics software, provide additional guidance and requirements for achieving functional safety in these domains.

Testing plays a critical role in ensuring the functional safety of a system or equipment. The purpose of testing is to verify that the safety measures that have been implemented are working as intended and to identify any potential defects or failures that could compromise the system's safety.

Functional safety testing typically involves a combination of different types of testing:

1. Integration Testing: This involves verifying that the different components of the system are working together correctly.
2. Unit Testing: This involves ensuring that individual software or hardware components function correctly.
3. System Testing: This involves examining the system to ensure it meets safety requirements and operates as intended.
4. Acceptance Testing: This involves testing the system with end users or stakeholders to meet their needs and expectations.

In addition to these standard testing types, functional safety testing often includes additional types of testing:

1. Fault Injection: This involves testing the system's ability to detect and respond to simulated faults or failures.
2. Safety Analysis: This involves testing the system's ability to detect and respond to potential hazards or safety-critical situations.
3. Environmental: This involves testing the system's ability to operate safely and reliably in different environmental conditions, such as extreme temperatures or vibrations.

Functional safety testing is critical to ensuring that a system or equipment can operate safely and reliably. It requires careful planning, execution, and documentation to be effective.

Functional Specification

A functional specification is a detailed document that outlines the requirements and specifications for a software application or system from an operational perspective. It defines the software's features, functions, behavior, and any specific business rules and processes the software must follow.

The functional specification document typically includes information on user interface design, input and output requirements, data processing and storage, system performance, and security. It also describes any external interfaces or integrations the software must support and any regulatory or compliance requirements it must meet.

The functional specification is created during the product development process as a collaboration between the software and hardware development team and the client or business stakeholders. Ensuring that all parties clearly understand the software's requirements and expectations is essential.

The functional specification serves as a blueprint for the software development process, providing a clear set of guidelines and requirements for the development team to follow. It is also used as a reference throughout the development process to ensure the software meets all desired functional requirements.

The functional specification is a critical component of the software development process, providing a clear and comprehensive description of the software's operational requirements and specifications. It helps to ensure that the software meets the desired functionality and performance standards and helps to minimize the risk of errors and defects during the development process [55, 56, 95].

Functional Testing

Functional testing focuses on the functionality of a software application or system. It involves evaluating the individual functions and features of the software to ensure that they work as expected and meet all desired functional requirements.

Functional testing typically involves the following steps:

1. Requirements Analysis: The first step in functional testing is to analyze the functional requirements of the software and identify the test cases that need to be executed to ensure that the software meets those requirements.
2. Test Planning: Once the test cases have been identified, a test plan is developed that outlines the testing strategy, testing objectives, and the resources required for testing.
3. Test Design: Test cases are designed and documented based on the identified requirements in this step. Test cases should be designed to cover all possible scenarios and edge cases to ensure maximum test coverage.
4. Test Execution: Once the test cases have been designed, they are executed and the results are recorded. Any defects or issues that are identified during testing are reported and tracked until they are resolved.

5. Test Reporting: Finally, a test report is generated that summarizes the results of the testing process, including the number of tests executed, the number of defects found, and any other relevant information.

Functional testing is performed using a combination of manual and automated testing tools. Automated testing tools can help to speed up the testing process and improve test coverage. In contrast, manual testing helps identify more complex issues and edge cases that can be missed by automated testing. Functional testing is an essential component of the software development process, helping to ensure that the software meets applicable requirements and performs as expected.

Fungus

Fungi can harm automotive products such as paint, upholstery, and other materials. Fungal growth can cause staining, discoloration, and deterioration of these materials, leading to reduced performance, decreased durability, and increased maintenance costs.

In particular, interior materials such as carpeting, seat covers, and headliners can be vulnerable to fungal growth in areas with high humidity or moisture, such as the floorboards or trunk of a vehicle. Fungal growth can also occur in areas with water intrusion on carpets, seat covers, and headliners, as well as in the case of a leaky windshield or sunroof.

To prevent fungal growth in automotive products, manufacturers may use fungicides or other antimicrobial agents in the production process. Proper maintenance and cleaning of the vehicle components can also help prevent fungal growth, such as regularly vacuuming and drying out the interior and promptly addressing any water leaks or damage [85, 86].

The most common way to determine the effect of fungal growth on electronic equipment is to inoculate the test item with a fungal spore solution, incubate the inoculated component to permit fungal growth, and examine and test the item. Incubation normally takes place under cyclic temperature and humidity conditions that approximate environmental conditions and promote suitable fungal growth.

A typical fungal spore mixture comprises Aspergillus flavus, A. versicolor, and Penicillium funiculosum. A 30-day growth period is allowed.

Any fungal growth that does occur should be examined and its long-term effects calculated. If a clear determination is made that no degradation of performance will occur over the life of the product and that the fungal growth will not detract from the appearance of visible portions of the product, the fungal growth is allowed. Otherwise, it is not permitted. Note: Conductive solutions used as a spore media and growth accelerator may affect operational tests [20].

Fuzzing Inputs

Fuzzing inputs is a software testing technique that involves feeding invalid, unexpected, or random data as input to a software application to identify bugs and vulnerabilities. Fuzz testing, also known as fuzzing, is an effective way to identify security flaws, memory leaks, crashes, and other software defects that other testing techniques may miss.

Fuzzing inputs involves the following steps:

1. Input Generation: Fuzz testing tools generate random or invalid input data, such as malformed files, invalid user input, or unexpected network traffic.
2. Input Injection: The generated input is injected into the tested software application. The information can be injected through various channels, such as network protocols, file inputs, or user inputs.
3. Monitoring and Analysis: The software application is monitored for any unusual behavior or errors that the fuzzed input may cause. The results are then analyzed to identify any software defects or vulnerabilities.
4. Reporting: The results of the fuzz testing are reported to the development team, who can then use the information to fix any identified bugs or vulnerabilities.

Fuzzing inputs can be performed manually or with automated tools and customized to suit the specific needs of the software application tested. Fuzz testing can be performed throughout the software development life cycle, from the early stages of development to post-release testing.

Fuzzing inputs is a powerful testing technique that can help to identify software defects and vulnerabilities that other testing techniques may miss. It is an essential component of software testing, particularly for applications requiring high security and reliability levels.

G

"If you want to be a big company tomorrow, you have to start acting like one today."

—Thomas J. Watson, Jr., IBM

Gauge Repeatability and Reproducibility

Gauge repeatability and reproducibility (gauge R&R is a statistical method used to assess the variability of the measurement system from the measuring device and the individual's interpretation to determine whether it can produce accurate and consistent measurements. Gauge R&R is an estimate of the combined variation of repeatability and reproducibility. Stated another way, GRR is the variance equal to the sum of within-system and between-system variances (Figure G.1) [87, 88].

$$\sigma^2_{GRR} = \sigma^2_{reproducibility} + \sigma^2_{repeatability}$$

Gauge R, or gauge repeatability, is one of the components of gauge R&R. It refers to the variation in measurements taken by the same operator using the same instrument. The other component, gauge reproducibility, refers to the variation in measures taken by different operators using the same device. Factors that impact the capability include:

- Ability to damen and instrument
- The skill of the operator
- Repeatability of the measuring device
- Ability to provide drift-free operation in the case of electronic and pneumatic gauges
- Conditions under which the instrument is being used, including ambient, light, dirt, and humidity [98]

Gauge R&R is a vital tool for ensuring the quality of manufacturing processes. By evaluating the measurement system, manufacturers can identify sources of variation and take corrective action to improve the accuracy and consistency of their measurements.

FIGURE G.1 An example of the gauge R&R samples of three operators.

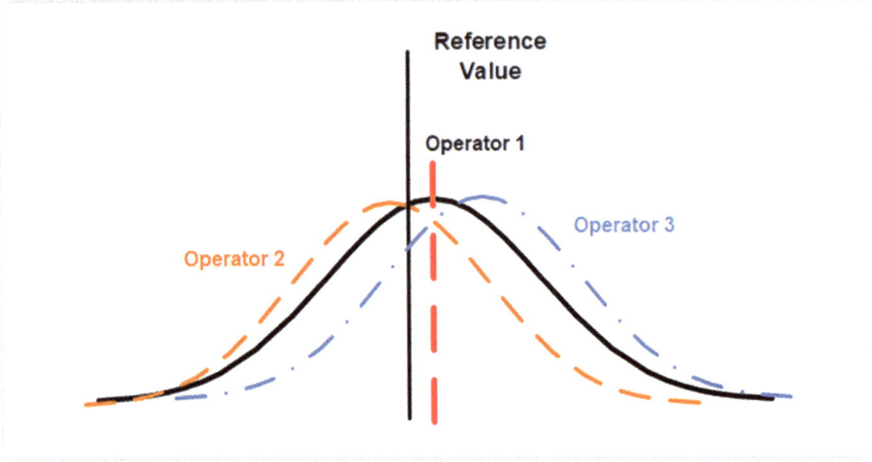

Gaussian Distribution

The Gaussian distribution, also known as the normal distribution or bell curve, is a continuous probability distribution that describes the distribution of a random variable (see Figure G.2). It is commonly used to model data distribution in a population.

FIGURE G.2 An illustration of the Gaussian distribution.

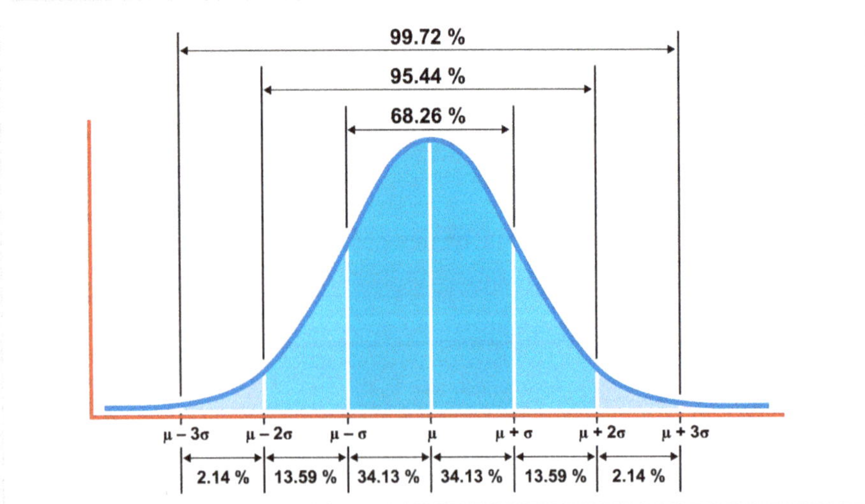

The Gaussian distribution is defined by its mean, representing the center of the distribution, and standard deviation, which represents the spread of the distribution. Finally, the distribution is defined by its shape, which is characterized by a single peak and a symmetrical curve.

The Gaussian distribution is commonly used in a variety of fields, including finance, economics, and engineering, to model and analyze data. It is beneficial for modeling symmetrical data that follows a normal distribution.

The Gaussian distribution has several important properties, including symmetry, unimodality, and asymptotic behavior. It is also characterized by the central limit theorem, which states that the sum of a large number of independent random variables is approximately normally distributed, regardless of the distribution of the individual variables.

Genetic Algorithm

A genetic algorithm is a type of optimization algorithm inspired by natural evolution. It uses principles of genetics and natural selection to search for the best solution to a problem. In the context of testing, a genetic algorithm is used to evaluate different combinations of inputs or variables to find the optimal solution to a problem.

For example, a genetic algorithm could be used to test different configurations of a product design to find the most efficient and cost-effective solution. The algorithm would generate a population of possible solutions and then evaluate their performance based on predetermined criteria. It would then use this evaluation to select the best solutions and breed them together to create a new generation. This process is repeated until the algorithm converges on the optimal solution.

Genetic algorithms can be helpful for testing and optimizing complex systems where traditional optimization methods may not be effective. However, they can be time-consuming and resource-intensive to implement and may require specialized expertise to set up and run.

Golden Unit

A golden unit is a standard or reference implementation of a system or component used as a benchmark for testing and comparison. It is typically a fully functional and tested version of the system that has been deemed to meet all the required specifications and standards.

In testing, the golden unit is used as a baseline to compare the performance and functionality of other versions or instances of the system. It is used to identify any discrepancies or deviations from the expected behavior and ensure that the system operates correctly. The development team typically creates and maintains the golden unit, which is considered the gold standard for the system.

It is used as a reference point for testing and debugging, and any changes to the system are typically compared against the golden unit to ensure that they do not introduce any new defects or problems.

Gorilla Testing

Gorilla testing involves sending many concurrent requests or transactions to a system to test its performance and stability under a heavy load. It is designed to simulate real-world conditions in which a system may be required to handle many requests simultaneously, such as during peak usage or when handling a sudden influx of traffic.

Gorilla testing uses specialized testing tools or software that can generate and send high requests to the system to identify bottlenecks or other performance issues that may not be detectable under normal usage conditions. Gorilla testing is a valuable tool for ensuring that a system can handle the expected level of traffic and usage without experiencing failures or performance issues. However, it can be resource-intensive and require careful planning and coordination to ensure it is conducted safely and effectively.

Gray-Box Testing

Gray-box testing is an approach that combines elements of black-box testing and white-box testing. In black-box testing, the tester does not know the system's internal workings and only focuses on the input and output of the system. In white-box testing, the tester has complete knowledge of the system's internal structure and can thoroughly test individual components and functions.

Gray-box testing falls somewhere in between these two approaches. The tester has some knowledge of the system's internal structure and can test specific components or functions, but they are not necessarily testing every aspect of the system in detail. Instead, the focus is on trying the system as a whole, ensuring it functions correctly, and meeting desired performance standards.

Gray-box testing can be helpful when testing complex systems for which a complete white-box approach may not be practical or efficient. It can also benefit testing systems being developed or maintained by multiple teams or organizations where the tester may not fully know the system's internal structure.

Ground Truth Data

Ground truth data testing is verifying the accuracy and reliability of data by comparing it to a known, authoritative source. The term "ground truth" refers to using real-world observations or measurements as a reference point for determining the accuracy of a dataset.

In ground truth data testing, the data tested is compared to a set of ground truth data, considered the data's most accurate and reliable representation.

Ground truth data testing may involve manually reviewing the data, using automated tools to compare the datasets, or using statistical analysis to determine the level of agreement between the two datasets.

Ground truth data testing is used in fields such as mapping, remote sensing, and machine learning, where the accuracy and reliability of data can have significant consequences. It is crucial to ensure that the data used is accurate and can be trusted to make decisions or draw conclusions.

GUI

GUI stands for graphical user interface. It is a user interface that allows users to interact with a software application or operating system through graphical elements such as buttons, icons, menus, and windows. Examples include vehicle instrumentation (cluster) or human-machine interfaces (HMI) in the vehicle [49, 89].

Software testing that focuses on an application's GUI aims to ensure that the GUI meets the application's functional and usability requirements that it behaves as expected in different scenarios. It typically involves automated testing tools that simulate user interactions with the GUI, such as clicking buttons, entering data into fields, and navigating menus. These tools can help identify missing or incorrect labels, broken links, layout problems, and functionality errors.

In addition to automated testing, manual testing may also be used to supplement GUI testing. Manual testing involves human testers interacting with the GUI and documenting any issues or defects encountered.

H

"Honestly is the single most important factor having a direct bearing on the final success of an individual, corporation, or product."

—Ed McMahon

Hacker

Hackers use their technical knowledge and skills to gain unauthorized access to computer systems, networks, or other digital devices. The term "hacker" can refer to both ethical and malicious hackers. Ethical hackers, also known as "white hat hackers," use their skills to improve security systems and protect against cyber threats. Malicious hackers, also known as "black hat hackers," use their skills to steal data, spread malware, or cause other types of harm (see Figure H.1). Other hackers consider themselves "grey hat hackers," who may engage in ethical and unethical hacking activities. It is important to note that hacking without proper authorization is illegal and can result in severe legal consequences.

FIGURE H.1 Hacker breaking into a vehicle.

Melnikov Dmitriy/Shutterstock.com.

Hacking

Hacking refers to gaining unauthorized access to a computer system or network to disrupt or steal data. Hackers use various techniques and tools to exploit vulnerabilities in computer systems, including social engineering, malware, and brute force attacks.

Modern testing uses various techniques, including vulnerability scanning, penetration testing, and code analysis, to identify exploitation possibilities for remediation based on risk assessments. Therefore, organizations must conduct regular testing to ensure their systems and networks are secure and identify potential vulnerabilities before exploitation. This can help prevent data breaches, financial loss, and other negative consequences associated with hacking. Additionally, organizations should implement proactive security measures such as firewalls, intrusion detection systems, and employee training to minimize the risk of successful hacking attempts.

Halo Effect

The halo effect is a cognitive bias in which a positive impression of a person or entity in one area leads to positive perceptions in other areas, even if those perceptions are not warranted. In the testing context, the halo effect can manifest when testers form positive impressions of a system or product based on previous positive experiences rather than objective evaluation of its merits.

The halo effect is particularly problematic in testing because it may lead testers to overlook or downplay potential issues or vulnerabilities in a system. This, in turn, can result in inadequate testing and potentially leave the system vulnerable to exploitation by malicious actors. Testers need to approach each testing effort with a fresh and objective mindset to minimize the negative impact halo effect bias. This means evaluating the system or product on its merits rather than being influenced by previous positive experiences or biases. Additionally, testers should use various testing techniques and tools to thoroughly evaluate the system or product and identify potential vulnerabilities. By minimizing the halo effect's impact and conducting thorough and objective testing, organizations can improve the security and reliability of their systems and products and reduce the risk of successful exploitation by malicious actors.

Halstead Number

The Halstead number measures the complexity of a software system or function. It is commonly used in the automotive industry to evaluate the complexity of software systems and functions used in vehicles or in the support systems that enable vehicle development, testing, and operation.

The Halstead number calculation uses a set of metrics based on the number of unique operators and operands used in a software system or function. In the automotive industry, the Halstead number evaluates the complexity of software systems and functions used in vehicles or the support systems that enable vehicle development, testing, and operation. It is a tool for evaluating the complexity of these systems and functions and can help to identify potential issues or areas for improvement.

The Halstead number is commonly used with other software quality metrics, such as code coverage, test case count, and defect density, to comprehensively analyze the quality and complexity of software systems and functions. As a result, it is an important tool for ensuring that software systems and functions used in the automotive industry are reliable, efficient, and meet all required specifications and standards.

HALT

Highly accelerated life testing (HALT) is used to evaluate the reliability and durability of a system or product over an extended period. It is used to identify any issues or weaknesses that may arise during the life of the system or product and to ensure that it meets required specifications and standards. It is used in aerospace, defense, automotive, and manufacturing industries, where complex systems must be highly reliable and precise.

HALT involves subjecting the system or product to extreme conditions and stresses beyond its normal operating range. These stresses replicate the conditions that the system or product will experience over its lifetime and are applied in a controlled and incremental manner [6, 101].

HALT involves subjecting a product or system to various stress tests, including thermal, vibration, and electrical stress. The goal is to identify potential failure points or weaknesses in the design by subjecting the system to extreme conditions beyond which it would typically experience. The equation is based on the premise that exposing a product or system to extreme stress can accelerate aging and reveal potential issues that might not surface under normal conditions.

The steps of the HALT equation are as follows:

1. Step Stress: The system is subjected to increasingly severe stress levels until a failure point is reached.
2. Margin Testing: The system is subjected to various stress levels to determine the minimum and maximum operating ranges.
3. Failure Analysis: The failed components are analyzed to identify the root cause of the failure.

HALT is a step in developing and maintaining any system or product, as it helps ensure that it is reliable and meets the required specifications and standards. Manufacturers and suppliers commonly use it to ensure that their products are of high quality and meet the expectations of their customers.

Hardness Testing O-Rings

Hardness testing is used to measure the firmness of materials, including elastomeric O-rings. Elastomeric O-rings are used in various applications, including seals for fluid and gas transfer systems. Therefore, it is essential to ensure they have the appropriate hardness to perform their intended function.

Different methods are used for hardness testing, but the most commonly used method for elastomeric O-rings is the Shore A test. The Shore A test involves using a specialized instrument, called a durometer, to measure the hardness of the O-ring by pressing a specified size and shape indentor into the material and measuring the depth of the indentation. The result is a numerical value on a scale from 0 to 100, where higher values indicate greater hardness.

The appropriate hardness for an elastomeric O-ring depends on the specific application and the requirements of the system in which it will be used. For example, O-rings used in high-pressure applications generally require higher hardness to resist deformation and maintain their seal. On the other hand, O-rings used in low-pressure applications generally require lower hardness for ease of installation and removal.

By conducting hardness testing on elastomeric O-rings, manufacturers and users can ensure that the O-rings have the appropriate hardness for their intended use and reduce the risk of system failure or leakage due to inadequate sealing performance. For more information, see **J417** [90].

Hardware-Based Fault Injection Testing

Hardware-based fault injection testing in the automotive industry is used to evaluate the reliability and robustness of a vehicle or its components. It is used to identify any vulnerabilities or weaknesses that may arise during the vehicle's life and to ensure that it meets required specifications and standards.

Hardware-based fault injection testing involves intentionally introducing faults or errors into a vehicle's hardware or its components to evaluate their behavior and performance. This may involve the use of specialized tools and equipment, such as fault injection boards or probes, to introduce the faults or errors.

The steps involved in hardware-based fault injection testing in the automotive industry typically include the following:

1. Identify the hardware components that are critical to the operation of the vehicle or component.
2. Develop a plan for injecting faults into these hardware components in a controlled and incremental manner.
3. Use specialized tools and equipment to inject faults into the hardware components such as simulators or tools.
4. Observe the behavior and performance of the vehicle or component in the presence of the injected faults, and record any issues or failures that may occur.
5. Analyze the data collected during the testing process to identify any vulnerabilities or issues that may arise in the event of a hardware failure and to provide recommendations for improving the vehicle's or component's robustness and reliability.

Hardware-based fault injection testing is essential in developing and maintaining any vehicle or component. It helps ensure that vehicles and parts are robust and reliable in the presence of hardware faults or failures. In addition, automotive manufacturers and suppliers commonly use it to verify that their products meet all required specifications and standards for reliability and safety.

Hardware Platform

An automotive hardware platform is a set of hardware components used to support the development, testing, and operation of a vehicle or its components. It is typically composed of a combination of hardware components, such as microcontrollers, processors, memory, sensors, and communication interfaces, designed to work together to enable the functions and features of the vehicle.

An automotive hardware platform may include a variety of components:

1. Microcontrollers: These are small, specialized computers used to control the functions and features of the vehicle.
2. Processors: These are central processing units (CPUs) used to execute software programs and perform computations.
3. Memory: This includes various types of memory, such as ROM, RAM, and flash memory, used to store data and software programs.
4. Sensors: These devices are used to measure and detect various data types, such as temperature, pressure, and position.
5. Communication interfaces: These are devices that are used to enable communication between different components and systems within the vehicle.

An automotive hardware platform is an essential component of any vehicle or its components, as it provides the foundation for the development and operation of the vehicle. Automotive manufacturers and suppliers commonly use it to develop and maintain complex vehicles and systems that support vehicle development, testing, and operation.

Hardware-Software Integration Testing

Hardware-software integration testing is used to evaluate the integration and compatibility of hardware and software components in a system or product. It is used to ensure that the hardware and software components work together as intended and to identify any issues or conflicts that may arise between them.

Hardware-software integration testing typically involves testing the interactions and dependencies between hardware and software components in a controlled and simulated environment. This may include using test equipment, simulators, specialized testing tools, and software and diagnostic systems.

Hardware-software integration testing is a step in the development and maintenance of any system or product that involves the integration of hardware and software components, as it helps to ensure that the components work together as intended and meet required specifications and standards. It is commonly used in aerospace, defense, automotive, and manufacturing industries, where complex systems must operate with high reliability and precision.

Hardware-software integration testing is typically performed in conjunction with other types of testing, such as system testing and acceptance testing, to provide a more comprehensive analysis of the system or product. It is an essential tool for ensuring the quality and reliability of systems and products that involve integrating hardware and software components.

HASS

Highly accelerated stress screening (HASS) in the automotive industry is used to evaluate the reliability and durability of a vehicle or its components under extreme conditions and stresses. It is used to identify any issues or weaknesses that may arise during the life of the vehicle or component and to ensure that it meets all required specifications and standards. It is an accelerated reliability screen that can identify hidden faults not picked up by traditional test methods like burn-in and ESS (environmental stress screening). The stresses used in HASS testing are over specification but yet within the design's capacity as verified by the HALT.

HASS involves subjecting the component to extreme conditions and stresses beyond its normal operating range. These conditions and stresses are designed to replicate the circumstances that the component will likely experience over its lifetime and are typically applied in a controlled and incremental manner.

HASS is used in aerospace, defense, automotive, and manufacturing industries, where complex systems must be highly reliable and precise. HASS is essential in developing and maintaining any vehicle or component, as it helps ensure that it is reliable and meets the required specifications and standards.

High-Order Tests

High-order tests, also known as integration tests, are designed to evaluate the interactions and dependencies between different components or modules of a system. They are typically used to test how well a system functions as a whole rather than just testing individual components or functions in isolation.

High-order tests are usually conducted after lower-level tests, such as unit or component tests, have been completed and any issues are resolved. They involve testing the system in a more realistic or representative environment and may include simulating different scenarios or conditions to evaluate its behavior and performance.

High-order tests are essential in the testing process as they help to identify issues that may not be detectable at the component level and to ensure that the system functions correctly and meets desired performance standards. However, they can be resource-intensive and require careful planning and coordination to ensure they are conducted effectively.

Hijacking

Hijacking and testing are two different activities with different goals and methodologies. Hijacking is done with malicious intent, while testing is performed to identify and address vulnerabilities to improve security.

However, some ethical hackers and security testers may conduct hijacking activities as part of their testing efforts. This is known as ethical hacking or penetration testing, and it is done with the permission of the system owner to identify vulnerabilities and improve security.

Organizations must conduct regular testing and implement proactive security measures to minimize the risk of successful hijacking attempts. This can help prevent data breaches, financial loss, and other negative consequences associated with hijacking. Additionally, organizations should ensure that ethical hacking and testing activities are controlled and responsible to avoid any unintended damage or disruption to the system or network.

HIL Testing

Complex real-time embedded systems are developed and tested using hardware-in-the-loop (HIL) simulation, also known as HWIL or HITL. HIL testing involves using physical hardware components in conjunction with a simulated environment to evaluate the performance and functionality of a system. It is commonly used to test control systems, such as those used in automotive, aerospace, and industrial applications.

In HIL testing, the physical hardware components, such as sensors, actuators, and controllers, are connected to a computer-based simulation of the system's environment and behavior. The hardware components are then controlled and monitored through the simulation, allowing the tester to evaluate the system's performance and behavior in a controlled and repeatable manner.

HIL testing evaluates the performance of hardware components in a real-world environment without the need for physical testing facilities or equipment. It can also be beneficial in testing complex systems that may be difficult or impractical to test in a real-world setting. However, it can be resource-intensive and require specialized software and hardware.

Hindsight Bias

Hindsight bias is a cognitive predisposition in which individuals believe that events that have occurred were more predictable than they actually were. In the testing context, hindsight bias can lead testers to believe that they would have predicted a particular defect or issue if they had known about it beforehand, even if the issue was not predictable based on the available information.

Hindsight bias is problematic in testing because it can lead to overconfidence in predicting and preventing issues, resulting in inadequate testing and a false sense of security. This can leave a system vulnerable to exploitation by malicious actors.

Testers must maintain an objective and open-minded approach, consider multiple perspectives, and experiment with testing to overcome hindsight bias. This means focusing on the available information and conducting thorough testing to identify potential issues and vulnerabilities rather than relying on assumptions or past experiences.

Testers can also use techniques such as exploratory testing and scenario-based testing to simulate real-world conditions and identify potential issues that may not be immediately apparent based on prior experiences or assumptions.

Histogram

A histogram is a graphical representation of the distribution of a dataset (see Figure H.2). A bar chart shows a dataset's frequency or number of occurrences of different values. The x-axis of a histogram represents the range of values in the dataset. The y-axis represents the frequency occurrence of those values. The bars of the histogram are drawn to touch each other. The height of each bar represents the number of data points that fall within the range of values represented by the x-axis.

FIGURE H.2 Example of five histograms.

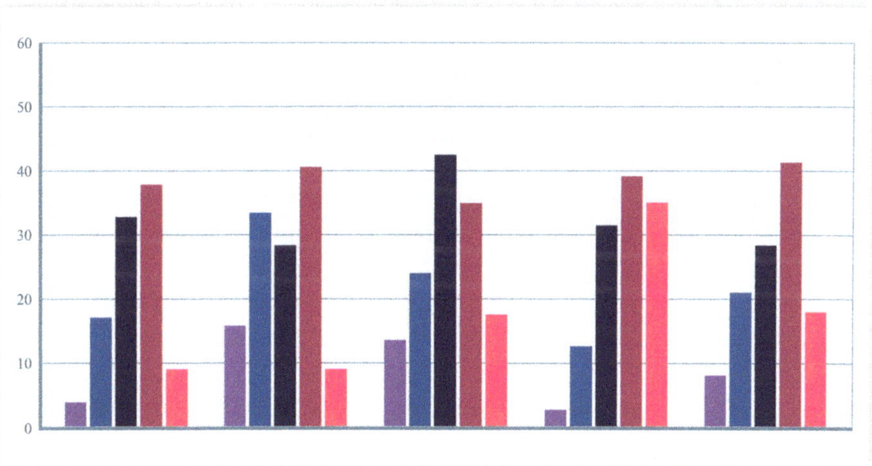

Nazarii M/Shutterstock.com.

For example, consider a dataset containing the heights of a group of people. A histogram of this data might have bins (or ranges) of heights on the x-axis and the frequency of people in each bin on the y-axis. The histogram would show how many people fall within each range of heights, allowing you to see the overall distribution of heights in the group.

Histograms are helpful in understanding the distribution of a dataset and identifying patterns or trends within the data. They can also identify outliers or unusual values in the data.

To create a histogram, follow these steps:

1. Collect data for the histogram. Identify and collect data of interest. For example, measurements such as the length of a key dimension on a collection of parts or the current consumption of a set of parts.
2. Divide the data into intervals or bins. These intervals should be of equal size, and they should cover the range of the data.
3. Count the number of data points that fall into each bin.
4. Plot the bins on the x-axis and the frequencies on the y-axis.
5. Draw bars to represent the frequencies in each bin. The height of each bar should be proportional to the number of data points in the bin [40].

Homologation Testing

Homologation testing certifies that a product, system, or vehicle meets specific standards and regulations for use or sale in a particular market. In addition, homologation testing ensures that products are safe and reliable and that they meet relevant legal and regulatory requirements.

Homologation testing is commonly used in the automotive industry, where vehicles must meet specific safety and emissions standards before being sold in a particular market. In addition to safety and emissions, homologation testing may cover other areas, such as performance, durability, and noise levels.

Homologation testing typically involves a series of tests and inspections conducted by authorized testing organizations or government agencies. These tests may include laboratory tests, on-road testing, and real-world simulations to assess the product's safety, performance, and durability under various conditions.

The specific requirements for homologation testing vary depending on the product and the market in which sold. For example, in the European Union, vehicles must comply with the European Community Whole Vehicle Type Approval (ECWVTA) regulations, which cover a wide range of safety, environmental, and technical requirements.

Once a product has completed homologation testing and meets all the required standards and regulations, it can be certified for use or sale in the relevant market. Homologation testing helps ensure that products are safe and reliable and meet the legal and regulatory requirements essential for their use or sale in a particular market.

Horns Effect

The horns effect is a cognitive bias that refers to the tendency to perceive someone or something as negative based on a single negative trait or characteristic, even if that trait or characteristic is irrelevant to the overall evaluation. For example, in the testing context, the horns effect can lead to negative assessments or assumptions about a system or product based on a single defect or issue, even if the flaw is minor and the system performs well in other areas.

The horns effect is problematic in testing because it can lead to an overly negative evaluation of a system or product, resulting in overtesting, overengineering, and wasted resources. It can ultimately impact the project's cost, schedule, and quality.

To avoid the horns effect, testers must evaluate a system or product based on performance and capabilities rather than focus solely on individual defects or issues. Testers can also use risk-based and exploratory testing techniques to identify potential problems and vulnerabilities based on their overall impact on the system rather than their severity. By minimizing the impact of the horns effect and conducting objective and balanced testing, organizations can improve the accuracy and effectiveness of their testing efforts and reduce the risk of adverse impacts on the cost, schedule, and quality of their projects.

Horsepower Testing

Horsepower testing refers to the process of measuring and evaluating the power output of an engine or motor, typically expressed in horsepower (hp). This testing helps determine the engine's performance, efficiency, and suitability for specific applications. This testing helps determine the engine's performance, efficiency, and suitability for specific applications. Here's an overview of the horse power testing process:

1. Test Setup:
    - The engine or motor is connected to a dynamometer, which acts as a load to simulate real-world operating conditions.
    - Instrumentation, such as sensors and data acquisition systems, is installed to measure various parameters, including torque, rotational speed, and fuel consumption.
2. Load Testing:
    - The engine is subjected to controlled loads and operated across a range of speeds and load conditions.
    - The load is varied using the dynamometer, and measurements are taken at different operating points to create a power curve.

3. Power Calculation:
   - Torque is measured using a torque transducer, while rotational speed is typically obtained from a tachometer or encoder.
4. Data Analysis and Reporting:
   - The collected data is analyzed to determine the engine's power output at various operating points.
   - Power curves are generated to visualize the relationship between horsepower, torque, and rotational speed.
   - Test results are reported, including peak horsepower, torque values, and any relevant performance parameters.
5. Performance Evaluation:
   - The horsepower testing results are compared against the engine's specifications and performance targets.
   - Performance metrics, such as power-to-weight ratio or specific power output, may be calculated to assess the engine's efficiency and performance relative to similar engines.

FIGURE H.3 A vehicle on the dynamometer testing horsepower.

A dynamometer, also known as a "dyno," is a device used to measure force, torque, or power (see Figure H.3). Dynamometers test the performance of engines, motors, and other mechanical systems by simulating real-world loads and conditions.

There are several types of dynamometers, including absorption dynamometers and motoring dynamometers. Absorption dynamometers, also known as load cells, measure the force generated by a system and then absorb that force using resistance. Motoring dynamometers, on the other hand, supply power to the system and measure the resistance that the system provides. Dynamometers are commonly used in the automotive industry to tune and optimize engine performance, as well as in other industries such as aerospace, defense, and power generation to test and evaluate the performance of mechanical systems. By measuring the force, torque, or power generated by a system, a dynamometer can provide valuable information about the system's performance and help identify improvement areas.

Hostile Attribution Bias

Hostile attribution bias occurs when individuals interpret ambiguous or neutral situations as intentionally hostile or antagonistic. Hostile attribution bias can lead testers to perceive defects or issues in a system as intentional or malicious, even if they are simply the result of unintentional errors or oversights. It is problematic in testing because it can lead to overzealous testing efforts, excessive suspicion of legitimate system (and team member) behavior, and the introduction of unnecessary security measures or restrictions.

Testers need to maintain an objective and evidence-based approach to testing to minimize the impact of hostile attribution bias on testing. This means avoiding assumptions about the system's or its users' intent or motivations and focusing on the available evidence to identify potential issues and vulnerabilities. Testers can also use risk-based and scenario-based testing techniques to simulate real-world conditions and identify potential problems that may not be immediately apparent based on assumptions or biases.

By minimizing the impact of hostile attribution bias and conducting objective and evidence-based testing, organizations can improve the accuracy and effectiveness of their testing efforts and reduce the risk of negative effects on the cost, schedule, and quality of their projects.

Hot Hand Fallacy

The hot hand fallacy is a cognitive bias that refers to the belief that a person who has experienced success or good fortune is more likely to continue experiencing success in the future. In the context of testing, the hot hand fallacy can lead testers to assume that a system or product will continue to perform well simply because it has performed well in the past.

The hot hand fallacy is problematic in testing because it can lead to complacency, overconfidence, and a lack of thoroughness. For example, if a system has performed well in previous tests, testers may be less likely to rigorously test it in future iterations, assuming that it will continue to perform well.

Minimizing the impact of the hot hand fallacy on testing is essential. Testers need to maintain an objective and evidence-based approach to testing. This means evaluating a system or product based on its current capabilities and performance rather than assuming that past performance will predict future success.

Testers can also use techniques such as regression testing and exploratory testing to ensure that changes or updates to a system do not introduce unexpected issues or vulnerabilities, regardless of its past performance. By minimizing the impact of the hot hand fallacy and conducting objective and evidence-based testing, organizations can improve the accuracy and effectiveness of their testing efforts and reduce the risk of negative impacts on the cost, schedule, and quality of their projects.

Humidity

Humidity testing measures the ability of a product or system to withstand exposure to high levels of environmental humidity or moisture. This type of testing is performed on electronic components, consumer products, and industrial equipment that may be exposed to high levels of moisture in their intended use environment.

Humidity testing typically involves exposing the product or system to a controlled environment with high humidity levels for a specified period. During the test, the product or system is monitored for signs of damage or malfunction, such as corrosion, electrical shorts, or changes in performance.

Exposure to high levels of humidity can cause damage to sensitive components and materials, leading to reduced performance, premature failure, and safety hazards. By performing humidity testing, manufacturers can identify potential issues and vulnerabilities in their products or systems and take steps to improve their durability and reliability in humid environments.

Several standards and test methods are commonly used for humidity testing, including ASTM D2247, IEC 60068-2-3, and MIL-STD-810. These standards provide guidelines for conducting humidity testing, including the test conditions, duration, and performance criteria.

Humidity testing is an essential aspect of product and system testing that can help manufacturers ensure that their products are reliable and durable in humid environments and can ultimately improve the safety and performance of their products.

Hybrid Fault Injection Testing

Hybrid fault injection testing combines two or more fault injection techniques to identify and assess the robustness and reliability of a software system. Fault injection is a technique used in testing to intentionally introduce faults or errors into a system to evaluate its response and behavior.

Hybrid fault injection testing involves combining different types of fault injection techniques, such as code-level fault injection, hardware fault injection, and network fault injection. For example, a hybrid fault injection test might simulate a hardware fault by introducing a voltage spike while also simulating a network failure by injecting a delay in network traffic.

The benefits of hybrid fault injection testing include identifying complex and hard-to-detect faults and vulnerabilities in a system and evaluating how the system responds to multiple types of defects and errors. This can provide a more comprehensive assessment of the system's overall robustness and reliability and help identify areas for improvement.

However, hybrid fault injection testing can also be complex and resource-intensive, requiring specialized tools and expertise to design and execute tests effectively. It may also be challenging to accurately reproduce real-world scenarios and conditions, which can limit the effectiveness of the testing.

Hybrid fault injection testing can be a valuable tool for identifying and assessing the robustness and reliability of software systems, particularly in complex and high-risk applications where system failures can have significant consequences.

Hyper D

Hyper D (or hyper defect detection) is a software testing technique that uses machine learning and data analytics to identify defects and vulnerabilities in software systems. This technique involves analyzing the large volumes of data a software system generates, such as logs, user behavior, and system performance metrics, to identify patterns and anomalies that may indicate defects or vulnerabilities.

Hyper D is a relatively new approach to software testing that leverages advancements in machine learning and big data analytics to improve the accuracy and efficiency of defect detection. By analyzing large volumes of data, Hyper D can identify defects and vulnerabilities that may be difficult or impossible to detect using traditional testing techniques.

Hyper D is particularly useful with complex and dynamic software systems, such as those used in financial services, healthcare, and transportation, where defects and vulnerabilities can have significant consequences. By detecting flaws and vulnerabilities earlier in the development process, Hyper D can help improve software systems' overall quality and reliability and reduce the risk of costly and disruptive system failures.

However, Hyper D is still a relatively new and evolving technique, and there are challenges and limitations associated with its use. For example, Hyper D requires large volumes of high-quality data to be effective, which may not always be available or easy to obtain. Additionally, the accuracy and effectiveness of Hyper D may depend on the quality and relevance of the data used for analysis and the quality of the machine learning models used to analyze the data.

Hyper D is an innovative and promising approach to software testing that has the potential to improve the accuracy and efficiency of defect detection and reduce the risk of system failures. First, however, it is essential to carefully evaluate this approach's benefits and limitations and consider its suitability for specific testing scenarios and applications.

I

"Innovation has never come through bureaucracy and hierarchy. It's always come from individuals."

—John Scully, CEO, Apple Computer

ICE

See In-Circuit Emulator.

Illusion of Control

The illusion of control is a cognitive bias in which people overestimate their ability to control events or outcomes that are determined by chance or external factors. In the testing context, the illusion of control can lead testers to believe they have more control over the testing process and outcomes than they actually do.

For example, if testers believe that they can prevent defects or vulnerabilities from occurring by carefully planning and executing tests, even though there may be factors outside of their control that can influence the behavior of the software tested, they may overestimate their testing efforts' effectiveness and overlook potential risks or vulnerabilities. The illusion of control can also make testers overly confident in their ability to identify and address defects and vulnerabilities, leading them to overlook subtle or complex issues that may be present in the software tested.

Testers must maintain a realistic and objective perspective on the testing process and outcomes. They should acknowledge the limitations and uncertainties of testing and recognize that defects and vulnerabilities may be present despite best efforts to identify and address them. Testers can also benefit from using techniques such as risk analysis and peer reviews to help identify potential areas of concern and ensure that testing efforts focus on the software's most critical and high-risk areas.

The illusion of control has important implications for testing, but by being aware of this bias and taking steps to mitigate its effects, testers can help ensure that testing efforts are effective, efficient, and focused on the software's most critical areas.

Illusion of Validity

The illusion of validity is a cognitive bias in which people overestimate the accuracy and reliability of their judgments or predictions based on limited or incomplete information. In the context of testing, the illusion of validity can lead testers to overestimate the accuracy and reliability of their test results and the conclusions they draw from those results.

For example, if testers believe that a particular test accurately represents the software's behavior, even though the test may have limitations or not fully capture all relevant aspects of the software's behavior, they may develop a false sense of confidence in the validity of the test results and the conclusions drawn from those results.

Testers should use multiple testing techniques and carefully evaluate each test's limitations and assumptions to help eliminate the illusion of validity. This can involve using techniques such as test coverage analysis and boundary value analysis to identify areas where testing may be incomplete or invalid assumptions are present.

Testers can also benefit from collaborating with other team members and stakeholders to validate test results and interpretations and to ensure that testing efforts are focused on the most critical areas of the software tested.

Illusory Correlation

The illusory correlation is a cognitive bias that occurs when people perceive a relationship or correlation between two variables when none exists. This can lead to incorrect or unsupported conclusions based on flawed or incomplete data.

In the testing context, the illusory correlation is problematic when it leads to incorrect assumptions about the relationship between different variables or factors tested. This can result in errant test results or inaccurate conclusions about the system tested.

To avoid the illusory correlation in testing, it is essential to ensure that the data collected and analyzed is accurate and representative of the tested system. This may involve using statistical analysis to confirm the existence of a correlation and carefully considering any potential biases or confounding factors that could influence the results. It is also essential to be cautious about making conclusions or decisions based on incomplete or limited data.

Impact Analysis

Impact analysis and testing are closely related, as impact analysis helps to identify the potential impact of changes to a software system, which can inform the testing process. During the testing process, impact analysis prioritizes the test cases and focuses on the areas of the system that are most likely to be affected by the changes. This can help ensure that testing focuses on the areas of the system that are most critical and most likely to have defects or issues.

In addition, impact analysis can help to identify additional test cases that may need to be added to the testing plan to cover the areas of the system impacted by the changes. This analysis can help to ensure that the testing is comprehensive and that all potential issues are identified and addressed.

Finally, impact analysis can be used to evaluate the testing process's effectiveness and determine whether additional testing is needed to verify the changes' impact. This can help to ensure that the software system is thoroughly tested and that any potential issues are identified and resolved before the changes are released to users.

Impact analysis and testing are complementary processes that work together to ensure the stability and reliability of software systems. By carefully analyzing the impact of changes and incorporating this information into the testing process, software development teams can minimize the risks associated with modifications to the system and ensure that the changes are implemented successfully.

Impact Bias

Impact bias is evident when people overestimate the emotional impact of future events, both positive and negative. In the context of testing, the impact bias can lead testers to overestimate the importance of certain defects or issues or to become overly anxious about the potential impact of those issues.

For example, testers may become overly concerned about a particular defect or issue, even though it may only have a minor impact on the overall functionality or usability of the tested software. This can lead testers to focus too much on resolving that issue, potentially at the expense of other more critical defects or issues.

In mitigating the effects of the impact bias in testing, testers must maintain a balanced and objective perspective on the importance of defects and issues. This can involve using techniques such as risk analysis and prioritization to identify the most critical and high-impact defects and issues and to focus testing efforts on those areas. Testers can also benefit from collaborating with other team members and stakeholders to validate the impact of defects and issues and to ensure that testing efforts focus on the most critical areas of the software.

By being aware of impact bias and taking steps to mitigate its effects, testers can help ensure that testing efforts are effective, efficient, and focused on the software's most critical areas.

Incident Logging

Incident logging in testing is the process of recording and tracking issues or incidents that may occur while testing a system or product. It is used to ensure that problems or incidents are identified, tracked, and resolved promptly and effectively and to minimize the impact of these issues or incidents on the testing process.

Incident logging in testing typically involves the following steps:

1. Identify and report an issue or incident during the testing process.
2. Evaluate the impact and severity of the problem or incident and determine the appropriate course of action.
3. Document the issue or incident, including the details, the actions taken to resolve it, and any relevant information or data.
4. Track the issue or incident through to resolution and update the status of the issue or incident as it progresses.
5. Review the issue or incident to identify any lessons learned or areas for improvement and implementing any necessary changes to prevent similar problems or incidents from occurring in the future.

Manufacturers and suppliers commonly use incident logging to ensure the quality and reliability of their products and that they meet the required specifications and standards.

Incident Management Life Cycle

The automotive incident management life cycle is a process used to manage and resolve issues or incidents that may arise while testing a vehicle or its components. It is used to ensure that problems or incidents are identified, tracked, and resolved promptly and effectively and to minimize the impact of these issues or incidents on the testing process.

The automotive testing incident management life cycle typically includes the following steps:

1. Incident Identification: This involves identifying and reporting an issue or incident that has occurred during the testing process.

2. Incident Triage: This involves evaluating the impact and severity of the incident and determining the appropriate course of action.
3. Incident Resolution: This involves identifying and implementing a solution to the issue or incident and verifying resolution.
4. Incident Documentation: This involves documenting the issue or incident and the actions taken to resolve it and updating any relevant documentation or records.
5. Incident Review: This involves reviewing the issue or incident to identify any lessons learned or areas for improvement and implementing any necessary changes to prevent similar problems or incidents from occurring in the future.

Automotive manufacturers and suppliers commonly use the incident management life cycle to ensure that their products are safe and reliable and that they meet all applicable specifications and standards.

Incident Report

An incident report documents any issues or problems that arise during the testing process. These issues include software bugs, hardware malfunctions, or other unexpected results or behaviors observed during testing.

When an incident occurs during testing, it's important to document it in an incident report. This report should include a detailed description of the incident, including any steps taken to reproduce it, its impact on the system or application tested, and any relevant screenshots or error messages.

By documenting incidents in this way, testing teams can more easily track and manage issues that arise during testing and ensure that they are addressed promptly and effectively. Incident reports can also be disseminated to other development team members or stakeholders outside of the testing team. Incident reports help to ensure that everyone is aware of any issues identified and how they are being addressed. Incident reports are essential for managing testing processes and ensuring that issues are identified and resolved before software or systems release into production.

In-Circuit Emulator

In-circuit emulation (ICE) is debugging embedded system software using a hardware device or in-circuit emulator. It works by utilizing a CPU that can perform the system's primary function in addition to supporting debugging procedures. See Emulator.

Incremental Integration Testing

Incremental integration testing is used to evaluate the integration and compatibility of different components or modules in a system or product. It is used to ensure that the components or modules work together as intended and to identify any issues or conflicts that may arise between them.

Incremental integration testing typically involves testing the interactions and dependencies between different components or modules in a controlled and simulated environment. This may include using test equipment, simulators, specialized testing tools, and software and diagnostic systems.

Incremental integration testing is performed gradually, with different components or modules added and tested step-by-step. This allows issues or conflicts to be identified and resolved early in the development process and helps ensure the final product's quality and reliability.

Incremental integration testing is essential in developing and maintaining any system or product that integrates different components or modules. It helps ensure the components or modules work together and meet the required specifications and standards. It is used in aerospace, defense, automotive, and manufacturing industries, where complex systems are needed to operate with high reliability and precision.

Incremental Testing

Incremental testing is a software testing technique in which tests are designed and implemented in small increments, each building on the previously completed work. This can involve testing individual functions or components of a system and testing the system as a whole at various stages of development. Incremental testing can be contrasted with regression testing, in which the entire system is tested after each change or addition to ensure that the change has not introduced any new defects. The main goal of incremental testing is to catch defects early in the development process while they are still easy to fix. Incremental testing can help to reduce the overall cost of testing and improve the quality of the final product.

Independent Test and Verification

Independent test and verification (also known as independence of testing) is a process in which a team of testers who are independent of the development team performs testing on a software system. This independence ensures that the system has been implemented correctly and meets the specified requirements. The testers use various techniques and tools, including manual testing, automated testing, and static analysis, to thoroughly evaluate the system and identify any defects or issues.

The main benefit of independent testing and verification is that it provides an unbiased evaluation of the system, as the testers are not involved in the development process and do not have any vested interest in the outcome of the testing. This can help increase the credibility and reliability of the test results and ensure that defects are identified and addressed before the system is released. As a result, independent test and verification is often used in critical or safety-critical systems, where the consequences of defects could be severe.

Independent Test Group (ITG)

An independent test group (ITG) is a team of testers who are independent of a software system's development team. The main goal of an ITG is to ensure that the system has been implemented correctly and meets all specified requirements. The ITG uses various techniques and tools, including manual testing, automated testing, and static analysis, to thoroughly evaluate the system and identify defects or issues.

The ITG typically works closely with the development team to understand the system and its intended requirements. They may also work with other stakeholders, such as users or customers, to understand their needs and expectations for the system. The ITG is responsible for creating and executing a comprehensive test plan, documenting the test results, and reporting any defects or issues discovered.

The main benefit of having an ITG is that it provides an unbiased evaluation of the system, as the testers are not involved in the development process and do not have any vested interest in the outcome of the testing. This can help increase the credibility and reliability of the test results and ensure that defects are identified and addressed before the system is released.

Infant Mortality

Product infant mortality is a term used in the context of quality control and reliability engineering to refer to the likelihood of a product failing or malfunctioning within a short period after it is put into use. It measures the product's reliability during its early life stages, similar to the infant mortality rate among newborns.

During the early stages of a product's life cycle, the product is relatively new and has not yet been extensively tested by users. Therefore, it is more likely to experience failures or malfunctions, resulting in lower customer satisfaction and increased costs for the manufacturer due to warranty claims and product returns (Figure I.1).

FIGURE I.1 Reliability (bathtub) curve of product failures.

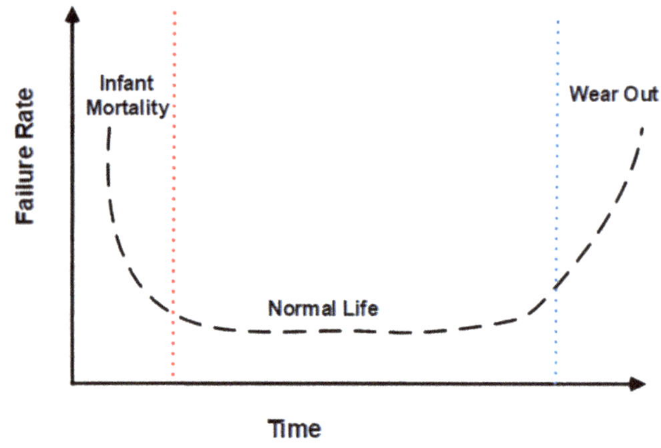

To minimize product infant mortality risk, manufacturers must prioritize product quality and reliability throughout the product life cycle. This includes design, development, testing, and production stages. Manufacturers must also conduct thorough quality control tests and validation procedures to ensure their products meet the necessary quality standards and regulations before being released to the market.

By addressing product infant mortality and ensuring high product reliability and quality, manufacturers can improve customer satisfaction, reduce warranty claims and returns, and enhance their brand reputation.

Information Bias

Information bias in testing occurs when there are errors or limitations in the information collected or used to make decisions. This bias can result in inaccurate or incomplete information, leading to biased or incorrect conclusions.

Several types of information bias can occur in testing:

1. Selection Bias: This occurs when the participants or subjects in a study are not representative of the target population or are selected in a way that systematically favors one group over another. This can lead to biased results that do not accurately reflect the population.

2. Measurement Bias: This occurs when the measurement tools or methods used in testing are inaccurate or imprecise, leading to incorrect or inconsistent results. This can happen when the testing instruments are incorrectly calibrated or the test procedures are not standardized.
3. Reporting Bias: This occurs when the information reported or recorded is incomplete, inaccurate, or biased in some way. This bias can arise when participants do not report all relevant information or researchers selectively report only specific results or outcomes.
4. Recall Bias: This occurs when participants in a study do not accurately recall past events or experiences, leading to incorrect or biased responses. This can happen when participants have faulty memories or current beliefs or attitudes influence them.

Researchers can take several steps to minimize the impact of information bias in testing:

1. Ensure that study participants are representative of the target population.
2. Use validated and standardized measurement tools and procedures.
3. Collect and analyze all relevant information rather than selectively report specific results or outcomes.
4. Use objective measures and avoid reliance on participant recall.

By minimizing information bias in testing, researchers can ensure that the results are accurate, reliable, and valid, allowing for informed decision-making based on sound information.

Infrared Testing

The American Society for Nondestructive Testing has classified passive thermographic inspection techniques, often known as infrared and thermal testing, as a class of nondestructive testing. The science of measuring and mapping surface temperatures is known as infrared thermography.

Infrared (IR) testing is a nondestructive testing method that uses electromagnetic radiation in the IR spectrum to inspect an object's surface or internal structure (see Figure I.2). IR testing is commonly used in a variety of industries, including aerospace, automotive, and manufacturing, to identify defects or imperfections in materials or components.

FIGURE I.2 An example of an IR view of the engine compartment.

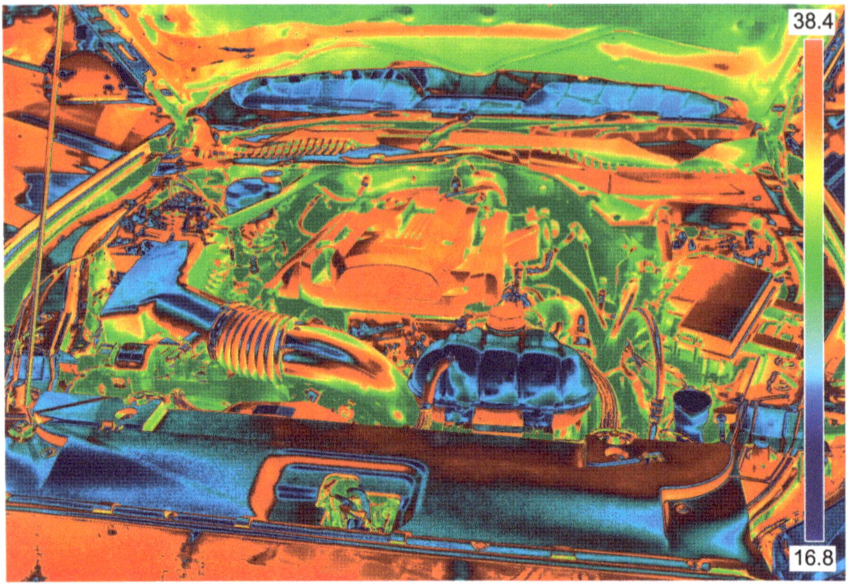

Ivan Smuk/Shutterstock.com.

In IR testing, an IR camera or thermal imaging device captures images or temperature readings of the object being tested. These readings are then analyzed to identify any differences in temperature or other characteristics that may indicate the presence of a defect.

IR testing is a quick and noninvasive way to identify defects in materials or components. It can detect a wide range of defects, including cracks, delamination, and voids, and can be performed on a wide range of materials and surfaces. However, it may not be suitable for detecting defects in objects with low thermal conductivity or complex internal structures. IR testing also is used to test wire harness temperatures when under load or extreme conditions; for example, an IR camera can be coupled with short-circuit testing to ensure the temperatures do not reach unacceptable levels.

Ingroup Bias

Ingroup bias occurs when individuals tend to favor and show preference toward people who are part of their social or identity group. This bias can affect testing, particularly in team-based testing environments where testers are part of different groups or departments.

When ingroup bias is present in testing, testers are more likely to favor their team or department, potentially overlooking or downplaying issues or problems identified by testers from other groups. This bias can ultimately lead to a biased testing process and result in the release of software or systems with unresolved issues or defects.

To combat ingroup bias in testing, it's essential to establish clear and objective testing criteria and protocols that apply equally to all testers, regardless of their team or department. In addition, testers should be encouraged to collaborate and share information openly, and efforts should be made to ensure all testers have access to the same information and resources.

Insensitivity to Sample Size

This cognitive bias occurs when people assess the likelihood of obtaining a sample statistic without considering the sample size. Insensitivity to sample size refers to the property of specific statistical methods or techniques that remain stable and produce consistent results regardless of the sample size. This means that changes do not significantly influence the results obtained from these methods in sample size and that increasing the sample size does not significantly change the results.

For example, some commonly used statistical techniques, such as the t-test and chi-squared test, are relatively insensitive to sample size and produce similar results regardless of whether the sample size is large or small.

In contrast, some statistical methods, such as simple regression analysis, are sensitive to sample size and can produce different results depending on the sample size. However, it is important to note that insensitivity to sample size is not always desirable, as it may result in reduced power or decreased accuracy in certain circumstances. In such cases, it may be necessary to use a larger sample size or a different statistical technique to obtain accurate results.

Inspection

Inspection (a static testing technique) sometimes refers to examining or evaluating a product, system, or service to ensure that it meets specified requirements or standards. Inspection can be performed at various stages of production, from raw materials to the final product, to ensure that the product meets the desired quality and safety standards.

Inspection can be performed via visual, manual, and automated means (see Figure I.3). As the name suggests, visual inspection involves examining the product to identify any defects or deviations from the specified requirements. The manual assessment involves physically manipulating the product to test its performance and functionality. Automated inspection uses machines and technology to inspect the product, such as machine vision or X-rays.

FIGURE I.3 Inspections can be conducted manually or through automation.

Gorodenkoff/Shutterstock.com.

Inspection is essential to quality control to prevent defects and ensure that products meet customer expectations. It also helps identify and correct production process problems and improve overall quality. The manufacturer, the customer, or a third-party inspector can perform an inspection. The frequency and type of inspection will vary depending on the product and industry. Still, the goal of the examination is always to ensure that the product meets the desired quality and safety standards [12].

Installation Testing

Installation testing focuses on verifying that software can be installed, configured, and uninstalled correctly and without issues. Installation testing ensures end users can quickly and successfully update, install, and set up the software on their computers or devices.

During installation testing, testers typically perform the following tasks:

1. Installation Verification: Testers verify that the installation process is smooth and error-free and that the software is installed in the expected location with the correct configuration.
2. Configuration Verification: Testers verify that the software has been properly configured and is ready for use, including any necessary customizations or settings.

3. Compatibility Testing: Testers verify that the software is compatible with the target operating system, hardware, and other software or applications that it may interact with.
4. Uninstallation Testing: Testers verify that the software can be uninstalled without any issues or adverse effects on the system or other applications.
5. Error Handling Verification: Testers verify that the software handles errors and exceptions appropriately during the installation and configuration process.

By performing installation testing, software developers and testers can ensure that end users have a positive experience when installing and setting up their software, ultimately leading to greater user satisfaction and fewer support requests or issues related to installation or configuration.

Instrumentation

Instrumentation in testing refers to adding specialized software or hardware to a system or application to measure or monitor its performance during testing. Instrumentation includes tools such as debuggers, profilers, and performance monitors, which are used to gather data on system behavior and identify any issues or bottlenecks impacting performance.

Developers and testers use instrumentation to gain greater visibility into the inner workings of a system or application, allowing them to identify and diagnose issues more effectively. This can be particularly useful for performance testing, where testers may need to measure and analyze system performance under different load conditions or in response to specific user actions.

Some common types of instrumentation used in testing include:

1. Code Profilers: These tools are used to monitor the execution of code to identify performance bottlenecks and areas of inefficiency.
2. Debuggers: These tools allow developers and testers to step through code to identify and diagnose issues such as logic errors, memory leaks, and other bugs.
3. Load Testing Tools: These tools simulate user traffic and activity to measure system performance under different load conditions.
4. Logging and Tracing Tools: These tools are used to capture and record system events and interactions, providing a detailed record of system behavior that can be used for analysis and diagnosis.

Instrumentation allows developers and testers to gain deeper insights into system behavior and performance and to more effectively identify and diagnose issues impacting system reliability and user experience.

Integration Coverage

Integration coverage refers to the extent to which a software system's different components are integrated and tested to ensure they work correctly. Integration coverage is a critical aspect of software testing, as it helps to ensure that the entire system, rather than individual components, functions correctly.

$$\text{Integration Coverage} = \frac{\text{Modules}_{tested}}{\text{Modules}_{total}}$$

There are several different types of integration testing, each of which focuses on testing different levels of integration:

1. Component Integration Testing: This type of testing focuses on testing individual components of the software system and their interactions with one another.
2. API Integration Testing: This type of testing focuses on testing the interfaces between different system components, such as APIs or web services.
3. System Integration Testing: This type of testing focuses on testing the system as a whole, including all of its components and subsystems.
4. End-to-End Integration Testing: This type of testing focuses on testing the entire system, from end to end, to ensure that all components and subsystems work together correctly.

High integration coverage is achieved by ensuring that all system components and subsystems are properly integrated and thoroughly tested, and that all potential failure points are identified and addressed before the software is released. Ultimately, high integration coverage can help to ensure that the software is reliable, functional, and meets the needs of its end users.

Integration Testing

Integration testing focuses on verifying that the individual components of a software system work correctly when integrated. In addition, integration testing aims to identify and address any issues or defects arising from interactions between different system components.

Integration testing typically follows unit testing, which involves isolating each system component. Once each component has been tested, it is integrated and tested to ensure it functions correctly.

There are several different types of integration testing:

1. Big Bang Integration Testing: In this approach, all system components are integrated and tested together at once.

2. Incremental Integration Testing: In this approach, the system is built up and tested incrementally, with new components added and tested one at a time.
3. Top-Down Integration Testing: In this approach, the higher-level components of the system are tested first, with lower-level components added and tested as the testing process progresses.
4. Bottom-Up Integration Testing: In this approach, the lower-level components of the system are tested first, with higher-level components added and tested as the testing process progresses.
5. Hybrid Integration Testing: This approach combines top-down and bottom-up integration testing elements, with testing proceeding upward and downward through the system.

During integration testing, testers typically ensure that all system components function, including verifying that data passes correctly between components, that system interfaces function as expected, and that any dependencies between components are appropriately managed. By performing thorough integration testing, developers and testers can help to ensure that software systems function correctly and reliably, ultimately leading to greater user satisfaction and fewer issues or defects in production.

Intermittent

Intermittent issues refer to defects or bugs that only occur sporadically or under certain (unknown) conditions, making them difficult to reproduce and diagnose. These issues can be particularly challenging for software testers and developers, as they may be difficult to replicate and not show up during standard testing processes.

There are several strategies that testers and developers can use to identify and address intermittent issues:

1. Collecting Detailed Data: Testers should collect detailed data on the conditions under which the issue occurs, including any error messages or system logs that may provide additional insights into the cause of the problem.
2. Using Automated Testing Tools: Automated testing tools can help to reproduce the conditions under which the issue occurs, making it easier to identify and diagnose the underlying cause.
3. Implementing Monitoring and Analytics Tools: Monitoring and analytics tools can help to identify patterns and trends in system behavior that may be contributing to intermittent issues.

4. Testing with Real-World Scenarios: Testers should try to replicate real-world scenarios and user workflows, as these may reveal issues not evident in more controlled testing environments.
5. Collaborating with Development Teams: Testers should work closely with development teams to identify and diagnose intermittent issues, leveraging the expertise of developers who are more familiar with the underlying code and architecture of the system.

Addressing intermittent issues requires persistence, attention to detail, trial and error, and a willingness to collaborate and explore multiple potential sources of sporadic performance. Using various testing techniques and collaborating closely with development teams, testers can help identify and resolve intermittent issues, ensuring that software systems are reliable and perform as expected.

Invalid Partitions

Invalid partitions refer to situations in which the data or test cases used in software testing do not represent the conditions the software will encounter in the real world. This disparity can result in inaccurate or incomplete testing results, leading to software defects and issues in production.

Invalid partitions can arise for several reasons:

1. Incomplete or Outdated Test Data: If the test data used in software testing is incomplete or obsolete, it may not accurately represent the conditions the software will encounter in the real world.
2. Overgeneralization of Test Cases: Overgeneralized test cases may not account for the specific conditions that the software will encounter in different environments or particular uses.
3. Incorrect Assumptions about User Behavior: If testers make false assumptions about how users will interact with the software, they may not adequately test all potential scenarios that could arise in real-world use.
4. Inadequate Consideration of Edge Cases: Edge cases, or scenarios where the input or conditions fall outside the expected range, are often overlooked in testing but can be critical in identifying potential defects and issues.

To avoid invalid partitions, testers should ensure that their test data and scenarios accurately reflect the conditions the software will (can potentially) encounter in the real world. Avoidance may involve using real-world data or simulating realistic scenarios that reflect how users interact with the software. Testers should also ensure that they test a wide range of scenarios, including edge cases and unexpected conditions, to ensure that the software is robust and reliable. Finally, testers should collaborate closely with development teams and other stakeholders to ensure they have a thorough understanding of the software and its intended use cases so that they can design effective testing strategies that reflect the needs and expectations of end users.

IRR

Inter-rater reliability helps ensure that the assessments and judgments made by different testers are consistent and reliable. It can be applied to product technical testing, especially when multiple testers or experts are involved in evaluating the performance, functionality, or quality of a product.

When conducting technical testing, such as evaluating the performance of software, hardware, or other technical products, it's important to establish clear criteria or standards for evaluation. These criteria serve as a guideline for the testers and help promote consistency in their assessments. Training and calibration sessions can also be conducted to align the testers' understanding and interpretation of the evaluation criteria.

To assess inter-rater reliability in product technical testing, complete the following steps:

1. Define the Evaluation Criteria: Clearly define the aspects or parameters that need to be evaluated, including performance metrics, functionality, usability, reliability, or any other relevant factors.
2. Provide Training and Guidelines: Train the testers on the evaluation criteria and provide them with clear guidelines and instructions on how to conduct the tests and record their observations.
3. Conduct Pilot Testing: Before the actual testing phase, perform a pilot test where all testers evaluate a set of sample products or test cases. This helps identify any discrepancies or areas of confusion in the evaluation process.
4. Calculate Inter-Rater Reliability Measures: Once the testing phase is complete, calculate inter-rater reliability measures such as Cohen's kappa, ICC, or Fleiss' kappa, depending on the type of data and the number of testers involved. These measures provide insights into the level of agreement or consistency among the testers.

5. Address Discrepancies and Reevaluate: If there are low inter-rater reliability scores, review the evaluation process, clarify ambiguous criteria, or provide additional training to improve consistency. Reevaluate the product or conduct further testing as needed.

High IRR indicates a high level of agreement between different raters or evaluators, indicating that the testing process is reliable and consistent. On the other hand, low IRR suggests a lack of understanding between different raters or evaluators, indicating that the testing process may be unreliable or inconsistent.

IRR is a vital test measure that can provide valuable insights into the reliability and consistency of test results and the effectiveness of different testing techniques and methodologies. By assessing IRR, testers can ensure that their testing efforts are reliable and consistent and identify areas for improvement and optimization in the testing process.

Irrelevant Failures

Irrelevant failures refer to test failures that do not indicate a problem with the software tested but are caused by issues with the test environment, test data, or other factors unrelated to the software itself. These failures can be frustrating for testers and developers, as they can consume valuable time and resources and distract from genuine software defects that require attention. There are two categories of irrelevant failures:

1. An actual defect causes no observable malperformance on the product features.
2. A product performance malady was observed due to testing interactions.

Irrelevant failures due to test interactions occur for several reasons:

1. Issues with the Test Environment: If the test environment is not configured correctly or does not accurately reflect the real-world environment in which the software will be used, it may produce irrelevant test failures.
2. Inadequate Test Data: If the test data used in software testing does not represent the real-world data that the software will encounter, it may produce irrelevant test failures.
3. Poorly Designed Test Cases: If the test cases used in software testing are poorly designed or do not accurately reflect the expected use cases of the software, they may produce irrelevant test failures.

4. System or Network Issues: If there are issues with the underlying systems or network infrastructure that the software relies on, it may produce irrelevant test failures.

Testers should design effective test environments to avoid irrelevant failures, test data, and test cases that accurately reflect the real-world conditions the software will encounter. Testers should also collaborate closely with development teams and other stakeholders to ensure that they have a thorough understanding of the software and its intended use cases so that they can design effective testing strategies that reflect the needs and expectations of end users.

In addition, testers should identify and address irrelevant failures as quickly as possible so that they do not consume unnecessary time and resources and do not distract from genuine software defects that require attention. This may involve implementing automated testing tools or other testing techniques that can help to identify and diagnose the underlying causes of test failures quickly.

Iterative Development

Iterative development is a software development methodology that involves breaking down a project into smaller, more manageable parts or "iterations," each designed, developed, and tested before being incorporated into the larger project. This process is repeated until the project is completed.

The iterative development process involves the following steps (see Figure I.4):

1. Planning: The project is divided into smaller iterations and the goals and requirements for each iteration are established.
2. Design: The design for the current iteration is created, including any necessary changes or updates to the previous iteration.
3. Implementation: The software is developed according to the design specifications for the current iteration.
4. Testing: The software is tested to ensure it meets the current iteration's requirements and specifications.
5. Review: The results of the testing and implementation are reviewed, and any necessary changes or updates are made.
6. Repeat: The process is repeated for the next iteration, building on the work completed in the previous iterations.

FIGURE I.4 Example of iterative development steps.

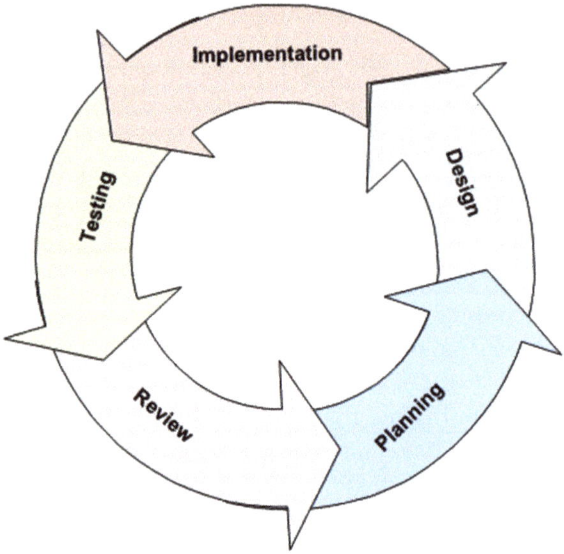

Iterative development allows for greater flexibility and adaptability in the software development process, as changes and updates can be made at each stage of development. This approach is beneficial for complex or large-scale projects where it is difficult to define all requirements upfront.

One of the key benefits of iterative development is the ability to receive feedback early and often from users or stakeholders, allowing for adjustments to be made in real time. That feedback is in the form of reviews and testing. This approach can help to identify potential issues or challenges early on in the development process, ultimately leading to a more efficient and effective end product.

K

"Stumbling is not failing."

—Portuguese proverb

Key Control Characteristics

According to the Automotive Industry Action Group (AIAG), key control characteristics refer to features that are critical to the quality of a product or service and that must be closely monitored and controlled throughout the production process to ensure compliance with required specifications and standards [91].

These key control characteristics are identified through a process known as statistical process control (SPC), which involves analyzing data from the production process to identify any patterns or trends affecting the quality of the product or service. By closely monitoring and controlling these key control characteristics, manufacturers can ensure that their products and services consistently meet all required standards and specifications, resulting in higher quality and greater customer satisfaction.

Key Product Characteristics

According to the Automotive Industry Action Group (AIAG), a key product characteristic (KPC) is a product feature or attribute critical to its performance, safety, or compliance with regulatory requirements. These characteristics are identified through a process known as design failure mode and effects analysis (see DFMEA) [92].

The purpose of identifying and defining KPCs is to ensure that they receive special attention during the design and manufacturing process. This involves setting specific requirements and limits for each KPC and developing tests and inspections to ensure that they are met.

Examples of KPCs in the automotive industry include engine power, braking distance, and emissions levels. By closely monitoring and controlling these characteristics throughout the design, development, and production process, manufacturers can ensure that their products are safe and reliable and that they meet all applicable standards and specifications.

KLOC

A "thousand lines of code," or KLOC, is a standard unit of measurement for the size, complexity, and human effort required to build a computer program. Typically, source code is what is being measured.

L

"The only person who behaves sensibly is my tailor. He takes new measurements every time he sees me. All the rest go on with their old measurements."

—George Bernard Shaw

Labcar

Labcars are advanced automotive testing tools that allow engineers to perform complex vehicle tests in a controlled laboratory environment. Labcars typically consist of a vehicle cockpit placed on a motion platform that simulates driving conditions, along with a range of sensors and monitoring equipment that capture data on the vehicle's performance.

Labcars are commonly used in the automotive industry for a range of testing activities:

1. Vehicle Dynamics Testing: Labcars can simulate various driving conditions and test the vehicle's response to speed, acceleration, and braking changes.
2. Durability Testing: Labcars can be used to simulate extended periods of driving and test the vehicle's performance over time.
3. Safety Testing: Labcars can simulate various crash scenarios and test the vehicle's safety features and crashworthiness.
4. Emissions Testing: Labcars can measure a range of emissions, including exhaust emissions, and test the vehicle's compliance with regulatory standards.

Labcars are typically used with other testing techniques, including computer simulations and real-world testing, to ensure that the vehicle meets performance and safety standards.

Using labcars can significantly accelerate the testing process, allowing engineers to perform various tests in a controlled environment without needing real-world testing. This can help reduce costs and improve testing accuracy while providing valuable insights into the vehicle's performance under various conditions.

Law of the Instrument

The cognitive bias known as the "law of the instrument" refers to the tendency to rely too heavily on a familiar tool or approach. The law of the instrument has important implications for testing. If testers rely too heavily on a particular testing tool or approach, without considering alternative and potentially more effective options, they can develop blind spots and potentially miss essential defects or issues.

To avoid falling prey to the law of the instrument testers need to remain open-minded and consider a range of testing approaches and tools. Furthermore, testers should be willing to experiment with new tools and techniques and regularly evaluate their testing approach's effectiveness to ensure that it achieves accurate outcomes.

It is also essential for testing teams to have a culture of continuous improvement, where feedback is actively sought and used to inform improvements to the testing process. Feedback can help identify areas where the team may be relying too heavily on a particular tool or approach, and can support adopting new and more effective testing strategies.

Ultimately, by remaining vigilant to the law of the instrument and actively seeking out new approaches and techniques, testing teams can ensure that they deliver high-quality software that meets end users' needs.

Level of Effort

Testing's level of effort (LOE) refers to the amount of time, resources, and energy required to perform a specific testing activity. The level of effort needed for testing can vary depending on factors such as the software's complexity, the testing team's size, the testing methodology used, and the types of tests being performed.

Factors that can impact the level of effort required for testing include the following:

1. Test Coverage: The more comprehensive the testing, the more effort is needed to cover all possible scenarios and edge cases.
2. Test Automation: Automated tests can save time and effort by reducing the need for manual testing but require an upfront investment in setting up the automation framework.
3. Test Environment: The complexity of the test environment, including the number of configurations and dependencies, can impact the effort required for testing.

4. Test Data: Generating or collecting test data can be time-consuming, particularly for complex or large datasets.
5. Test Documentation: Creating and maintaining test plans, cases, and reports can require significant effort.

By carefully estimating the effort required for testing, testing teams can develop realistic project timelines and budgets and ensure that testing activities are completed on time and to a high standard. Accurately estimating the effort required for testing can also help identify areas where improvements can be made, such as through test automation or more efficient testing methodologies.

Life Cycle Model (Product)

The life cycle model describes the stages a product goes through in its market existence. The model is represented by a curve that starts at the product's launch and ends at its eventual discontinuation (see Figure L.1). The various stages of the life cycle model for a product introduction and maturation are as follows:

1. Development Stage: This is when the product is developed, including designing, prototyping, and testing the product to meet the desired quality standards, evoking the design's weak points, and refining the features. Testing escalates up to the introduction phase of the development.
2. Introduction Stage: This is when the product is first introduced to the market. It is often characterized by low sales volume and high marketing costs as companies work to create awareness and generate interest in the new product. Testing explores latent failures in the field to determine the root cause and corrective action.
3. Growth Stage: This is when the product gains acceptance in the market and starts seeing increased sales volume. Companies may focus on expanding distribution channels, improving product features, and reducing costs to maximize profits. Adapting the product for other markets will require testing consideration.
4. Maturity Stage: This is when sales growth begins to slow down as the market becomes saturated with competitors and the product reaches its peak sales potential. Companies may focus on differentiating their product, reducing costs, or finding new markets to continue growth. Product cost improvement changes, including manufacturing processes, may necessitate testing to confirm that the changes have no adverse impact.
5. Decline Stage: This is when sales decline as the product becomes outdated or replaced by newer products. Companies may discontinue the product or attempt to extend its life by repositioning or rebranding it.

FIGURE L.1 An example of the product life cycle.

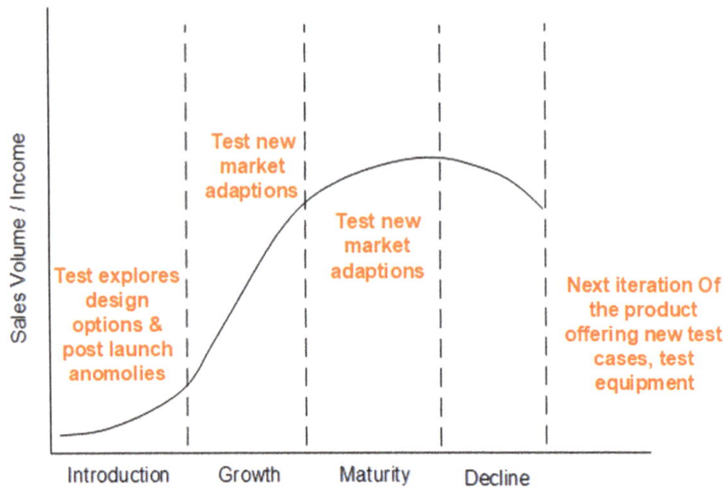

Understanding the life cycle model of a product is essential for companies to make informed decisions about product development, marketing, and sales strategies. By identifying a product's stage, companies can better allocate resources and make decisions to maximize its profitability throughout its life cycle.

Load Dump

Load dump refers to the disconnection of a high-current load from a vehicle's electrical system, resulting in a rapid voltage increase. Load dump testing is conducted to evaluate the resilience of automotive electrical systems and components to sudden voltage surges that can occur in a vehicle's electrical system during load dump events.

Load dump events can happen when a high-current load, such as the alternator or other electrical components, is suddenly disconnected while the battery remains connected. This disconnection can occur due to faulty connections, component failure, or switching off the engine. The load dump generates a surge of voltage that can reach several times the normal operating voltage of the vehicle's electrical system.

Load dump testing ensures that automotive electrical systems and components can withstand these voltage surges without experiencing damage or malfunction. In addition, it helps manufacturers verify the robustness and reliability of their products under such extreme conditions.

During load dump testing, a surge generator simulates the load dump event by introducing a controlled voltage surge into the electrical system or component under test. As relevant standards and industry specifications define, the surge generator generates a high-voltage pulse with specific characteristics, including amplitude, duration, and waveform shape.

The system or component being tested is subjected to the simulated load dump event, and its performance and behavior are observed. This includes monitoring voltage levels, current flow, functionality, and any deviations or abnormalities during and after the load dump event.

Load dump testing is typically performed in compliance with international standards and specifications specific to the automotive industry, such as J1455 (see Figure L.2). These standards define the test procedures, voltage characteristics, and performance criteria for evaluating the resistance of automotive electrical systems and components to load dump events.

FIGURE L.2 Load dump test parameters as defined in the J1455 standard [20].

System Voltage	Waveform Definition (Volts)	Rise Time (ms)	Source Resistance (ohms)	Pulse Rate	# of Pulses	Energy
12 V	$14+86e^{(-t/0.4s)}$	<5	0.4	10 second interval	5	70 J
24 V	$28+122e^{(-t/0.4s)}$	<5	0.8	10 second interval	5	100 J

Load Testing

Load testing is used to evaluate the performance and determine the stability of a software application under normal and peak usage conditions. It aims to identify the maximum number of concurrent users or demands on the system or transactions a software application can handle without experiencing performance issues or system failures.

Load testing involves simulating real-world usage scenarios and applying a load to the application to determine how it will perform under different stress levels. The loads generated are through simulating concurrent demands of the system or by simulating heavy usage over a specific period.

Common types of load testing include the following:

1. Volume Testing: This involves testing the application's ability to handle a large amount of data, such as user accounts, product listings, or other data types.
2. Stress Testing: This involves testing the application's ability to handle heavy usages, such as a sudden increase in user traffic or a surge in transactions.

3. Endurance Testing: This involves testing the application's ability to handle sustained usage over an extended period, such as several hours or days.
4. Spike Testing: This involves testing the application's ability to handle sudden spikes in usage, such as a sudden surge in traffic due to a marketing campaign or other event.

By performing load testing, developers can identify potential performance bottlenecks, such as slow response times, server crashes, or other user experience issues. Load testing can also help developers optimize the application's performance by identifying areas for improvement, such as optimizing database queries, caching data, or increasing server capacity. Overall, load testing is an essential component of software testing to help ensure that an application performs well under normal and peak usage conditions.

Localization Testing

Localization testing ensures that a software product is adapted to meet a specific geographic region or target market's language, culture, and other regional requirements. Localization testing aims to ensure that the software is easy to use and understand by the target market. In addition, it helps to identify and correct any issues related to language, formatting, or cultural differences: for example, instrument clusters with displays in a variety of languages.

Localization testing involves a variety of activities:

1. Language Testing: This involves verifying that all text displayed in the software product (e.g., instrument cluster) has been translated correctly and is free of spelling, grammar, and punctuation errors.
2. Formatting Testing: This involves verifying that the software product's formatting, such as date and time formats, currency symbols, and other regional formatting requirements, are correct and consistent with local standards.
3. Cultural Testing: This involves verifying that the software product is sensitive to local cultural norms and practices, such as using specific colors, images, or symbols.
4. Functionality Testing: This involves verifying that localization changes do not impact the software product's functionality and that all features and functions work as expected.
5. Usability Testing: This involves verifying that the software product is easy to understand and use by the target market and that the user interface is intuitive and well-designed.

By performing thorough localization testing, developers can ensure that their software products meet target market needs, improve user satisfaction, and increase product success in the marketplace.

Loop Testing

Loop testing evaluates the functionality and behavior of loops in software programs. Loops are a fundamental programming construct used to execute instructions repeatedly until a specific condition is met.

Loop testing involves designing test cases that exercise different scenarios involving the loop, such as testing the loop's execution when the loop condition is true, false, or invalid. The goal of loop testing is to ensure that the loop is executed the correct number of times and produces the expected results, as well as to identify and correct any issues related to the loop's behavior.

Common types of loop testing include the following:

1. Simple Loop Testing: This involves testing loops with a fixed number of iterations to ensure that they execute the expected number of times and produces the desired results.
2. Nested Loop Testing: This involves testing loops containing other loops to ensure that the nested loops are executed correctly and produce the expected results.
3. Infinite Loop Testing: This involves testing loops with no exit condition to ensure that the program does not get stuck in an endless loop, which can cause the program to crash or become unresponsive.
4. Boundary Testing: This involves testing loops that operate on arrays or other data structures to ensure that they execute correctly when the array or data structure is empty, contains a single element, or includes the maximum number of elements.

By using loop testing techniques, developers can ensure that their software programs function correctly and produce expected results, which can improve the overall quality and reliability of the software.

M

"Good judgment comes from experience, and experience comes from bad judgment.

—Rita Mae Brown

Machine Learning

Machine learning and testing are closely related because testing plays an essential role in ensuring the accuracy and reliability of machine learning models. In addition, testing can help identify potential issues or errors in the model and provide valuable feedback for improving its performance and functionality.

Some ways in which testing can improve machine learning models include the following:

1. Verifying that the model is performing as expected and producing accurate results.
2. Detecting and correcting errors or biases in the training data that may affect the model's accuracy.
3. Testing the model under various conditions and scenarios ensures it remains stable and reliable.
4. Evaluating the model's performance against established benchmarks or industry standards.
5. Identifying areas where the model may be overfitting or underfitting the data and adjusting the model accordingly.
6. Testing the model's ability to generalize to new and unseen data.
7. Identifying and mitigating potential security or privacy risks associated with the model.

In addition to these benefits, testing can also help ensure that machine learning models are of high quality and meet the needs of their users, which contributes to their overall success and impact. By prioritizing testing throughout the machine learning development life cycle and using testing as a tool for continuous improvement, developers can create models that are more accurate, reliable, and useful in real-world applications.

Maintainability

Product maintainability and testing are closely related because testing is critical in ensuring that a product can be easily maintained, updated, and repaired over time. Testing can help identify potential issues or areas of the product that may be difficult to maintain and can also provide valuable feedback for improving product maintainability.

Some ways in which testing can improve product maintainability include the following:

1. Detecting defects and errors early in the development process can prevent them from becoming more complex and difficult to fix later on.
2. Identifying areas of the product that are particularly complex or difficult to maintain and working to simplify or streamline them.
3. Verifying that changes or updates to the product do not have unintended consequences or introduce new issues.
4. Testing the product under various conditions and scenarios ensures it remains stable and reliable.
5. Providing feedback on the product's usability and user experience can inform future updates and improvements.

In addition to these benefits, testing can also help ensure that a product is of high quality and meets the needs of its users, which can contribute to its overall maintainability. By prioritizing testing throughout the product development life cycle and using testing as a tool for continuous improvement, developers can create easier products to maintain, update, and repair over time, ultimately leading to higher customer satisfaction and increased profitability.

Mean

The mathematical mean, also known as the average, measures the central tendency of a set of numbers. Calculate the mean by adding the numbers in the set and dividing that by the number of elements in the dataset. For example, consider the set of numbers {1, 2, 3, 4, 5}. The sum of these numbers is 15, and there are 5 numbers in the set, so the mean of the set is as follows:

$$Avg = \frac{15}{3} = 5.$$

The mean is a useful measure of central tendency because it considers all the values in the set and gives an idea of its "typical" value. It is sensitive to every value in the set, so it is affected by extreme values (outliers) more than other measures of central tendency, such as the median.

Note that the mean is not always a good measure of the central tendency for skewed distributions, where the values are not evenly distributed around the center. In these cases, the median may better measure central tendency.

Mean Time between Failure

In the automotive industry, the mean time between failures (MTBF) measures the reliability of a vehicle or its components. It is defined as the average time a vehicle or component is expected to operate without failing. MTBF is typically measured in hours and calculated by dividing the total operating time of a vehicle or part by the number of failures it experiences.

MTBF is an essential consideration for automotive manufacturers, as it can impact a vehicle's overall quality and reliability. It can also significantly impact a vehicle's ownership cost, as higher MTBF typically translates to lower maintenance and repair costs. Several factors impact the MTBF:

- Material selection
- Quality of the materials
- Manufacturing processes used
- Operating conditions
- Environment in which the product is deployed
- Level of maintenance and care that the vehicle receives

Automotive manufacturers typically strive to design and build vehicles with high MTBF to reduce the cost of ownership for their customers and improve customer satisfaction.

The MTBF of a product or system is calculated with the following equation:

$$MTBF = \frac{Total\ Working\ Time - Total\ Breakdown\ Time}{Number\ of\ Breakdowns}$$

Where:

- Total working time corresponds to the number of hours the machine would have been operating had it not failed.
- Total breakdown time is the unplanned downtime of the sum of the samples (excluding scheduled maintenance, i.e., inspections, periodic revisions, or preventive replacements).
- Number of breakdowns equals the number of failures.

For example, if a product has been in operation for 1,000 hours and has experienced two failures (one at 300 hours and the other 200 hours) during that time, its MTBF can be calculated as follows:

$$MTBF = \frac{1000 - (150 + 200)}{2}$$

$$MTBF = \frac{1000 - (350)}{2} = 325$$

This means that, on average, the product can expect to operate for 325 hours between failures.

It is important to note that MTBF is a statistical measure based on the assumption that the failure rate of a product or system is constant over time. In reality, the failure rate of a product or system may vary over time due to various factors, such as wear and tear, changes in operating conditions, or the effects of aging. As a result, the actual MTBF of a product or system may differ from the calculated value.

Mean Time to Failure

The mean time to failure (MTTF) is a metric that measures the expected time between the initial use of a product or system and its first failure. It is commonly used in reliability engineering to evaluate the performance and reliability of products and systems.

The MTTF is calculated using the following equation:

$$MTTF = \Sigma \frac{(\mathrm{T_i})}{N}$$

Where:

T_i is the time until the i-th failure occurs.

N is the total number of failures.

In other words, the MTTF is the sum of the time between each failure divided by the total number of failures. This equation assumes that the failures occur randomly and independently of each other.

For example, suppose a system has experienced five failures, with the following times of failure (in hours): 200, 400, 600, 800, and 1,000. Using the MTTF equation, we can calculate the expected time between failures as follows:

$$MTTF = \frac{200+400+600+800+1000}{5}$$

$$MTTF = 400\,hours$$

This means we can expect the system to function for an average of 400 hours before experiencing a failure. However, it is important to note that the MTTF does not guarantee that the system will last for this amount of time, as failures can occur at any point during the system's operation. It is simply a statistical measure of the system's expected reliability.

Measurement

Automotive testing often involves the use of various types of measurements to assess the performance and safety of a vehicle. Some standard measures taken during automotive testing include the following:

1. Vehicle Performance: Acceleration, braking, handling, and other performance characteristics can be measured using specialized equipment, such as acceleration sensors and lap timers.
2. Emissions: Vehicles are often tested to ensure that they meet emissions standards. This can involve measuring the levels of various pollutants, such as carbon monoxide and nitrogen oxides.
3. Fuel Efficiency: Testing can measure a vehicle's fuel efficiency, which can be expressed in miles per gallon (MPG) or liters per 100 kilometers (L/100 km).
4. Safety: Vehicles are often tested to ensure that they meet safety standards, which involves testing things like the strength of the frame, the effectiveness of the brakes, and the performance of the airbags.
5. Noise: The noise levels of vehicles can be measured using specialized equipment, such as decibel meters.
6. Durability: Automotive testing can also be used to measure the durability of a vehicle, which can involve subjecting it to extreme conditions, such as high temperatures or rough terrain, to see how it holds up over time.

The measured test results will be compared to the expectation defined in specifications and other product description documentation to determine pass or failure.

Measurement System Analysis (MSA)

Automotive measurement system analysis (MSA) evaluates the reliability and accuracy of measurement systems used in the automotive industry. Here are some of the common MSA techniques used in the automotive industry [87, 88]:

1. Gauge R&R (Repeatability and Reproducibility) Analysis: This statistical method determines a measurement system's accuracy and repeatability. It measures the variation due to the gauge (repeatability) and the variation due to the operator (reproducibility).
2. Bias Studies: This technique is used to evaluate the accuracy of the measurement system by comparing the measured values to a known reference value or standard. Bias is the difference between the reference value and the measurement value.
3. Linearity Studies: This technique is used to evaluate the ability of the measurement system to measure over a range of values. Linearity is the ability of a measurement system to provide measurements that are proportional to the actual values.
4. Stability Studies: This technique is used to evaluate the consistency of a measurement system over time. It measures the variation in measurements taken over an extended period.
5. Precision-to-Tolerance Ratio: This technique is used to evaluate the ability of the measurement system to distinguish between parts that are within the specified tolerance limits and those that are not. In addition, it compares the measurement system variation to the specified tolerance limits.
6. Attribute Agreement Analysis: This technique is used to evaluate the reliability of the measurement system for categorical data. It measures the agreement between operators or observers for categorical data, such as pass/fail or good/bad.
7. Capability Studies: This technique is used to evaluate the ability of a process to produce parts that meet the specifications. In addition, it measures the capability of the measurement system to detect variation in the process.

Mechanical Shock

Mechanical shock testing is used to determine the strength and durability of automotive components under extreme conditions (Figure M.1). This type of testing is often used to simulate the effects of rough terrain, high speeds, and the sudden impacts a vehicle may encounter during regular use.

FIGURE M.1 An example of mechanical shock stimulus [20].

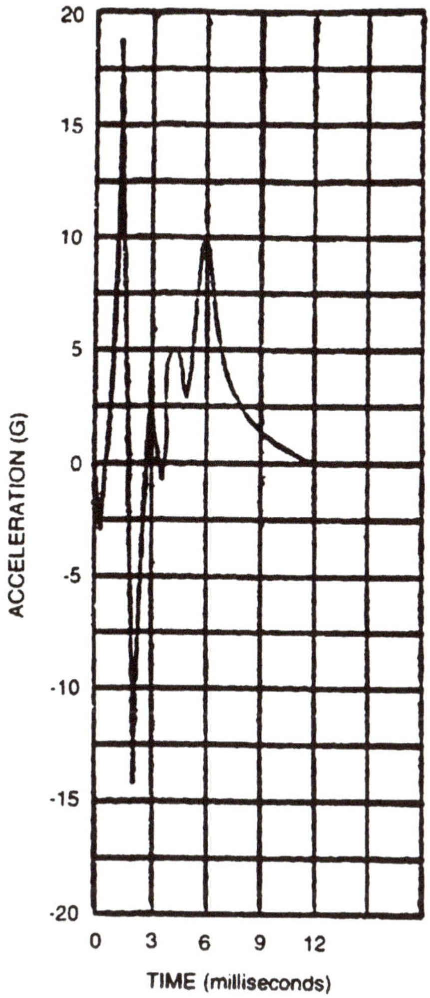

During mechanical shock testing, an automotive component is subjected to a series of mechanical shocks or impacts, typically using a shaker table or other mechanical device. The part is subjected to these shocks at various frequencies, amplitudes, and durations to mimic the stresses and strains it may experience in the real world.

Mechanical shock testing is an integral part of the development process for new vehicles and components, as it helps manufacturers ensure that their products can withstand the rigors of daily use. It also helps identify potential failure points or weaknesses in a design, allowing manufacturers to make necessary modifications before releasing a product.

Mechanical Vibration

Mechanical vibration testing evaluates the durability and reliability of automotive components and systems. It involves subjecting the parts to controlled vibration levels and simulating the stresses and strains they may experience during regular operation or extreme driving conditions. The testing is conducted using specialized equipment that can generate specific frequencies and amplitudes of vibration (see Figure M.2) [20].

FIGURE M.2 An example of measured vehicle vibration [20].

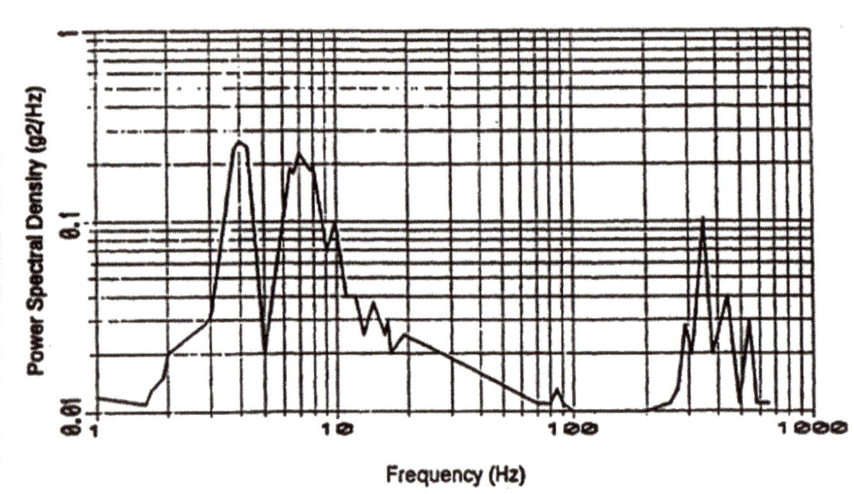

Mechanical vibration testing aims to identify weaknesses or vulnerabilities, such as cracks, deformations, or other signs of damage, in the component or system being tested. By identifying these issues early on, manufacturers can make necessary design changes or adjustments to improve the durability and reliability of the component.

Mechanical vibration testing is an integral part of the automotive industry, as it helps ensure that vehicles are safe, reliable, and perform well under various conditions (see Figure M.3). It is also an essential part of the product development process, allowing manufacturers to identify and address potential issues before releasing the product.

FIGURE M.3 An illustration of a PCB under mechanical vibration.

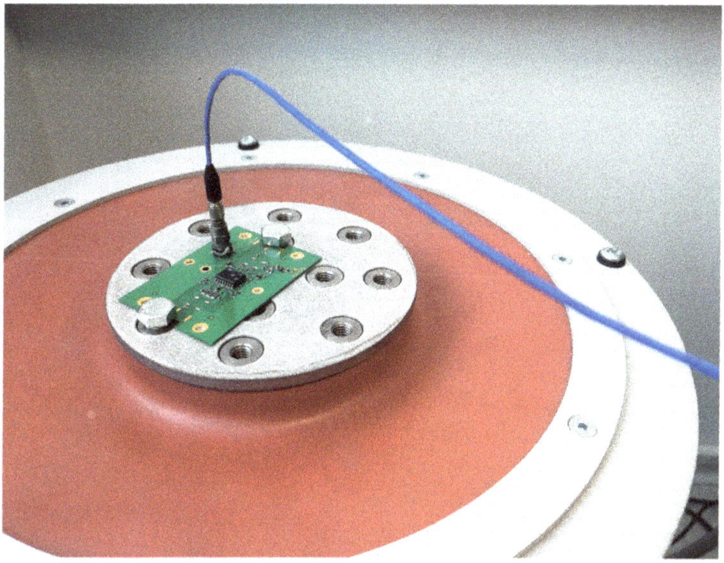

Audrius Merfeldas/Shutterstock.com.

Median

The median is a statistical measure representing a dataset's middle value. It is calculated by organizing the dataset in numerical order and finding the middle value. If the dataset has an odd number of values, the median is simply the middle value. If the dataset has an even number of values, the median is the average of the two middle values. The median helps understand the central tendency of a dataset, as it is less affected by extreme values or outliers. It is often used in place of the mean (average) when the dataset has many outliers or is skewed in distribution.

For example, consider a dataset of five values: 1, 2, 5, 10, and 20. The median value would be 5, the middle value when the dataset is organized in numerical order. If we added a value of 100 to the dataset, the median would still be 5, as it is unaffected by the outlier value of 100. The mean, however, would be significantly affected by the outlier value of 100, as it would be much higher than the other values in the dataset.

Microcontroller

A microcontroller is a small, self-contained computer system on a single integrated circuit (IC) that controls and manages devices or systems. It typically consists of a central processing unit (CPU), memory, input/output interfaces, and sometimes additional peripherals such as timers, analog-to-digital converters, and communication interfaces (see Figure M.4).

FIGURE M.4 An illustration of a microcontroller and peripherals.

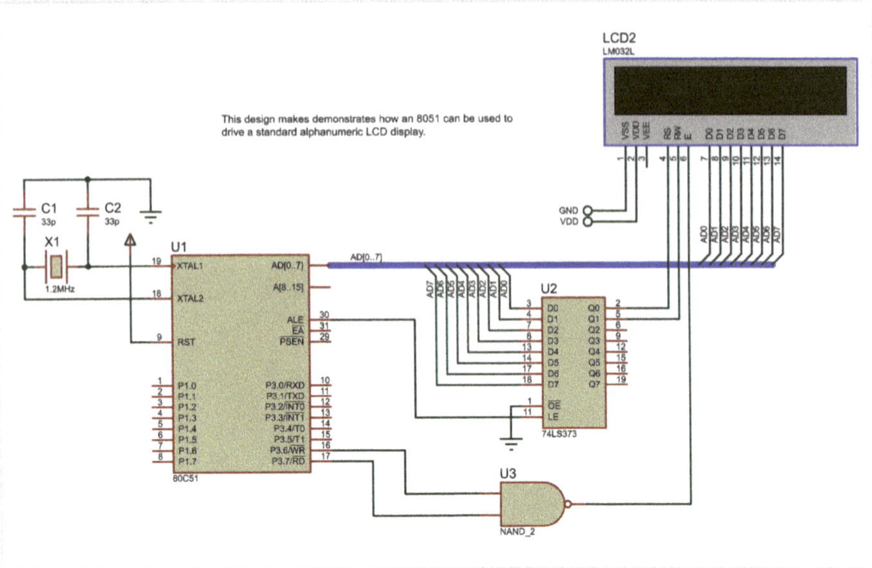

Microcontrollers are used in many applications, from simple control systems, such as thermostats and lighting controls, to more complex systems, such as automotive engine management, medical devices, and industrial automation.

One of the critical advantages of microcontrollers is their compact size and low power consumption, which makes them well-suited for applications where space and power are limited. They are also relatively inexpensive and easy to program, which makes them accessible to a wide range of users.

Microcontrollers are programmed using software, typically consisting of high-level programming languages like C or C++ and low-level assembly language instructions specific to the microcontroller hardware. In addition, microcontrollers are often combined with sensors, actuators, and other electronic components to create complete embedded systems. These systems can be programmed to perform a wide range of functions, from simple on/off control to complex data acquisition, processing, and communication.

Mileage Accumulation

Mileage accumulation testing involves gathering a specified number of miles or kilometers on a vehicle under controlled conditions to evaluate its performance, durability, and reliability over various driving conditions, tire types, and inflation. This type of testing is often used in the automotive industry to validate the performance and reliability of new vehicles before releasing them to the market.

During mileage accumulation testing, the vehicle is driven under various conditions, including different speeds, road surfaces, and temperature and humidity levels, to simulate real-world driving conditions. The testing is performed on a dynamometer or public road, depending on the requirements and objectives of the test.

The data collected during mileage accumulation testing can be used to evaluate the performance of various vehicle components, such as the engine, transmission, suspension, brakes, and tires. In addition, mileage accumulation testing helps identify any potential long-term issues or latent defects in the vehicle, such as wear and tear, vibration, noise, or fuel consumption.

Mileage accumulation testing is often used with other types of vehicle testing, such as durability testing, crash testing, and emissions testing, to achieve a comprehensive evaluation of the vehicle's performance and safety.

Overall, mileage accumulation testing is vital for vehicle testing and development. It helps ensure that new vehicles are safe, reliable, and perform well under real-world driving conditions. In addition, it is a critical step in bringing new vehicles to market and can help prevent costly recalls and warranty claims.

Milestone

A milestone is a significant point or achievement in a project or process. It is a specific event or moment representing a critical accomplishment or progress toward completing a testing goal. Milestones are often used to track and measure progress and ensure a project stays on schedule.

Testing is evaluating a product, system, or process to determine its quality or performance. It is a critical component of the product development process, as it helps identify defects, errors, and other issues that can impact a product's performance, safety, or usability. Testing can be performed at various stages of the product development process, including during design, development, and production. In addition, milestones and testing are closely linked, as testing validates progress toward project milestones, such as the following:

- Test plan completed date
- Test cases completed date
- Product available for a testing date
- Testing start date
- Test completion date
- Test result report date

Mode

The mode of a dataset is the value that appears most frequently. For example, the value of X at which the probability mass function reaches its highest value is known as the mode if X is a discrete random variable. In other words, the value will be sampled most frequently.

The mathematical mode is the most frequently occurring value in a dataset. It is used to represent the central tendency of the data and is used in statistical analysis. All values in the dataset are counted to find the mode, and the value that occurs most frequently is identified as the mode. If multiple values occur the same number of times, the dataset is considered to have various modes. If no value occurs more than once, the dataset is deemed to have no mode.

Model-Based Development

Model-based development and testing is an approach to software development that involves creating a mathematical model of the system developed and using this model to guide the design, implementation, and testing (Figure M.5) [93].

FIGURE M.5 An example of the distribution of model and real environment testing [93].

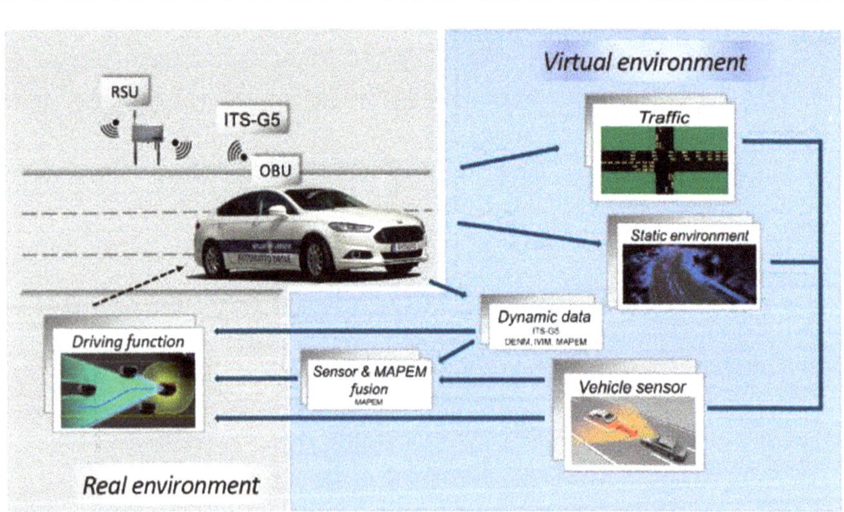

The model serves as a specification for the system, defining its behavior and interactions with other components. It can be used to generate code automatically, which can help to reduce errors and improve the efficiency of the development process. In addition, the model can be used to simulate the system's behavior under different conditions, which can help identify potential issues before software deployment.

Testing is an integral part of model-based development, and several testing techniques are commonly used in this approach. One technique is equivalence class testing, which involves dividing the input space into equivalence classes and selecting test cases from each category to ensure the software is tested under various conditions.

Another technique is boundary value analysis, which involves selecting test cases on the boundaries between equivalence classes. This technique helps ensure the software testing occurs under conditions likely to cause errors or unexpected behavior. Finally, model-based testing can also involve formal verification techniques, such as model checking, which automatically verify that the software satisfies specific properties or requirements specified in the model.

Model-based development and testing can help improve software development's efficiency and effectiveness by providing a rigorous and systematic approach to software systems' design, implementation, and testing. It can also help reduce the risk of errors and ensure the software meets its specified requirements.

Model-Based Testing

Model-based testing uses a model of the system under test to generate test cases. The model represents the system's expected behavior and serves as a guide to design test cases that verify the system's functionality (Figure M.6).

FIGURE M.6 An example of a model layout in Simulink [94].

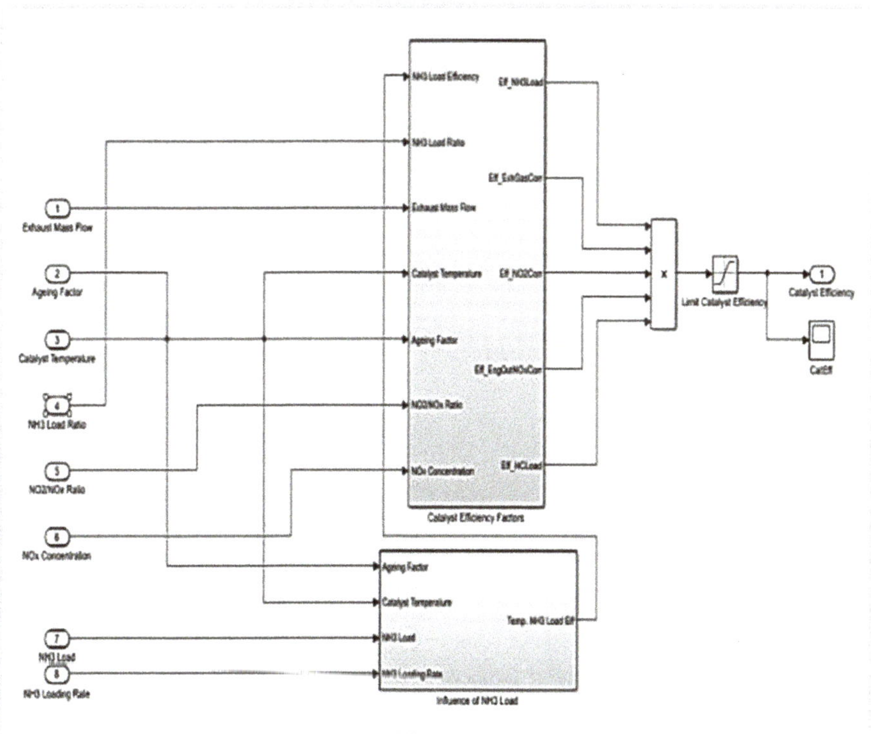

There are several benefits to using model-based testing. It can help identify potential defects in the system early in the development and process and provide a structured approach for prioritizing, creating and executing test cases. It also allows for automatically generating test cases, saving time and resources.

Model-based testing begins with the tester creating a model of the system using a modeling language or tool. This model represents the system's behavior and defines each function's or feature's input and output values. The tester then uses the model to generate test cases, which are executed to verify that the system behaves as expected.

Model-based testing is a helpful technique for ensuring the quality and reliability of software systems. It provides a structured approach for creating and executing test cases and can help to identify defects early in the testing process [94].

Modeling Tool

A modeling tool is a software application or physical device to create, manipulate, and analyze digital representations of physical or abstract systems. These tools can be used in various fields, including engineering, architecture, finance, biology, and computer science. Examples of modeling tools include computer-aided design (CAD), financial modeling, and simulation software. These tools allow users to create and analyze models of real or hypothetical systems to gain insights and make informed decisions.

Modeling tools are categorized as test design tools since they are essentially "model-based testing tools," which create test inputs or test cases from recorded information about a particular model (such as a state diagram). They aid in the validation of software or system models. For more information, see **J2812** [95].

Moderator

A moderator in a technical review is the person responsible for facilitating the review process and ensuring that it runs smoothly, including organizing and scheduling meetings, distributing materials and documents to the review team, and managing any conflicts or issues arising during the review process. The moderator may also be responsible for keeping track of the review team's progress and feedback and providing guidance and support to the team as needed. Generally, the moderator plays a key role in ensuring that the technical review process is thorough, efficient, and effective in identifying and addressing any issues or problems with the product or system being reviewed [12].

Module Testing

Module testing is a sort of software testing that examines certain classes, methods, subprograms, or subroutines within a program. Module testing advises testing the program's smaller building components before the entire program.

In module testing, individual units or components of a software application are tested in isolation. The goal is to validate that each software unit or component of the application works as intended, meets specified requirements, and performs as desired. The development team usually does module testing as part of the implementation phase of the software development life cycle. It is done before system testing, which tests the entire application.

Monkey Testing

Monkey testing involves randomly inputting data or performing actions on a software system to see how it reacts. For example, in the automotive industry, monkey testing could involve randomly pressing buttons or manipulating controls to see if they respond appropriately or causes any issues. This type of testing helps identify potential bugs or problems in the system not identified during more traditional testing methods.

Monte Carlo Simulation

Monte Carlo simulation is a computational method that uses random sampling to model and analyzes the probability of different outcomes in a system or process. It is named after the city of Monte Carlo in Monaco, famous for its casinos and use of randomness in games of chance. Monte Carlo simulations are used in finance, engineering, operations research, and research and development.

A computer program generates many random samples from a statistical distribution in a Monte Carlo simulation. These samples estimate the probability of different outcomes and the uncertainty associated with those outcomes.

One of the main advantages of a Monte Carlo simulation is its ability to model complex systems with many variables and dependencies, which can be challenging to analyze using traditional methods. It is also a valuable tool for evaluating the impact of risk and uncertainty on decision-making. However, Monte Carlo simulation can be time-consuming and requires high computational power. Moreover, the results are only as accurate as the assumptions and input data used in the simulation [96, 97].

Mutation Testing

Mutation testing in the automotive industry involves introducing small changes or "mutations" to the code of a software system and then testing to see if the system can still function correctly. This type of testing identifies code weaknesses or vulnerabilities and ensures that the system can handle unexpected changes or errors. In the automotive industry, mutation testing ensures the reliability and safety of software systems that control critical functions such as brakes, steering, and engine management.

Mutation testing begins with a tool automatically generating a set of mutated code versions, each with a single fault. The test suite is then run against each mutated version to determine whether it detects the fault. If the test suite passes for a mutated version, it does not cover the specific defect introduced by that mutation, and the transformation is considered "killed." On the other hand, if the test suite fails for a mutated version, it means that the test suite has successfully detected the introduced fault, and the mutation is considered "alive."

Mutation testing can be time-consuming and resource-intensive, but it can provide valuable insights into the test suite's quality. It can also help identify areas of the code that are not being adequately tested, leading to improvements in the overall quality of the software.

N

"There is only one thing more painful than learning from experience, and that is not learning from experience."

—Archibald MacLeish

Negative Testing

Negative testing in the automotive industry refers to testing a product or system to ensure it does not fail or malfunction under certain conditions. This includes testing for extreme temperatures, shock or vibration, collision, rollover, and other stressors that may cause a component or system to fail. Negative testing is essential in the automotive industry as it helps to ensure the safety and reliability of vehicles and prevent accidents or malfunctions.

Negativity Bias

Negativity bias refers to focusing more on negative than positive information. This bias can affect product testing in several ways. First, testers may be more likely to focus on finding faults and weaknesses in a product rather than highlighting its strengths. This hyperfocus can result in a disproportionate emphasis on negative results, which may not accurately reflect the product's overall performance.

Second, negativity bias can also lead testers to overlook a product's positive features or dismiss them as insignificant. Again, this can result in an incomplete or biased product performance evaluation. It is good to mitigate the effects of negative bias in product testing through standardized testing protocols, defining clear criteria for evaluating performance, and involving a diverse group of testers with different backgrounds and perspectives. However, it is also essential to remain open-minded and objective when assessing test results and to avoid making premature judgments based on early findings.

Minimize the effects of negativity bias through rigorous and standardized testing protocols, copious experience-based testing, and the involvement of a diverse group of testers.

Neglect of Probability

Neglect of probability is a form of cognitive bias that refers to the propensity to ignore a probable possibility when one is faced with an uncertain decision. It is one way that people reject the normative guidelines for making choices. As a result, small dangers are often entirely ignored or vastly overstated.

The neglect of probability in automotive engineering can lead to negative consequences. For example, if testers have neglected the probability of an event during the design and testing of a vehicle, it may be prone to unexpected failures or accidents due to unforeseen variables or circumstances that have gone unexplored and untested. This failure can lead to severe injury or death for the driver and passengers and damage to the vehicle and other property. In addition, neglecting probability can also lead to costly recalls and legal issues for the automotive company. Therefore, automotive engineers need to consider event probability to ensure the safety and reliability of vehicles.

Network Testing

Like software testing, network testing involves using a performance test to find flaws and performance issues, assess significant network modifications, and gauge network performance [98].

Network testing in the automotive industry involves testing the communication systems within a vehicle, such as the infotainment system, navigation system, and telematics system. This testing includes connection and functionality of the components, for example, sensors, cameras, and displays, as well as the data transmission and communication between these components. Network testing also involves checking for security vulnerabilities and protecting the network against cyber threats. This testing is essential as the increasing reliance on connected and autonomous vehicles has made them more vulnerable to cyberattacks. For more information, see **J2602-2_202110** [98].

Neural Networks

In artificial intelligence, neural networks are machine learning algorithms trained on large datasets to recognize patterns and make predictions. Therefore, testing neural networks is essential to ensure accurate and reliable predictions.

One common approach to testing neural networks is to split the dataset into training and testing sets. The network is trained on the training set and then confirmed on the testing set to measure its accuracy. This approach helps to ensure that the network has not simply memorized the training set but can generalize to new data.

Another approach is cross-validation, where the dataset is divided into several subsets. Then, the network is trained on each subset while being tested on the others. This approach can help to reduce overfitting, which occurs when the network becomes too specialized to the training set and performs poorly on new data. In addition to these approaches, various other techniques are used in testing neural networks, such as regularization, dropout, and early stopping, which can help prevent overfitting and improve the network's performance.

Testing is critical to developing and deploying neural networks, as it ensures that they are accurate and reliable in making predictions. Therefore, it is essential to use a variety of testing approaches and techniques to ensure that the network is robust and can generalize to new data [99].

Nonfunctional Testing

Nonfunctional testing in the automotive industry evaluates a vehicle's or system's nonfunctional requirements, such as performance, reliability, security, and usability. This type of testing ensures that a vehicle or system performs as expected under different conditions and scenarios [100, 101].

Examples of nonfunctional testing in the automotive industry include the following:

1. Performance Testing: This type of testing evaluates the speed and efficiency of a vehicle or system under different workloads. It helps to ensure that the vehicle or system can handle high usage levels without experiencing any issues.
2. Reliability Testing: This type of testing assesses the durability and longevity of a vehicle or system. It helps to ensure that the vehicle or system can withstand normal wear and tear and unexpected conditions such as extreme temperatures or rough terrain.
3. Security Testing: This type of testing evaluates the security of a vehicle or system, including its ability to protect against unauthorized access or tampering.
4. Usability Testing: This type of testing assesses the ease of use of a vehicle or system. It helps ensure the vehicle or system is intuitive and easy for users of all skill levels.

Normal Distribution

The normal distribution is a symmetrical probability distribution about the mean, with a bell shape, a mean (μ), and a standard deviation (σ) define it (see Figure N.1).

FIGURE N.1 An illustration of normal distribution and deviations.

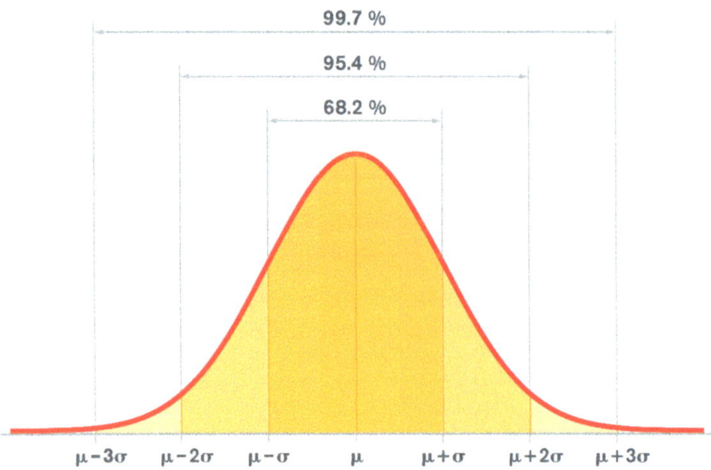

Peter Hermes Furian/Shutterstock.com.

A normal distribution can determine the probability of a given value occurring within a specific mean range. For example, if the norm is 50 and the standard deviation is 10, the probability of a value between 40 and 60 is 68%, as it falls within one standard deviation of the mean.

The normal distribution is often used in statistical analysis to model real-world phenomena, such as height, weight, and intelligence. It is also commonly used in hypothesis testing and to estimate the confidence intervals of a sample.

The normal distribution has several important properties:

- The area under the normal distribution curve equals 1, representing the probability of an event occurring.
- The mean, median, and mode are equal at the bell curve's peak.
- Approximately 68% of the values fall within one standard deviation of the mean, 95% within two standard deviations, and 99.7% within three standard deviations.
- The normal distribution is continuous, meaning that it can take on an infinite number of values.

Normalcy Bias

The normalcy bias can affect product testing in any industry, especially with technology. Product testers may become accustomed to specific test results and fail to consider the possibility of unexpected results or failures. For example, a company may conduct product testing on a new feature and find that it performs well under certain conditions. However, if the testers are affected by normalcy bias, they may fail to consider the possibility of the product failing under other conditions, such as extreme temperatures or high humidity.

This bias can result in insufficiently tested products with latent defects found after release. The effects of normalcy bias in product testing can be mitigated through diverse teams of testers with varying backgrounds and experiences, conducting tests under various conditions, and continually reviewing and updating test protocols.

O

OBD Testing

On-board diagnostics (OBD) testing analyzes and diagnoses the performance of a vehicle's onboard systems and components. OBD testing aims to identify problems or malfunctions that could affect the vehicle's emissions output or overall performance.

OBD testing is typically required by law for vehicles to be registered and operate on public roads. The testing is performed at authorized inspection centers or garages. During the test, a technician connects a diagnostic tool to the vehicle's OBD port to retrieve information about the engine's performance and emissions output (see Figure O.1). The device analyzes the data and reports any issues that need to be addressed.

FIGURE O.1 OBD diagnostics tool is used to view the state of the emissions systems.

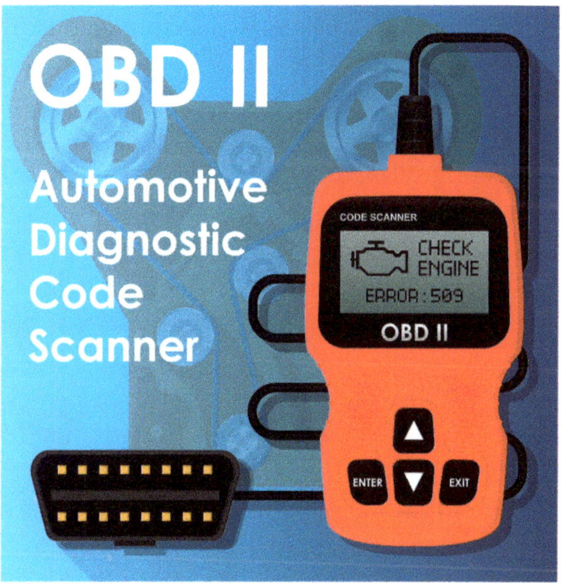

Pepermpron/Shutterstock.com.

Specific OBD requirements and standards vary by country and jurisdiction. For example, the Environmental Protection Agency (EPA) sets vehicle emission standards and mandates OBD testing to ensure compliance in the United States. The European Union (EU) has established similar emissions standards and OBD requirements in Europe.

OBD testing ensures vehicles' safety and environmental compliance on public roads. By identifying and addressing potential issues, OBD testing can help improve vehicles' overall performance and longevity while reducing harmful emissions. For more information, see **J1699-1**, **J1699-2**, **J1699-3**, and **J1699-4** [102, 103, 104, 105].

Object-Oriented Applications

Object-oriented applications can be used in embedded systems, which are computer systems built into other devices, such as cars, appliances, and medical devices. Embedded systems often have limited resources, such as memory and processing power, so object-oriented programming (OOP) can be used to create more efficient and modular software. Modular software offers possibilities for improved quality through use of pretested software assemblies, which are also quicker to market.

Rules in object-oriented programming (OOP) serve as guidelines and principles that help developers design and implement effective, maintainable, and organized code. (see Figure O.2) These rules exist to promote best practices and to harness the benefits of the OOP paradigm. Here are some reasons why rules are important in OOP:

1. Classes: The blueprint or template for creating objects. They define the properties (attributes) and behaviors (methods) that things of that class will possess.
2. Objects: Instances of classes. They represent real-world entities or concepts and encapsulate data (attributes) and behavior (methods) associated with that entity.
3. Encapsulation: Bundles data and methods within a class. It allows for data hiding, as the internal implementation details are hidden from external access and only the defined public interfaces are exposed.
4. Inheritance: Allows for the creation of new classes based on existing classes. The derived or child class inherits the properties and methods of the base or parent class, and it can also add or modify its own unique characteristics.
5. Polymorphism: Allows objects of different classes to be treated as objects of a common parent class. It allows methods to

be implemented differently in various subclasses while adhering to a common interface.

6. Abstraction: Simplifies complex systems by breaking them down into smaller, more manageable parts. It focuses on essential features and hides unnecessary details, allowing developers to work with high-level concepts.
7. Modularity: Promotes code organization into smaller, self-contained modules or units. Modularity improves code readability, maintainability, and reusability, as modules can be easily modified or replaced without affecting other parts of the program. These modules are unencumbered by other components and systems entanglements to test.
8. Message Passing: In OOP, objects communicate by sending messages. A message is a request for an object to perform a specific method. Objects can invoke methods on other objects to perform tasks or exchange information.

FIGURE O.2 The elements of object-oriented programming.

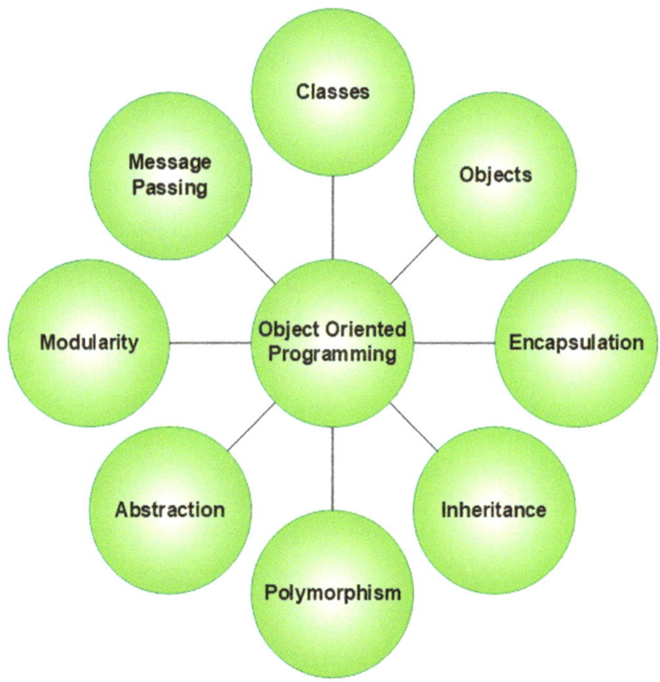

Testing is crucial to developing embedded software, as these systems are often used in safety-critical applications where software failures can have serious consequences. Testing begins with the smallest element of the software. Several techniques can be used to test object-oriented embedded software:

1. Unit Testing: This involves isolating individual code units such as specific software modules such as methods or functions. Unit testing can help ensure that each unit behaves correctly and can be used to identify and fix bugs early in the development process.
2. Integration Testing: This involves testing how different units of code work together. In OOP, integration testing can include testing how objects interact with one another and how messages are passed between them.
3. System Testing: This involves testing an entire system, including the hardware and software components. System testing can be particularly important in embedded systems, where the software must interact with sensors, actuators, and other physical components.
4. Automated Testing: This involves using software tools to automatically run tests and check for errors. can help save time and reduce the risk of human error in testing.

Ontology-Based Simulation

Ontology-based simulation is used to model complex systems by defining the behavior of objects and their relationships using an ontology, which is a formal description of the concepts and relationships within a domain. The ontology is created by vehicle experts, who carefully define each component and system within the vehicle and the relationships among them. This technique allows for the realistic simulation of various scenarios, such as a vehicle's performance under different driving conditions or the effects of additional maintenance and repair procedures on the vehicle's overall health, in a more accurate, detailed, and precise way.

Testing ontology-based simulations is a challenging task, as these simulations can be complex and involve many interconnected objects and relationships. However, several techniques can be used to test ontology-based simulations:

1. Verification: This involves checking whether the simulation accurately models the simulated system. Verification can include comparing the simulation results to real-world data or other simulation models of the same system.

2. Validation: This involves testing whether the simulation accurately captures the behavior of the objects and relationships within the system simulated. Validation can include comparing the simulation results to expected outcomes or to expert knowledge about the system.
3. Sensitivity Analysis: This involves testing how changes to different parameters within the simulation affect the overall results. Sensitivity analysis can help identify which parameters have the most significant impact on the simulation and can help refine the simulation model.
4. Scenario Testing: This involves testing the simulation under different scenarios or conditions. Scenario testing can help identify how the simulation behaves in different situations and can help ensure that the simulation is robust and can handle a wide range of scenarios.

This type of simulation is advantageous in developing new vehicles, allowing designers to test different configurations and components to see how they will perform under various conditions. For the maintenance and repair of vehicles, it helps technicians to predict the impact of different repair procedures on a vehicle's overall health.

Open-Loop Testing

Open-loop testing in the automotive industry refers to evaluating a vehicle or component without external feedback or control. Open-loop testing is a system that does not adjust or correct itself based on the input or output of the test.

For example, a standard open-loop test in the automotive industry checks a vehicle's fuel efficiency by driving it for a set distance and measuring the fuel used. The vehicle does not receive any external feedback or control to adjust its fuel usage during the test. Other examples of open-loop testing include testing an engine's or transmission's performance without external control or adjusting the brake system without external feedback.

Open-loop testing helps identify essential performance characteristics and identify problems. However, it is not as effective as closed-loop testing, which involves external feedback and control to simulate real-world conditions and identify potential issues more accurately. For more information, see **2017-01-1403** [106].

Operating System

An operating system is software that manages computer hardware and software resources and provides common services for computer programs (see Figure O.3). Testing an operating system is a complex and critical task, as it involves ensuring that it functions correctly and reliably in various environments and use cases.

FIGURE O.3 The operating system of a product is the mechanism by which the product functions are deployed.

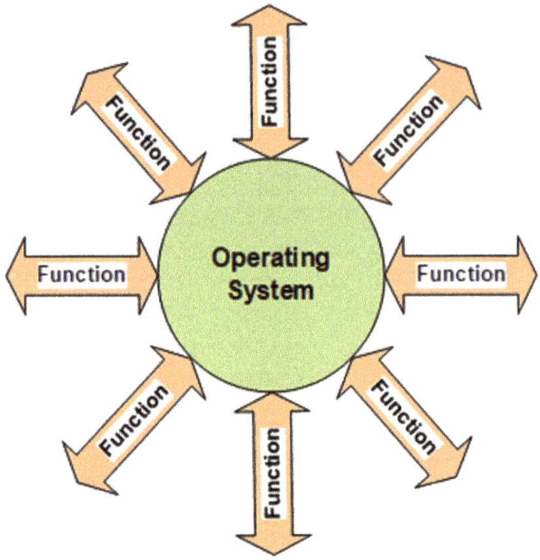

Several types of testing can be used to test operating systems:

1. Functional Testing: This involves testing the operating system to ensure that it performs the functions it is supposed to perform, such as managing hardware resources, running software applications, and providing system services.
2. Performance Testing: This involves testing the operating system's performance, such as speed, efficiency, and scalability, under different workloads and usage scenarios.
3. Security Testing: This involves testing the operating system's security features, such as access controls, authentication, and encryption, to protect the system and its data from unauthorized access and attacks.
4. Compatibility Testing: This involves testing the operating system's compatibility with different hardware and software configurations and other operating system versions.
5. Usability Testing: This involves testing the operating system's user interface and user experience to ensure that it is easy to use and understand for a wide range of users.
6. Regression Testing: This involves retesting the operating system after changes or updates to ensure that the system still functions correctly and reliably.

Testing an operating system requires a combination of manual and automated testing techniques and a wide range of testing tools and environments. By thoroughly testing an operating system, developers can ensure that it functions correctly and reliably in a wide range of settings and use cases, which is critical for ensuring the stability and security of computer systems.

Operational Acceptance Testing

Operational acceptance testing (OAT) evaluates whether a system or application is ready for deployment and use in a real-world operating environment. The main goal of OAT is to ensure that the system or application meets the end users' operational requirements and performance expectations.

The primary goal of OAT is to verify that the system can operate under typical and peak loads and handle expected variations in data and usage patterns. OAT is typically performed after system integration testing and before user acceptance testing, and it focuses on ensuring that the system is ready to be deployed into production.

There are several key activities involved in OAT:

1. Environment Preparation: The operational environment must be set up and configured to replicate the production environment as closely as possible. This preparation includes setting up hardware, software, and network configurations to match the production environment.
2. Test Planning: A detailed test plan is created that outlines the testing objectives, the testing methods to be used, and the acceptance criteria.
3. Test Execution: The system is tested in the operational environment to evaluate its performance, functionality, and reliability. The testing may include load, stress, and security testing.
4. Issue Tracking: Any issues that arise during testing are documented, tracked, and resolved.
5. Reporting: The results of the testing are compiled and reported to stakeholders. The report should include an assessment of the system's readiness for deployment and any remaining issues that need to be addressed.

The testing process in OAT includes a range of activities:

1. Performance testing ensures that the system can handle the expected load and usage patterns of the end-users.
2. Security testing ensures that the system is secure and meets the organization's security requirements.
3. Usability testing evaluates the system's user interface and ensures that it is intuitive and easy to use.

4. Integration testing ensures that the system can integrate with other systems or applications that are part of the operational environment.
5. Recovery testing ensures that the system can recover from failures or outages and resume normal operation.

The results of the OAT are used to determine whether the system or application is ready for deployment and use in the operational environment. If any issues or defects are identified during the OAT, they are addressed and resolved before deployment.

OAT is a critical step in the software development process, as it helps ensure the system is ready for deployment into production. By thoroughly testing the system in its operational environment, developers can identify and address issues before the system goes live, which can help to minimize the risk of downtime or other problems. For additional information on acceptance testing, see the following:

- Parasitic Battery Drain Problems and AUTOSAR Acceptance Testing [9]
- Unsettled Topics Concerning User Experience and Acceptance of Automated Vehicles [107]
- J2944 Operational Definitions of Driving Performance Measures and Statistics [108]

Operational Environment

An operational environment refers to the real-world environment in which the system or application will be used. This operating environment includes the hardware, software, networks, and other resources used in the system's day-to-day operation. Testing in an operational environment allows organizations to evaluate a system's performance, reliability, and security under real-world conditions. It also provides an opportunity to identify and address any issues or defects arising during use.

The operational environment can vary depending on the type of system or application being used. For example, a vehicle's operational environment will depend on its application. The operational expectations for passenger vehicles, commercial vehicles, and military vehicles are quite different. The latter will include communications networks, satellite systems, environmental considerations that may require rugged hardware, and additional test cases.

The operational environment can pose unique challenges to testing. For example, there may be limitations on the availability or accessibility of specific resources, such as networks or hardware. There may also be limitations on the tests that can be performed in the operational environment due to security or operational constraints.

It is important to carefully plan and design the testing process to address these challenges to ensure it can be effectively conducted in the operational environment. This planning may involve using specialized testing tools and techniques, collaborating with subject matter experts, and taking steps to ensure the security and integrity of the operational environment during testing. For more information, see **J2944** [108].

Optimism Bias

Optimism bias occurs when individuals overestimate the likelihood of positive outcomes and underestimate the possibility of adverse outcomes. This bias can have an impact on testing in a number of ways:

- Optimism bias can lead testers to overlook or minimize the potential risks or defects in the tested system. This can result in incomplete or inadequate testing, leaving critical defects undetected.
- Optimism bias can influence the interpretation of testing results. Testers may be more likely to interpret ambiguous or inconclusive results in a positive way, leading to false or inaccurate conclusions about the quality or performance of the system being tested.

To mitigate optimism bias in testing, it is important to be aware of its potential impact and take steps to minimize its effects. Testers should approach testing objectively and unbiasedly, focusing on identifying and addressing potential defects or risks rather than assuming that everything is working as expected.

One effective way to mitigate optimism bias is to involve multiple testers or subject matter experts in the testing process. By soliciting diverse perspectives and feedback, testers can identify potential issues or defects that may have been overlooked or minimized due to optimism bias.

Additionally, implementing a robust quality assurance and testing process that includes multiple levels of testing, such as unit testing, integration testing, and system testing, can help to ensure that defects and risks are identified and addressed before deployment. By being aware of the potential impact of optimism bias and taking steps to minimize its effects, testers can improve the accuracy and effectiveness of their testing results.

Oracles

Oracle Automotive Quality refers to a range of Oracle software solutions in the automotive industry to improve quality and efficiency in manufacturing, supply chain management, and customer service, including Oracle AutoVue, Oracle Transportation Management, and Oracle Warranty and Service Contracts. These solutions are specially designed to help automotive companies streamline processes, improve collaboration and communication, and reduce the risk of product errors and defects. Overall, using Oracle Automotive Quality solutions helps automotive companies increase customer satisfaction, reduce costs, and improve their competitive advantage in the market.

A system known as a test Oracle assesses whether a product meets the requirements of a test case. According to our definition, a test Oracle must have two key components: Oracle information that reflects expected output and an Oracle technique that contrasts the Oracle information with the actual production.

Oscilloscope

An oscilloscope is a laboratory instrument used to measure and analyze electronic signals (see Figure O.4). Oscilloscopes are commonly used in electronics testing, design, and troubleshooting, as they allow engineers and technicians to observe the behavior of electronic signals in real time.

FIGURE O.4 Oscilloscope make visible those things that cannot readily be seen.

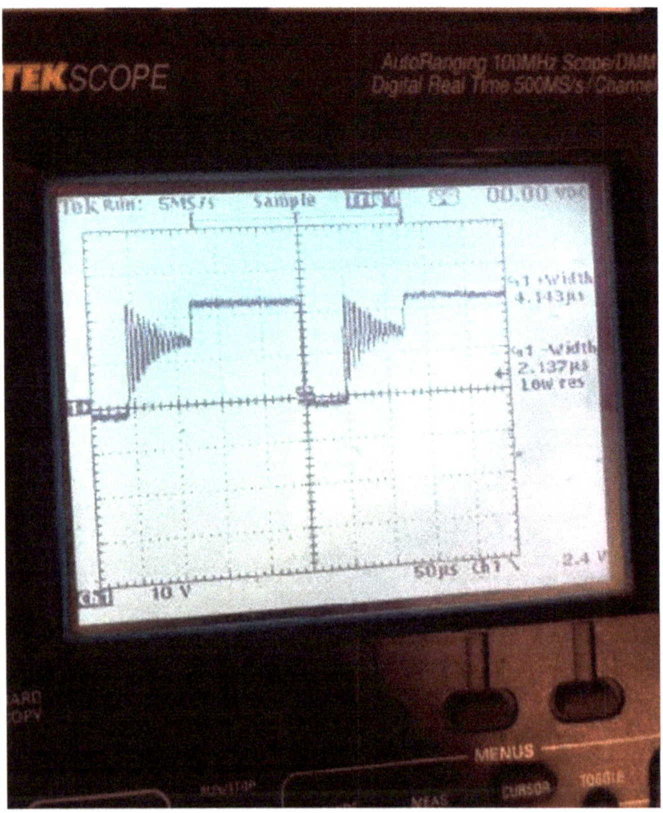

There are several ways in which oscilloscopes can be used in testing:

1. Signal Analysis: Oscilloscopes are used to analyze electronic signals to determine their frequency, amplitude, and other characteristics. This information can diagnose problems in electronic circuits and ensure that circuits are functioning as expected.
2. Timing Analysis: Oscilloscopes can be used to measure the timing of electronic signals, which is critical in applications where precise timing is required, such as in digital circuits and communication systems.
3. Waveform Capture: Oscilloscopes can capture and display waveforms of electronic signals in real time, allowing engineers and technicians to see how a signal changes over time. This can help diagnose problems in electronic circuits and verify that circuits are functioning as expected.
4. Debugging: Oscilloscopes are commonly used in debugging electronic circuits, as they allow engineers and technicians to observe electronic signals' behavior and identify and diagnose problems. For example, oscilloscopes can provide a visualization of electrical noise on the system or component under test that causes anomalous performance.
5. Performance Testing: Oscilloscopes can be used to evaluate electronic circuits' performance and ensure that they meet specified performance requirements.

Oscilloscopes are a versatile and powerful tool for electronic testing and design. By providing real-time analysis of electronic signals, oscilloscopes allow engineers and technicians to diagnose problems, verify performance, and ensure that electronic circuits are functioning as expected.

Ostrich Effect

The ostrich effect is a cognitive bias in which individuals ignore or avoid information or situations perceived as negative or threatening. This bias can have an impact on testing in several ways:

- The ostrich effect can lead testers to overlook or avoid testing certain areas or aspects of a system that are perceived as challenging or complex. This can result in incomplete or inadequate testing, which may leave critical defects undetected.
- The ostrich effect can also influence the interpretation of testing results. Testers may be more likely to interpret ambiguous or inconclusive results in a positive way, ignoring or minimizing potential risks or defects that may be present.

To mitigate the ostrich effect in testing, it is essential to be aware of its potential impact and take steps to minimize its effects. Testers should strive to approach testing objectively and unbiasedly, focusing on identifying and addressing possible defects or risks rather than avoiding them.

One effective way to mitigate the ostrich effect is to implement a structured and comprehensive testing process that includes defined testing objectives, test cases, and success criteria. By clearly understanding what needs testing and how to test it, testers can avoid the tendency to overlook or avoid challenging areas of the system.

Additionally, involving multiple testers or subject matter experts in the testing process can help to mitigate the ostrich effect. By soliciting diverse perspectives and feedback, testers can identify potential issues or defects that may have been overlooked or avoided due to bias.

Outcome Bias

Outcome bias occurs when individuals judge the quality of a decision or action based on its outcome rather than the decision-making process or information available when the decision was made. This bias can have an impact on testing in several ways:

- Outcome bias can lead testers to focus too heavily on testing results rather than the process used to conduct the testing. Testers may be more likely to view testing as successful or effective if no defects are found, even if the testing process is incomplete or inadequate.
- Outcome bias can also influence the interpretation of testing results. Testers may be more likely to interpret ambiguous or inconclusive results positively if the outcome is desirable, ignoring or minimizing potential risks or defects that may be present.

To mitigate outcome bias in testing, it is important to focus on the process used to conduct the testing rather than just the results. Testers should approach testing objectively and unbiasedly, identifying and addressing potential defects or risks regardless of the outcome.

One effective way to mitigate outcome bias is to establish clear and objective criteria for testing success. This can include specific testing objectives, success criteria, and key performance indicators that are defined and agreed upon before testing begins.

Additionally, involving multiple testers or subject matter experts in the testing process can help to mitigate outcome bias. By soliciting diverse perspectives and feedback, testers can identify potential issues or defects that may have been overlooked or minimized due to the bias.

Outgassing

Outgassing is the release of gases or vapors from materials or substances when exposed to certain conditions, such as heat or vacuum. Outgassing is important in testing rubber or plastic components, particularly in applications involving vacuum or low-pressure environments. Outgassing testing is an important consideration in a wide range of industries and applications, particularly in aerospace, electronics, and vacuum technology. Engineers and technicians who understand the outgassing properties of materials and components can ensure that their systems and products perform as expected and meet specified requirements.

Outgassing testing involves subjecting materials or components to a vacuum environment and measuring the amount of gas or vapor released. The testing is typically performed in a controlled environment, such as a vacuum chamber, and involves monitoring the pressure and composition of the gases released from the material.

There are several reasons why outgassing testing is important:

1. Contamination: The gases or vapors released during outgassing can contaminate other components or materials in a vacuum environment, affecting their performance or life span.
2. System Performance: Outgassing can affect the performance of vacuum systems or electronic components, particularly in applications where small amounts of gas can cause significant changes in pressure or conductivity.
3. Safety: Outgassing can pose safety risks in certain applications, such as in space missions, where the accumulation of gases can affect the performance of critical systems.
4. Regulatory Compliance: Outgassing testing is often required by regulatory agencies to ensure that materials or components used in certain applications meet specified requirements.

Outgassing can lead to issues such as reduced mechanical properties, changes in appearance, and decreased overall material quality. To minimize the effects of outgassing, it is important to consider the type of plastic or rubber used and the conditions under which it will be used. Then testing during the selection and design process will stress the product in ways to ensure the material selection meets the application demands. Some materials are formulated with low outgassing properties and can be used in applications where outgassing is a concern.

Overconfidence Effect

The overconfidence effect occurs when individuals overestimate their own abilities, knowledge, or judgments. This bias can have an impact on testing in several ways:

- The overconfidence effect can lead testers to underestimate the difficulty or complexity of the testing task at hand. Testers may be more likely to believe that they will be able to identify all potential defects or risks, even in situations where this may be challenging or unlikely.
- The overconfidence effect can also influence the interpretation of testing results. Testers may be more likely to interpret ambiguous or inconclusive results positively, assuming that their judgments and assessments are correct and accurate.

It is important to approach testing objectively to mitigate the overconfidence effect. Testers should strive to identify potential biases and limitations in their judgments and assessments and seek diverse perspectives and feedback to help identify potential issues or defects.

One effective way to mitigate the overconfidence effect is to establish clear and objective criteria for testing success. This can include specific testing objectives, success criteria, and key performance indicators defined and agreed upon before testing begins.

Additionally, involving multiple testers or subject matter experts in the testing process can help to mitigate the overconfidence effect. By soliciting diverse perspectives and feedback, testers can identify potential issues or defects that may have been overlooked or minimized due to bias.

Overstress

Automotive quality is of utmost importance, as it directly affects a vehicle's safety, reliability, and overall performance. Automakers and suppliers must prioritize and focus on producing high-quality components and systems to ensure the best possible experience for customers. Overstressing automotive quality means going above and beyond necessary measures to ensure that every aspect of a vehicle is of the highest standard. This involves additional testing, thorough inspections, and high-quality materials and processes. While it is important to prioritize quality, balancing it with cost and efficiency is essential to create a sustainable business model.

The overstress test successfully detects such flaws in both the product design and the manufacturing process. It ensures the overall robustness of the product by applying stresses over the design limit of the product. For additional information, see **2009-01-0294, 2019-28-2582** [109, 110].

P

Pairwise Testing

Pairwise testing, also known as all-pairs or combinatorial testing, is used to test various combinations of input parameters in software efficiently. In the automotive industry, pairwise testing can be used for multiple systems, such as engine management, climate control, and entertainment, among others [46].

Pairwise testing helps testers to identify defects and bugs in the system without examining all possible combinations of input parameters. Instead, testers can select a subset of varieties most likely to cause defects based on their knowledge of the system and its components.

For example, suppose an automotive manufacturer wants to test the climate control system in a car. The system has three input parameters: temperature, fan speed, and airflow direction. Each parameter has three possible values: low, medium, and high. Without pairwise testing, testers would need to test all 27 possible combinations of input parameters, which would be time-consuming and expensive. However, using pairwise testing, testers can select a subset of 9 combinations covering all input parameter pairs, significantly reducing the testing time and cost.

Pairwise testing is a valuable technique in the automotive industry, as it can help ensure the reliability and safety of various car systems while also saving time and resources.

Parameter

In software development, a parameter is a variable or input value that is used to define the behavior or output of a software program or function. Parameters define a software application's characteristics, constraints, or requirements and can be adjusted to achieve desired outcomes [46].

Testing is an essential part of software development, and it involves evaluating a software application or system to identify and resolve any defects, bugs, or errors before it is released to end users. Testing helps to ensure that the software works as expected, meets all required specifications, and is reliable, usable, and maintainable.

Software parameters are often used in testing to define the specific conditions and expected outcomes of the tests.

For example, in unit testing, individual components or modules of the software are tested to ensure that they work as intended. Parameters that may be tested in unit testing include input values, expected output, error handling, and performance.

Different software components are tested in integration testing to ensure they work as expected when integrated. Parameters that may be tested in integration testing include data flow, interoperability, and compatibility with other systems or components.

In system testing, the entire software system is tested to ensure that it meets the specified requirements and works as intended. Parameters that may be tested in system testing include functionality, usability, performance, security, and reliability.

Acceptance testing ensure that software meets the end users' requirements and expectations. Parameters that may be tested in acceptance testing include user experience, ease of use, and adherence to business rules or regulatory requirements.

Software parameters and testing are essential aspects of software development that help ensure software applications' quality, reliability, and usability.

Pareidolia

Pareidolia is a psychological phenomenon in which individuals perceive meaningful patterns or images in random or ambiguous stimuli. This can impact testing in specific contexts, particularly in evaluating images or visual data.

In the testing context, pareidolia can lead testers to perceive patterns or relationships in data or test results that may not exist. For example, a tester may see a pattern in a set of test results that is actually due to random variation or may identify a defect that is not present due to a perceived pattern or image.

To mitigate pareidolia in testing, it is essential to approach the analysis of test data and results objectively and without bias. Testers should be aware of the potential for pareidolia and take steps to confirm any perceived patterns or relationships before drawing conclusions or making decisions based on them.

One effective way to mitigate pareidolia is to establish clear and objective criteria for evaluating test results. This can include specific metrics or performance indicators defined and agreed upon before testing begins, as well as standardized methods for analyzing and interpreting test data.

Additionally, involving multiple testers or subject matter experts in the testing process can help to mitigate pareidolia. By soliciting diverse perspectives and feedback, testers can help to confirm any perceived patterns or relationships and identify potential issues or defects that may have been overlooked or minimized due to bias.

Pareto Chart

A Pareto chart is a graphical tool used to analyze and visualize the relative importance of different factors or issues in a dataset (see Figure P.1). It is based on the Pareto principle, which states that a small number of factors or issues typically account for the majority of the impact or occurrence of a problem [112].

FIGURE P.1 Pareto charts help us determine the most critical elements of an exploration.

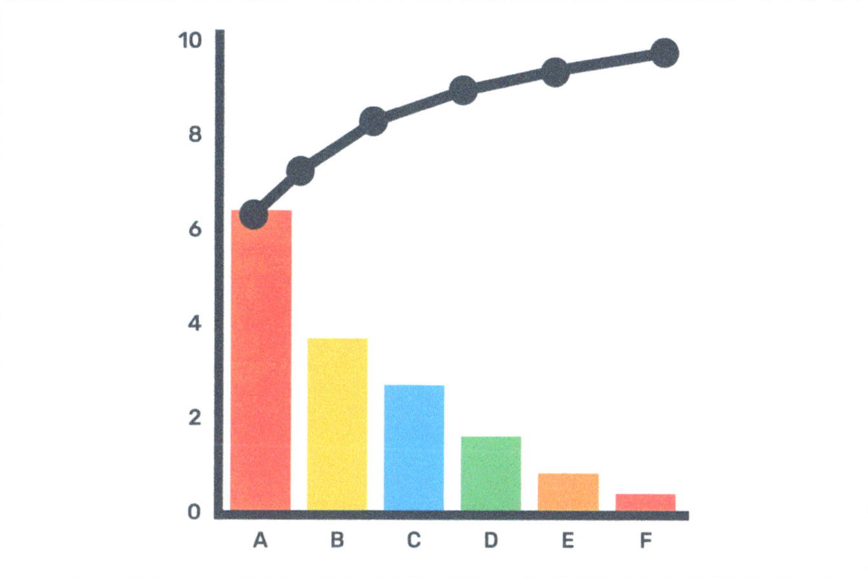

In the context of testing, Pareto charts can be used to identify and prioritize issues or defects that have the most significant impact on the system or application being tested. By analyzing the frequency and severity of different issues or defects, testers can identify the most critical problems and focus on addressing them first.

To create a Pareto chart for testing, testers can follow these steps [40]:

1. Identify the different types of issues or defects that have been identified during testing. This includes things like bugs, performance issues, security vulnerabilities, and usability problems.
2. Collect data on the frequency and severity of each type of issue or defect. This includes the number of occurrences, the impact on system performance, the severity of the issue, and any other relevant metrics.
3. Sort the data in descending order based on the frequency or severity of each issue or defect.

4. Plot the data on a bar chart, with the most significant issues or defects on the left side and the least significant on the right.
5. Overlay a line chart on the bar chart to show the cumulative percentage of the total issues or defects.

By analyzing the Pareto chart, testers can quickly identify the most significant issues or defects and focus their efforts on addressing those first. This can help to ensure that testing efforts are focused on the most important areas and can lead to more effective and efficient testing overall.

Parkinson's Law of Triviality

The law of triviality is the 1957 claim made by C. Northcote Parkinson that people frequently or usually assign disproportionate weight to trivial concerns inside an organization. Parkinson gives the example of a committee that was tasked with approving the plans for a hypothetical nuclear power plant and spent the majority of its time debating relatively unimportant but simple-to-understand matters, such as the type of material to use for the staff bicycle shed, while ignoring the proposed design of the plant itself, which is far more significant and a far more challenging and complex task.

In a testing context, testers may spend an excessive amount of time debating the color or layout of a button on a user interface, while neglecting to thoroughly test critical functionality or security features. This can lead to serious defects or vulnerabilities being overlooked or missed.

To mitigate the impact of Parkinson's law of triviality on testing, it is important to establish clear priorities and goals for the testing process. This includes identifying critical functionality, performance requirements, and security considerations that must be thoroughly tested and evaluated.

Testers should also be aware of the tendency to focus on trivial issues and take steps to ensure that they are allocating their time and resources appropriately. They should set strict deadlines for testing activities, establish clear criteria for prioritizing issues and defects, and involve subject matter experts and stakeholders in the testing process to provide guidance and perspective.

Additionally, testers can use automated testing tools and techniques to help prioritize and streamline the testing process. This includes automated testing scripts, regression testing tools, and performance testing tools that can help quickly identify critical issues and ensure they are evaluated.

Pass/Fail Criteria

Pass/fail criteria for testing automotive systems varies based on the specific system being tested and the requirements of the manufacturer or regulatory body. However, some common examples of pass/fail criteria in automotive testing include the following:

1. Performance Standards: The system must meet certain performance standards, such as speed, acceleration, braking, or fuel efficiency.
2. Safety Standards: The system must meet safety standards set by regulatory bodies, such as crash test ratings or airbag deployment.
3. Quality Standards: The system must meet standards set by the manufacturer, such as reliability, durability, or corrosion resistance.
4. Environmental Standards: The system must meet environmental standards, such as emissions levels or fuel efficiency.
5. Compatibility Standards: The system must be compatible with other systems in the vehicle, such as the electrical system or communication systems.
6. User Experience Standards: The system must meet user expectations, such as ease of use, comfort, or convenience.

Expressions called "pass/fail criteria" link data to inspection pass or fail standards. Mathematical evaluations can establish rules for true/false, integer, Boolean, and float data types.

Path Coverage

Path coverage is used in software testing to ensure that all possible paths through a program are tested. A path is a sequence of instructions a program executes from the start of a function to its end. Path coverage ensures or provides a method for evaluating the number of paths tested compared to every possible path through a program. The evaluation includes all branches, loops, and conditions considered for testing.

Path coverage aims to detect errors or bugs in less frequently executed paths, such as edge cases, corner cases, or error-handling scenarios. Path coverage can be used in both manual and automated testing.

$$Path\ Coverage = \frac{Paths_{tested}}{Paths_{total}}$$

Testers must identify all possible paths through a program to evaluate path coverage and create test cases to cover all or as many as risk and time permits. This process requires a thorough understanding of the program's logic, control structures, and data dependencies. Testers can use various techniques to identify paths, such as control flow graphs, decision tables, or state transition diagrams.

Path coverage is a rigorous testing technique that can help ensure the quality and reliability of software. However, achieving 100% path coverage is often impractical, especially for complex programs, as it requires many test cases and may not be feasible in terms of time and resources. Therefore, testers often use path coverage as one of several testing techniques, along with techniques such as boundary value analysis, equivalence partitioning, or error guessing, to ensure adequate test coverage.

Path Testing

Path testing aims to identify and test all possible paths through a software program to ensure that the program behaves as expected and does not contain bugs or errors. Once paths are identified, testers create test cases that exercise each path. This can be done using various techniques, such as control flow diagrams, decision tables, and finite state machines.

During path testing, testers follow each path through the software, executing the relevant test cases and observing the behavior of the software. This helps to identify any errors or defects that may be present in the software, which can then be fixed before the software is released to the public.

Path testing can be time-consuming and complex, especially for larger software systems. However, it is a valuable technique for ensuring the quality and reliability of the software, particularly for safety-critical applications where even a small error can have serious consequences.

Path testing typically involves three steps:

1. Path Identification: Testers must identify all possible paths, both normal and abnormal, through the program. This identification requires a thorough understanding of the program's logic, control structures, and data dependencies. Testers can use techniques such as control flow graphs, decision tables, or state transition diagrams to identify paths.
2. Path Selection: Testers then select a subset of paths most likely to cause defects or cover critical functionality in the program. This subset should include normal and abnormal paths, edge cases, and error-handling scenarios.
3. Test Case Creation: Finally, testers create test cases for each selected path, ensuring that the test cases cover all instructions, branches, loops, and conditions in the path. Test cases should also include both valid and invalid inputs.

P-Diagram

A P-diagram, or parameter diagram, is a helpful tool for identifying and documenting inputs, intended outputs, unintended outputs (also known as error states), noise factors, and control variables. Also known as P-chart, it is a statistical process control chart used to monitor the proportion of defective items in a process, identify trends in the process, and determine if it is in statistical control.

Automotive testing evaluates the performance and reliability of vehicles, including cars, trucks, and buses. This can include various testing types such as durability, safety, and emission testing. Automotive testing is crucial in ensuring that vehicles meet the necessary standards and requirements for use on public roads.

Pen Tester

Pen testing, or penetration testing, involves simulating an attack on a system or application to identify vulnerabilities and weaknesses that attackers can exploit. Pen testers use various techniques, tools, and methodologies to identify these vulnerabilities, including network scanning, vulnerability scanning, password cracking, and social engineering (see Figure P.2). Pen testing aims to identify and prioritize security issues so that they can be addressed and fixed before attackers can exploit them.

FIGURE P.2 Penetration testing takes multiple approaches [113].

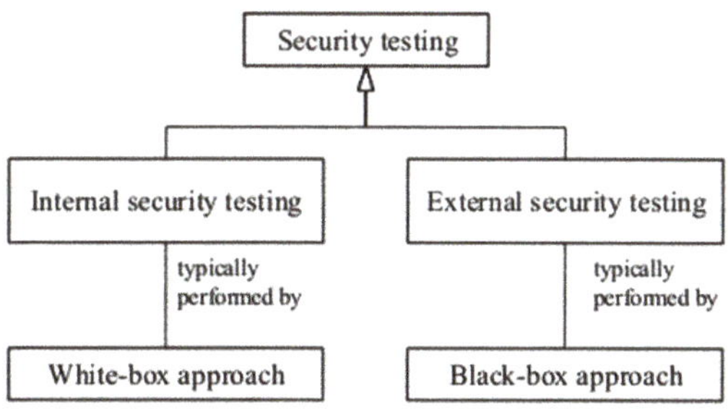

Penetration testing can also be applied to vehicles, specifically to their software and electronic systems. As vehicles become increasingly connected and autonomous, the potential attack options for hackers increases. Penetration testing can help identify vulnerabilities in these systems, such as remote code execution, denial-of-service attacks, and unauthorized access to sensitive information, that hackers could exploit.

Vehicle penetration testing typically involves testing various electronic control units (ECUs) and their communication protocols, along with the software and firmware that run on these ECUs. This includes infotainment systems, telematics systems, advanced driver assistance systems (ADAS), and autonomous driving systems. Penetration testers may also attempt to gain access to a vehicle's network and test the security of the wireless communication systems used by the vehicle.

The goal of vehicle penetration testing is to identify vulnerabilities and weaknesses that attackers could exploit and to provide recommendations for addressing these issues. This can help improve the overall security and safety of vehicles and reduce the risk of cyberattacks. As connected and autonomous vehicles become more widespread, penetration testing of vehicles is likely to become an increasingly important part of automotive testing and development.

Penetration Testing

See Pen Testing.

Perception Testing

Perception testing in the automotive industry refers to evaluating a vehicle's sensory systems, such as the visual and auditory systems, to ensure that they function correctly and provide accurate information to the driver. This type of testing is typically done in a controlled environment, such as a test track or simulation. Specialized equipment and sensors are used to measure the performance of the sensory systems. Perception testing is essential to vehicle safety and ensures drivers have the necessary information to make informed decisions while operating a vehicle.

Performance Testing

Performance testing in the automotive industry refers to evaluating a vehicle's performance and capabilities with respect to acceleration, braking, handling, fuel efficiency, and so on. This type of testing is typically done in a controlled environment, such as a test track or simulation. Specialized equipment and sensors are used to measure the vehicle's performance. Performance testing is essential to vehicle development and ensures that vehicles meet the necessary performance standards and requirements for use on public roads.

Performance testing is also often used in software quality assurance to assess how a system performs regarding responsiveness and stability under a specific workload. For more information, see **J3220_202301**, **J994_202306**, and **J2432_202111** [114, 115, 116].

Pessimism Bias

Pessimism bias involves overestimating the likelihood of adverse outcomes and underestimating the likelihood of positive results. This bias can impact software testing in several ways:

- Testers with a pessimism bias may be more likely to focus on negative scenarios and overlook positive ones. This can result in a testing strategy overly focused on finding defects and vulnerabilities rather than testing the software's features and functionality.
- testers with a pessimism bias may be inclined to assume that software defects and vulnerabilities are more severe than they actually are. This can result in a higher rate of false positives, where testers report defects that are not actually defects or that do not significantly impact the software's performance or security.
- Pessimism bias can lead to a lack of confidence in the software's performance and security, even if the software has been thoroughly tested and meets all requirements. This can result in unnecessary delays in software releases and a reluctance to adopt new technologies or features.

To address pessimism bias in testing, testers can take steps to ensure that they are testing the software in a balanced and objective way. This can include developing a testing strategy that covers positive and negative scenarios, using testing techniques such as exploratory and risk-based testing, and ensuring that all testing is based on clear and measurable criteria. Additionally, testers can work to build confidence in the software by thoroughly documenting testing results and disseminating the results for review.

PFMEA

Process failure mode and effects analysis (PFMEA) is used to identify potential failures in a process and assess the potential impact and likelihood of those failures. It is a proactive tool that helps organizations identify and address problems before they occur, improving a process's reliability and quality. In the automotive industry, PFMEA is commonly used to identify potential failures in the manufacturing process of vehicles and parts. It involves specifying the potential failure modes of each step in the process, evaluating the potential effects of those failures, and determining the likelihood of those failures occurring. It is then possible to implement corrective actions to prevent or mitigate potential failures [76, 117].

The items identified in the PFMEA documentation may require testing to explore if the estimated severity and likelihood of impact are accurately assessed. Testing is used to clarify, confirm, or refute the identified failure modes and explore alternative solutions that impact on the severity or risk priority number (RPN).

Planning Fallacy

The planning fallacy is evident when people underestimate the time, effort, and resources required to complete a task, even when they have experience with similar tasks. This bias can lead to unrealistic expectations, missed deadlines, and poor performance. It is also connected to the belief that the work requires a detailed, unalterable plan.

Testing can be affected by the planning fallacy if testers underestimate the time required to complete testing activities or fail to allocate enough resources for testing. This can lead to rushed testing, inadequate test coverage, and the potential for defects to go undetected.

To address the planning fallacy in testing, it's important to consider the system's complexity and carefully estimate the time and resources needed for testing activities. Testers can also use historical data and benchmarks to inform their estimates and adjust their plans as needed based on feedback and progress updates. Additionally, involving stakeholders in the testing process can help ensure that expectations are realistic and that testing is given the appropriate priority and resources.

Planning Poker

Planning poker is a collaborative technique Agile development teams use to estimate the effort or complexity of a software development or testing task. Planning poker is an approach like wideband Delphi, only faster. It involves team members, including developers, testers, and product owners, who estimate the work required to complete a task.

During a planning poker session, each team member is given a deck of cards with numbers representing the level of effort required for a given task. The numbers typically follow a modified Fibonacci sequence, such as 1, 2, 3, 5, 8, 13, 20, 33, and 53. The team then discusses the task and each team member privately selects a card that represents their estimate of the level of effort required for that task.

Once everyone has selected their card, the team members reveal their cards simultaneously. If there is a wide range of estimates, the team members discuss their reasoning and the task is reestimated. This process continues until the team reaches a consensus (or at least converges) on the required effort for the task.

Planning poker is effective because it encourages team collaboration and communication, helps identify potential issues early in the development process, and ensures that everyone understands the level of effort required for a given task.

Population

Population refers to the entire group of individuals, objects, or events to be studied (see Figure P.3). It can be any group of people or things, such as all the residents of a particular country, all the users of a specific model of vehicle, or all the cars produced by a particular manufacturer, or even parts from a specific supplier and manufacturing line [87].

FIGURE P.3 Population is the entire group of products or people; testing works with samples from that population to make inferences about the total population.

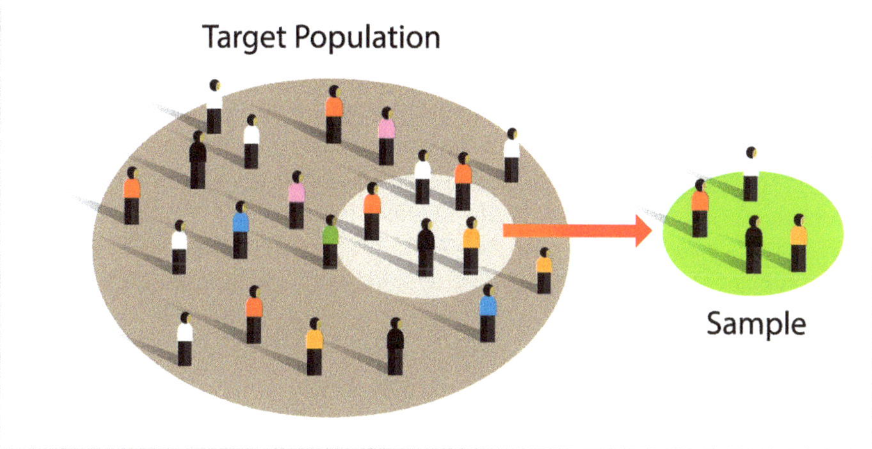

Bakhtiar Zein/Shutterstock.com.

Statistics is the branch of mathematics that deals with data collection, analysis, interpretation, presentation, and organization. It involves using quantitative methods to analyze data and draw conclusions about the population from a sample.

Testing in statistics refers to using statistical methods to test hypotheses about the population. Hypothesis testing involves formulating a null hypothesis, which is the hypothesis that there is no difference or no relationship between variables, and an alternative hypothesis, which is the hypothesis that there is a difference or relationship between variables. The null hypothesis is then tested against the alternative hypothesis using statistical tests. The test result helps to determine whether the null hypothesis can be rejected and whether there is evidence to support the alternative hypothesis.

Portability Testing

Portability testing evaluates a software application's ability to be transferred from one environment to another. It tests the software's compatibility with different hardware, operating systems, and network configurations. Portability testing aims to ensure that a software application can be easily installed, configured, and run on other platforms without any issues.

Product and portability testing are essential in automotive software development to ensure that the application is high quality and can be integrated into a vehicle. Portability testing helps to ensure that the software application can be easily moved from one vehicle platform to another, while product testing helps to ensure that the software application meets specified requirements and performs well in different vehicle incarnations, scenarios, and environments.

Positive Testing

Positive testing in the automotive industry refers to evaluating a vehicle's performance and capabilities under favorable or optimal conditions. This type of testing is typically done in a controlled environment, such as a test track or simulation. Specialized equipment and sensors are used to measure the vehicle's performance. Positive testing is essential to vehicle development and ensures that vehicles meet the necessary performance standards and requirements for use on public roads. It is typically done with negative testing, which evaluates a vehicle's performance and capabilities under unfavorable or challenging conditions.

A type of testing known as "positive testing" is carried out on a software program by using valid datasets as input. It determines whether the software application responds to positive inputs as predicted. Positive testing is done to see if the software application performs as it should.

Postcondition

In product testing, a postcondition is a statement that defines the state of a product, system, or software application after a particular function or operation has been executed. It describes the expected result or outcome of a function or operation. Postconditions are used to verify that the system's desired behavior has been achieved after the function or operation is executed [118].

After testing, postconditions are used to evaluate the correctness and completeness of a system or application. Test cases are designed to check whether the postconditions are satisfied. If the postcondition is satisfied, the system or application functions correctly and produces the expected result. On the other hand, if the postcondition is not satisfied, it indicates an error or defect in the system or application, requiring either a product correction or specification and test case updates.

Clearly defining the postconditions for each function or operation is vital for creating compelling test cases. This helps to ensure that the test cases cover all possible scenarios and that the system or application is thoroughly tested. Postconditions are also helpful for debugging and troubleshooting issues in the system or application. By examining the postconditions, developers can identify where the error occurred and take appropriate corrective action.

Pragmatic Testing

Pragmatic testing focuses on the practical, real-world application of a product or system. It evaluates how well a product or system performs in actual usage scenarios and can involve functional and nonfunctional testing. Actual users typically do pragmatic testing in a real-world environment, including user acceptance testing, field testing, and performance testing. The goal of practical testing is to identify any issues or limitations affecting the product or system's usability, functionality, or performance in the real world. It is an essential aspect of software development and is used to ensure that a product or system meets the needs and expectations of its intended users.

Precondition

Preconditions to testing in the automotive industry refer to the conditions that must be met before testing begins [113]. For example, a vehicle must be parked with the brake set (the precondition) to start the testing of the brake driving alarm test. Some common preconditions to automotive testing include the following:

1. Test Facilities and Equipment: Testing requires specialized facilities and equipment, such as test tracks, simulation environments, and measurement and monitoring equipment. These must be in place and operational before testing can begin.
2. Personnel: Testing requires trained and qualified personnel to operate the equipment, perform the tests, and collect and analyze the data.
3. Regulatory Requirements: Automotive testing is often subject to various regulatory requirements, such as safety standards and emissions standards. These requirements must be met before testing can begin.

Satisfying these preconditions is necessary to ensure that testing is controlled and conducted consistently, producing accurate and reliable results.

Probability

Probability measures the likelihood of an event occurring in a statistical experiment. It is a value between 0 and 1, where 0 represents an impossible event and 1 represents a specific event. For example, the probability of flipping a coin and getting heads is 1/2, or 0.5.

In statistics, probability is used to make predictions about the likelihood of certain events occurring based on statistical data. For example, suppose a study found that 60% of people who eat a certain type of food regularly develop a certain illness. In that case, we can use that probability to predict the likelihood of other people who eat that food regularly developing the illness.

Different ways to calculate probability include classical, empirical, and subjective probability. Classical probability involves estimating the likelihood of an event occurring based on the number of possible and favorable outcomes. Empirical probability is calculated based on observed data from a sample. Subjective probability involves judging the likelihood of an event based on personal belief or opinion.

Probe Effect

The probe effect in testing refers to the influence of testing on the test results. The interaction between a device under test and the test equipment can impact the test results.

Consider a scenario where a team is conducting performance testing on a new smartphone. The team wants to measure the battery life by running a specific application continuously and recording the time until the battery is drained. The testers start the application, set up the measurement equipment, and begin the test.

During the test, they notice that the battery life is significantly shorter than expected, and the device's performance seems subpar. This unexpected outcome puzzles the testers because the smartphone has undergone thorough pretesting and its battery life was deemed satisfactory.

Upon further investigation, they discover that the presence of the measurement equipment, which includes additional sensors and connections, caused the device's processor to work harder, leading to increased power consumption. In other words, the act of measuring the battery life was inadvertently affecting the battery performance itself [119].

Process Metric

Process metrics specify numerical and graphical measures of a process's performance and development. A process entity (such as an action, role, artifact, condition, or asset) or group of entities may be essential to process metrics [120].

Many metrics can be used to evaluate the testing progress and process in product testing. Some standard metrics include the following [121]:

1. Test Case Coverage: This is the percentage of the codebase evaluated by the test cases. A higher range indicates that more of the code has been tested.
2. Defect Density: This is the number of defects found per unit of functions. A lower defect density indicates that the product is of higher quality.
3. Time to Complete Testing: This is how long it takes to complete the testing process. A shorter time to complete testing indicates that the process is efficient.
4. Cost of Testing: This is the resources (e.g., time, personnel, equipment) required to complete the testing process. A lower cost of testing indicates that the process is efficient.

5. Number of Defects Found: This is the total number of defects discovered during testing. A lower number of defects indicates that the code is of higher quality.
6. Defect Discovery Rate: This is the number of defects found per unit of time. A lower defect discovery rate indicates that the testing process is efficient.
7. Pass/Fail Rate: This is the percentage of test cases that pass or fail. A high pass rate indicates that the code is of high quality.
8. Cost: Testing will have associated budgets; price is one of the reasons to terminate product testing and can be a test process metric.

These metrics evaluate the testing process's effectiveness and efficiency and identify improvement areas.

Processor Load Testing

Processor load testing is used to evaluate a system's behavior under a high load. This testing aims to determine how well the system can handle high levels of demand and identify any potential bottlenecks or issues.

During processor load testing, the system is subjected to a high processing activity, such as running many tasks or processing a large amount of data. The system's performance is then monitored to see how it handles the increased load. The testing may involve simulating the processing activity using specialized tools or running actual tasks on the system.

Processor load testing is commonly used to ensure systems can handle the expected demand levels in a production environment. It can help identify potential issues when the system is under high loads, such as slow performance or crashes. Identifying and addressing these issues before the system is deployed can help ensure that the system is stable and reliable.

The number of processes running on the CPU or awaiting its execution is known as CPU load. The average number of processes running or waiting to run over the previous 1, 5, and 15 minutes is thus CPU load average.

Product Metric

Product metrics are quantitative measurements to evaluate a software product's quality and performance, such as its complexity, reliability, maintainability, usability, and efficiency. Product metrics are essential for assessing the product's quality and are used to evaluate the test results compared to the expected product performance.

Product performance metrics are defined in the product specifications. From these specifications, test cases are generated to confirm or refute the specifications and to evaluate product performance via testing. Product performance metrics include such things as current (amperes) consumption, key dimensions, and processing speed and throughput.

Product metrics are tangible measurement points for testing. These product attributes will compare points and be part of the pass/fail evaluation.

Product Quality Predictions

Product quality predictions are estimations of the expected quality of a software product prior to its market release to end users. These predictions are based on data from previous projects, historical data, and metrics gathered during the development and testing.

There are several methods for forecasting the quality of a product. Regression analysis and decision trees are two statistical modeling techniques that may be used to assess historical data on product performance and to find trends that can be used to predict future performance. Another strategy is to examine massive datasets using machine learning methods, like support vector machines or random forests, and then generate predictions based on the patterns found in the data.

Testing is crucial in predicting product quality because it helps identify and eliminate defects and issues that can affect product quality. By thoroughly testing the product, testers can identify potential quality issues and address them before the product is released to the market.

During the testing process, various techniques are used to evaluate the quality of the software product, including functional testing, performance testing, security testing, usability testing, and compatibility testing, among others. Each of these techniques helps to identify different types of defects and issues that can affect the quality of the product.

By analyzing the results of these tests, testers can identify patterns and trends in the quality of the product and make predictions about its overall quality. For example, if the number of defects found during testing is decreasing over time, it may be a sign that the product quality is improving. Similarly, if the performance testing results indicate that the product meets the specified performance requirements, it may be a sign that the product is of high quality.

Some examples of product quality predictions that can be made based on the testing process include the following:

1. Defect Density: The number of defects found per code unit. By analyzing the defect density, testers can make predictions about the overall quality of the product. For example, if the defect density is high, it may indicate that the product has quality issues and requires further testing.
2. Test Coverage: The extent to which the product has been tested. By analyzing the test coverage, testers can predict the likelihood of defects and issues being found in untested areas of the product. For example, if the test coverage is low in a particular area, it may be a sign that potential quality issues need to be addressed.
3. Performance Metrics: Measure the product's performance, such as response time, throughput, and latency. By analyzing these metrics, testers can make predictions about the product's overall performance. For example, if the response time is consistently fast and meets the specified requirements, it may indicate that the product is of high quality.
4. Usability Metrics: Measure the product's usability, such as the ease of use and user satisfaction. By analyzing these metrics, testers can predict the overall user experience and satisfaction with the product. For example, if the usability metrics indicate that the product is easy to use and intuitive, it may indicate that it is of high quality.
5. Security Metrics: Measure the security of the product, such as vulnerability scans and penetration testing. By analyzing these metrics, testers can make predictions about the overall security of the product. For example, if the security metrics indicate that the product is secure and free from vulnerabilities, it may indicate that it is of high quality.

Product Requirements Analysis

Product requirements analysis and testing are closely related processes essential for ensuring a software product's quality and success. Testing is closely connected to requirements; correct and understandable requirements are a prerequisite for many testing approaches [122, 39].

Product requirements analysis involves identifying and defining the functional and nonfunctional requirements of a software product (see Figure P.4). This process involves gathering information about the stakeholders' needs and expectations, defining the product's scope, and documenting the requirements in a clear and unambiguous manner.

FIGURE P.4 Good requirements are influenced by many things.

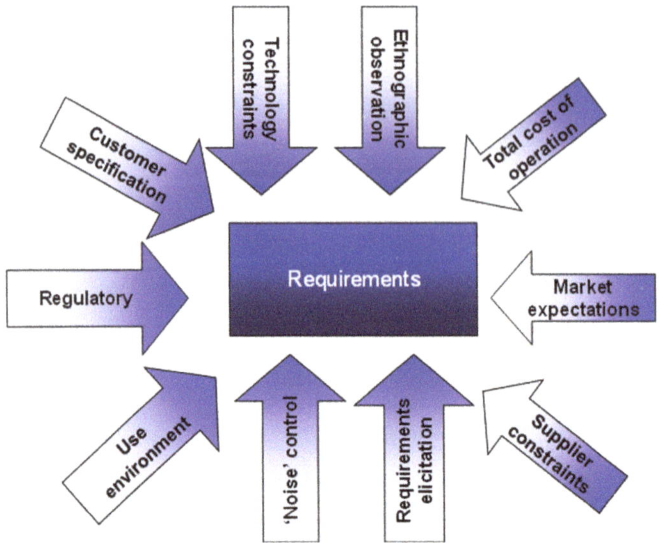

Product requirements analysis and testing are closely linked because the requirements define what the product should do and the testing ensures that it does it correctly. Testing is based on the requirements, and the requirements are used to define the test cases and scenarios that are used during testing (see Figure P.5). By ensuring that the product meets the specified requirements, which are used to define the test cases and scenarios through testing, testers can verify that a product is of high quality and meets the needs of the stakeholders.

FIGURE P.5 Requirements are gathered and evaluated to ascertain if indeed a requirement needs to be put into the requirements documents.

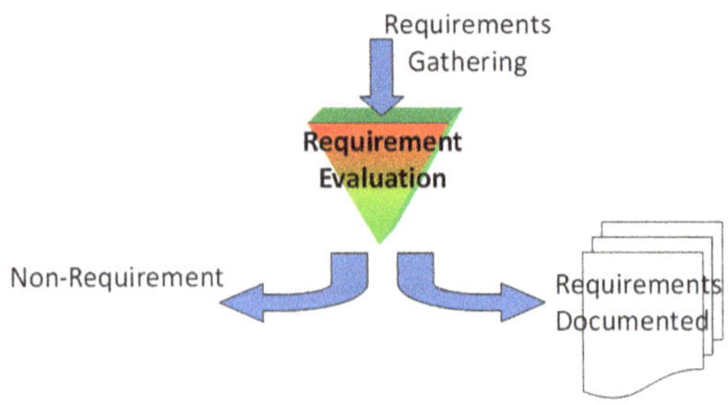

Product requirements analysis and testing also help ensure the product is delivered on time and within budget. Products are generally delivered in increments (see Iterative Development), each containing some of the requirements, until an entire product is produced or the funding is exhausted. By clearly defining the conditions delivered to testing, the product can be incrementally tested against those requirements, and stakeholders can sure that the product is delivered with minimal defects and meets their needs and expectations. This helps to reduce the risk of costly rework, delays, and other issues that can arise when a product does not meet the requirements.

Product requirements analysis and testing are critical for ensuring a software product's quality and success. In tandem, requirements help to ensure that a product meets the needs of the stakeholders, performs as expected, and is delivered on time and within budget.

Project Deafness

Project deafness occurs when project team members become so focused on their tasks and responsibilities that they fail to communicate effectively with each other. This lack of communication can lead to misunderstandings, missed deadlines, and other issues that can affect the quality of the software product.

Testing plays an important role in addressing project deafness by providing a common language and framework for communication. Additional testing can provide objective data to show how the product and by extension the project are performing—how the product is rather than how it should be at a given point in time. By defining clear test cases and test scenarios, testers can ensure that all team members have a shared understanding of the expected behavior of the software product. This can help to reduce misunderstandings and miscommunications that can arise due to project deafness. For example, if a developer is working on a particular module and fails to communicate changes to other team members, this can lead to conflicts and issues during testing. However, by testing the product thoroughly, testers can identify and address these issues before they become a problem.

Furthermore, testing can help to promote collaboration and communication among team members. For example, testers can work closely with developers to identify and address issues during testing, leading to better communication and collaboration. This can help to break down silos and promote a more collaborative and cohesive team environment.

Project Management

Project management involves planning, organizing, and overseeing the resources and tasks required to deliver a product or service. Testing ensures that the software product meets the required specifications and works as intended [123].

Effective project management requires clear goals, timelines, and communication channels. A project manager must ensure that the development team knows the requirements, deadlines, and milestones. The project manager should also identify and mitigate any risks arising during the project [124].

Testing is a critical part of the development process as it helps identify bugs and errors before a product is released to users. Different types of testing, such as unit testing, integration testing, and acceptance testing, ensure that the software product meets the required standards [39].

In modern software development methodologies, project management and testing are often integrated. Agile methods, for example, emphasize the importance of collaboration between developers, testers, and project managers to deliver high-quality software products. By involving testers early in the development process, teams can identify issues early on to ensure that software products meet all required standards.

Prototype

A prototype is a model of an instantiation of a product at a given point in the development of the product [125]. They are a mechanism for learning about the product. Prototype and testing go hand in hand in product development. Prototype parts are used to test and refine their design. If an organization uses simulation as part of product development, these prototype parts will assist in evaluating the simulation models. This way, simulations and prototypes work together in maturing the product design. Prototype parts can be created using 3D printing or other rapid prototyping techniques to test different aspects of the product, such as its form, fit, and function (see Figure P.6).

FIGURE P.6 Prototype parts are used to test product ideas and learn throughout development.

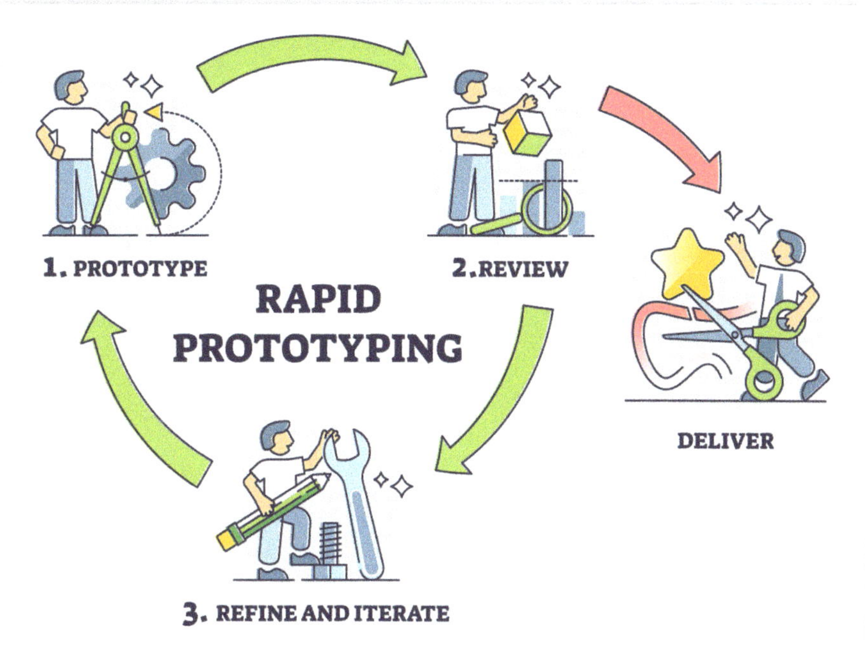

VectorMine/Shutterstock.com.

There are often varying levels of prototype parts capabilities defined by the organization (supplier or customer). The automotive industry uses different prototype parts (capability) levels, sometimes called A, B, C, and P samples, to develop new vehicles and automotive systems. The following are examples of the levels of prototype parts typically used in the automotive industry:

1. Level 1—Virtual Prototype: This level involves creating a digital model of a vehicle or automotive system using computer-aided design (CAD) software. This level allows engineers to test and refine the design before creating physical prototype parts.
2. Level 2—Foam Prototype: This level involves creating a physical model of a vehicle or automotive system using foam materials. Foam prototypes are typically used to evaluate the form and fit of the design and can be created quickly and cost-effectively.
3. Level 3—Alpha Prototype: This level involves creating a working vehicle or automotive system prototype using production-ready materials and components. Alpha prototypes are used to evaluate the performance and functionality of the design.
4. Level 4—Beta Prototype: This level involves creating a preproduction prototype of a vehicle or automotive system using production tooling

and processes. Beta prototypes are used to test and validate the manufacturing process and to identify and resolve any issues before production begins.

5. Level 5—Production Prototype: This level involves creating a final vehicle or automotive system prototype using production tooling and processes. Production prototypes are used to evaluate the final product before it is released to the market.

The level of prototype parts used in the automotive industry depends on the stage of development and the project's specific requirements. By using prototype parts, automotive manufacturers can identify and resolve issues early in the development process, which can help to reduce costs and improve the quality of the final product.

PUGH Matrix

Stuart Pugh developed the decision-matrix approach, often known as the Pugh method or Pugh idea selection, a qualitative method for ranking the multidimensional alternatives in an option set.

A Pugh matrix, a decision matrix or selection matrix, is a tool used to evaluate and compare the performance of multiple options or alternatives. It is often used in the early stages of product development to help identify the most promising design concept.

A Pugh matrix consists of a table with rows representing different design options or alternatives and columns representing evaluation criteria or performance characteristics. Each cell in the table contains a rating or score that reflects the relative performance of the option being evaluated on the corresponding criterion. The ratings can be based on objective data or subjective judgment.

Once all of the options have been rated, the scores are totaled and the option with the highest total score is identified as the preferred option. The Pugh matrix can be used to compare multiple options in a structured, systematic way and to identify the strengths and weaknesses of each option. It is often used in conjunction with other tools and techniques, such as brainstorming and prototyping, to support the decision-making process.

Q

"Quality is not an act, it is a habit."

—Aristotle

Quality Assurance

The manufacturing and service sectors use the term "quality assurance" to refer to the systematic measures taken to guarantee that a product delivered to the customer meets the contractual and other agreed-upon performance, design, reliability, and maintainability expectations of that customer.

Quality assurance (QA) in the automotive industry refers to the processes and activities used to ensure that vehicles are high quality and meet the required standards and specifications. These processes include testing, inspections, and audits to ensure that vehicles meet safety, reliability, and performance requirements [126].

In the automotive industry, QA is critical for maintaining customer satisfaction and trust, as well as for ensuring that vehicles are safe and reliable. Automotive manufacturers and suppliers have strict QA processes to ensure that the vehicles they produce meet the necessary standards and requirements.

QA in the automotive industry also involves continuous improvement efforts to identify and address any issues or defects that may arise during production. This can include using data analysis and other tools to identify trends and patterns in defects and implementing corrective actions to prevent similar issues from occurring in the future.

Quality Assurance Plan

A quality assurance plan (QAP) is a document that outlines the processes and procedures that will be used to ensure that a product or service meets specified quality standards. The QAP is a vital part of the overall quality management system and helps to ensure that the product or service is consistently high-quality. The latest applicable standard is ISO 9001:2015.

A QA plan typically includes the following components:

1. Quality Objectives: These are specific, measurable goals that the QAP aims to achieve.
2. Quality Standards: These are the specific quality criteria the product or service must meet to be considered acceptable.
3. Testing and Inspection Procedures: These methods are used to verify that the product or service meets the quality standards.
4. Training and Education: This section outlines the training and education programs that will be provided to ensure that employees have the necessary skills and knowledge to produce high-quality products or services.
5. Continuous Improvement: This section describes how the QAP will be reviewed and updated on an ongoing basis to ensure that it remains effective.

Overall, a QAP is a key tool for ensuring that a product or service is of consistently high quality, and helps to identify and address any issues that may arise during the development process.

Quality Audit

A quality audit is a procedure by which a team of auditors thoroughly examines a quality system.

A quality audit is a systematic, independent evaluation of an organization's quality management system (QMS) to determine whether it conforms to the requirements of a standard or other specified quality criteria. It is a crucial component of a company's quality management system. Quality audits are typically conducted by external auditors trained to evaluate the effectiveness of an organization's QMS [126].

There are several types of quality audits:

1. Compliance Audits: These audits determine whether the organization's QMS complies with a specific standard or regulation.
2. Process Audits: These audits focus on the organization's processes and procedures to ensure they are practical and efficient.
3. System Audits: These audits evaluate the overall QMS, including the policies, procedures, and processes that comprise the system.
4. Product Audits: These audits focus on the quality of the organization's products or services.

During a quality audit, the auditor will review the organization's quality policies, procedures, and records and may also observe the organization's processes and interact with employees to gather information. Based on the audit's findings, the auditor will provide a report with recommendations for improvement.

Quality audits are essential for ensuring that an organization's QMS is effective and for identifying areas for improvement.

Quality Circle

A quality circle is a small group of employees, typically from the same department or area of the organization, who perform the same or comparable tasks and regularly get together to discuss, evaluate, and resolve issues relating to their jobs. It is also known as quality control circle.

A quality circle has a minimum of three and a maximum of twelve members [40]. Quality circles are led by a facilitator who helps guide the group's efforts.

A quality circle aims to identify problems or opportunities for improvement and develop and implement solutions. Quality circles may focus on a specific process, product, or service or may focus more broadly on improving overall quality within the organization.

Quality circles are often used with other quality improvements tools, such as root cause analysis and process mapping. They can effectively involve employees in the quality improvement process and identify and address issues that may not be apparent to management.

Overall, quality circles are a valuable tool for organizations seeking to continuously improve the quality of their products, processes, or services.

Quality Control

Quality control (QC) in the automotive industry refers to the processes and procedures to ensure that vehicles and vehicle components meet the required quality standards. Automotive QC is significant because vehicle defects or failures can have serious consequences, including accidents, injuries, and even fatalities.

There are several steps involved in quality control in the automotive industry:

1. Setting Quality Standards: This involves establishing specific quality criteria that vehicles and components must meet.
2. Design and Development: During the design and development phase, quality considerations are integrated into the product to ensure it meets the required quality standards.
3. Testing and Inspection: Vehicles and components are subjected to various tests and assessments to ensure that they meet quality

standards. These include laboratory tests, field tests, and examinations at multiple stages of the production process.

4. Continuous Improvement: Automotive QQC is an ongoing process, and organizations use tools such as root cause analysis and statistical process control to improve the quality of their products on an ongoing basis.

Overall, the goal of automotive QC is to produce vehicles and components that are safe, reliable, and meet the customer's needs. For more information, see **J661_202110** [127].

Quality Management

Quality management and testing are closely related concepts. Testing is a critical component of quality management because it helps to ensure that products or services meet the required quality standards. Here are some ways in which quality management and testing are related:

1. Quality Planning: This is an essential component of quality management that involves defining quality objectives and establishing processes to achieve those objectives. Testing is a key aspect of quality planning, as it helps identify potential quality issues and establish procedures to prevent or mitigate them.
2. Quality Control: This involves monitoring and measuring quality throughout the product or service life cycle. Testing is a critical part of quality control, as it helps to identify defects and other quality issues that may arise during development or production.
3. Quality Assurance: This involves consistently establishing processes and procedures to meet quality standards. Testing is essential to quality assurance, as it helps verify that products or services meet the required quality standards.
4. Continuous Improvement: This is a key principle of quality management (examples are Total Quality Management and Kaizen) and testing is a critical component of this process. Testing helps identify improvement areas and provides valuable feedback to refine and improve products or services over time.

Testing is a key component of quality management, as it helps to identify and mitigate quality issues throughout the product or service life cycle.

R

"The roots of education are bitter, but the fruit is sweet."

—Aristotle

Ramp Testing

Ramp testing in the automotive industry is used to evaluate the performance of vehicles under different driving conditions, such as driving up a steep hill or accelerating onto a highway. In this context, ramp testing typically involves evaluating a vehicle's acceleration, braking, and handling on various inclines and declines. The testing may be performed on a test track or public roads, depending on the specific requirements of the test.

Ramp testing is an essential part of vehicle development and validation, as it helps to ensure that vehicles meet the required performance standards and are safe to operate under a wide range of driving conditions. The results of ramp testing can be used to optimize vehicle design and configuration and identify areas for improvement in vehicle performance.

Some of the benefits of ramp testing in the automotive industry include the following:

1. Evaluating Performance under Different Driving Conditions: Ramp testing helps to assess the performance of a vehicle under a variety of driving conditions, such as inclines, declines, and heavy loads.
2. Identifying Potential Issues: Ramp testing can help identify issues such as brake fade, overheating, or engine performance issues that may not be apparent under usual driving conditions.
3. Optimizing Vehicle Design: The results of ramp testing can be used to optimize vehicle design and configuration to improve performance and ensure safety.
4. Improving Vehicle Safety: Ramp testing is an integral part of ensuring that vehicles are safe to operate under a wide range of driving conditions, helping to improve overall vehicle safety.

Ramp testing is an important part of automotive development and validation, helping to evaluate vehicle performance under various driving conditions and ensure that vehicles are safe and reliable.

Range

In statistics, range refers to the difference between a dataset's largest and smallest values. The range is a simple measure of the variability or spread of the data. It is often used as a preliminary analysis to gain insight into the dataset's values field.

The range can be calculated by subtracting the smallest value from the largest value in the dataset. For example, if a dataset contains the values 2, 4, 6, 8, and 10, the range would be 10 – 2 = 8.

The range is a fundamental statistical measure that is often used in conjunction with other statistical measures such as the mean, median, and standard deviation. For example, the range can be used to identify outliers or extreme values in the dataset, which may significantly impact the mean and standard deviation.

While the range is a simple and easy-to-understand statistical measure, it has some limitations. For example, the range is sensitive to outliers or extreme values in the dataset. It may not accurately represent the variability in the data if there are significant outliers.

To overcome some of the limitations of the range, other statistical measures such as the interquartile range or standard deviation may be used to provide a more robust measure of the variability in the data.

Range Testing (Vehicle)

In the automotive industry, range testing evaluates the range or distance a vehicle can travel on a single charge or tank of fuel. Range testing aims to accurately estimate the vehicle's driving range under normal operating conditions.

Range testing for electric vehicles typically involves testing the vehicle's battery performance under various driving conditions, such as different speeds, terrain, and climate conditions. The testing may be performed on a test track or public roads, depending on the specific requirements of the test.

For gasoline-powered vehicles, range testing typically involves testing the vehicle's fuel efficiency under various driving conditions. This includes testing the vehicle's fuel consumption at different speeds, driving styles, and terrain conditions.

Some of the benefits of range testing in the automotive industry include the following:

1. Providing Accurate Range Estimates To Customers: Range testing ensures that customers receive accurate range estimates, helping to improve customer satisfaction and confidence in the vehicle's performance.

2. Evaluating Performance under Different Driving Conditions: Range testing helps to assess the performance of the vehicle under a variety of driving conditions, such as different speeds, terrain, and climate conditions.
3. Optimizing Vehicle Design: The results of range testing can be used to optimize vehicle design and configuration to improve performance and efficiency.
4. Improving Vehicle Efficiency: Range testing is an essential part of improving vehicle efficiency, helping to reduce fuel consumption and emissions and improve overall sustainability.

Range Testing (Wireless)

Range testing is used to evaluate the range or coverage of a wireless or communication system, such as a Wi-Fi network, Bluetooth device, or cellular network. The goal of range testing is to determine the maximum distance the system can effectively transmit data or signal without experiencing significant degradation in performance or signal quality.

Range testing is typically performed by measuring the signal strength or quality at various distances from the transmitter or access point and recording the data to create a range profile. The range profile can then be used to determine the maximum range at which the system can operate effectively.

Range testing is a part of wireless system testing and validation, as it helps to ensure that the system meets the required performance standards and is reliable under a wide range of operating conditions. Range testing also identifies potential interference issues or areas with poor signal quality that may impact system performance.

Some of the benefits of range testing include the following:

1. Ensuring Effective Wireless Communication: Range testing helps ensure that wireless communication systems can effectively transmit data or signal over the required range.
2. Identifying Areas with Poor Signal Quality: Range testing can help to identify areas with poor signal quality or interference issues that may impact system performance.
3. Optimizing System Configuration: The results of range testing can be used to optimize the configuration of wireless systems, such as adjusting antenna placement or power levels, to improve performance and range.
4. Improving User Experience: Range testing helps to ensure that wireless systems provide reliable and consistent performance, improving the overall user experience.

Rayleigh Distribution

The Rayleigh distribution is a continuous probability distribution for nonnegative-valued random variables in probability theory and statistics. It is consistent with the chi distribution with two degrees of freedom up to rescaling. The distribution bears the name of Lord Rayleigh, a British mathematician and physicist.

The Rayleigh distribution is a continuous probability distribution used to model the magnitude distribution of a two-dimensional vector whose components are independent and identically distributed Gaussian random variables. The Rayleigh distribution is commonly used in engineering and physics to model random vibrations or noise amplitude [39].

One use of the Rayleigh distribution is to test for noise or interference in a signal. In this application, the signal is first analyzed to determine the amplitude distribution and the resulting distribution is compared to a Rayleigh distribution. If the signal's amplitude distribution closely matches a Rayleigh distribution, this indicates that the signal is likely to be affected by random noise or interference.

To perform a Rayleigh distribution test, the following steps are taken:

1. Collect Data: Collect a sample of data from the signal to be tested. This sample should contain measurements of the signal amplitude.
2. Calculate the Rayleigh Distribution: Calculate the Rayleigh distribution parameters based on the collected data. The Rayleigh distribution has two parameters: scale and location. The scale parameter is related to the average amplitude of the signal, while the location parameter is related to the position of the distribution on the x-axis.
3. Compare the Distribution: Compare the collected data's amplitude distribution to the Rayleigh distribution. This can be done using statistical measures such as the Kolmogorov-Smirnov or Anderson-Darling tests.
4. Interpret the Results: If the amplitude distribution closely matches the Rayleigh distribution, the signal is likely to be affected by random noise or interference.

To calculate the Rayleigh distribution, you need to know the scale parameter, which is related to the average amplitude of the signal or the magnitude of the two-dimensional vector. The probability density function (PDF) of the Rayleigh distribution is determined by the following equation:

$$f(x)=\left(\frac{x}{\sigma^2}\right)\times\exp\left(\frac{-x^2}{\left(2\times\sigma^2\right)}\right)$$

Where:
x is the amplitude of the signal.
Σ is the scale parameter of the Rayleigh distribution.

The cumulative distribution function (CDF) of the Rayleigh distribution is determined by the following equation:

$$f(x) = 1\exp\left(\frac{-x^2}{2 \times \sigma \wedge 2)}\right)$$

To calculate the Rayleigh distribution for a given set of data, calculate the average amplitude of the signal by taking the mean of the amplitude measurements in your dataset. Then calculate the scale parameter, σ, using the following formula:

$$\sigma = (\sqrt{\frac{\pi}{2}}\left(\frac{\textit{Sum Of Squared Amplitudes}}{\textit{Number Of Amplitudes}}\right)^{0.5}$$

Where:
π = pi (approximately 3.14159).
sqrt = square root sum of squared amplitudes = sum of the squared amplitude measurements in your dataset.

The number of amplitudes is the number of amplitude measurements in your dataset.

1. Calculate the PDF of the Rayleigh Distribution: Using the calculated σ value, you can calculate the PDF of the Rayleigh distribution for a given amplitude x using the formula given above.
2. Calculate the CDF of the Rayleigh Distribution: Using the calculated σ value, you can calculate the CDF of the Rayleigh distribution for a given amplitude x using the formula given above.
3. Interpret the Results: The PDF and CDF values can be used to understand the distribution of the amplitudes in your dataset and make inferences about the signal's properties.

Reactance

Reactance is a measure of a component's opposition to changes in current or voltage due to the presence of capacitance or inductance. In electrical testing, reactance is often measured to characterize the properties of electrical components such as capacitors and inductors.

Several methods are available to test for reactance:

1. Impedance Testing: Impedance is the combination of resistance and reactance in a circuit. Impedance testing involves measuring the impedance of the component at different frequencies to determine the reactance of the component. This is done using an impedance analyzer or a network analyzer.
2. Phase Angle Testing: Phase angle is the difference in phase between the voltage and current in a circuit. The phase angle of a circuit can be used to determine the circuit's reactance. Phase angle testing involves measuring the phase angle of the component at different frequencies using an oscilloscope or a phase angle meter.
3. Resonance Testing: Resonance is a condition in which the reactance of a circuit is zero. Resonance testing involves measuring the frequency at which the reactance of the component is zero. This is done using a signal generator and an oscilloscope or spectrum analyzer.
4. Q-factor Testing: The Q-factor measures the damping of a circuit, which is related to the circuit's reactance. Q-factor testing involves measuring the Q-factor of the component at different frequencies using an impedance analyzer or a network analyzer.

Reactive Devaluation

Reactive devaluation can affect testing in various ways. If an opposing party, such as a competitor or a rival team, presents a test result, a receiver may be more likely to dismiss or devalue the result without a fair evaluation. Similarly, if an opposing party proposes a testing method or procedure, the receiver may be more likely to view it with suspicion or skepticism, even if the method is objectively valid and reliable.

This bias can lead to missed opportunities for collaboration and cooperation in testing. For example, if two companies are working on a joint testing project, reactive devaluation could lead to a breakdown in communication and a failure to reach mutually beneficial agreements. It can also lead to a lack of trust in testing results and methodologies, undermining the credibility of the testing process.

To overcome reactive devaluation in testing, focus on the objective merits of test results and methodologies rather than the identity of the proposer or tester. This can be accomplished by independently verifying and validating the results and methods, using neutral third parties or objective standards to evaluate them. It can also be helpful to build trust and rapport between testing parties to reduce perception of them as adversaries and promote collaboration and cooperation. Finally, it can be useful to establish clear and transparent testing procedures and protocols to increase the credibility and reliability of the testing process.

Real-Time Simulation

Real-time simulation is a powerful tool used in automotive testing to validate the performance of vehicle systems and components. Real-time simulation uses high-performance computing systems and software to simulate the behavior of a vehicle system in real time, allowing engineers to test and optimize the system's performance under various operating conditions.

Real-time simulation is beneficial for automotive testing because it allows engineers to evaluate the behavior of vehicle systems and components under realistic conditions, including extreme driving scenarios and environmental conditions. This can help identify potential issues and optimize the system's performance before deployment in a real-world setting.

Real-time simulation is used for a variety of automotive testing applications:

1. Powertrain Testing: Simulating the behavior of the engine, transmission, and drivetrain systems under various operating conditions allows engineers to optimize performance and fuel efficiency.
2. Vehicle Dynamics Testing: Simulating the vehicle's behavior under different driving scenarios, including acceleration, braking, and cornering, allows engineers to optimize handling and stability.
3. Autonomous Vehicle Testing: Simulating the behavior of autonomous vehicle systems allows engineers to test and optimize the system's performance in various stimulus scenarios.
4. Crash Testing: Simulating the vehicle's behavior during a crash allows engineers to evaluate the performance of the vehicle's safety systems and to optimize their design.

Recovery Testing

Recovery testing is a type of testing in which a system or application is intentionally subjected to failures or errors to test its ability to recover and resume typical operation. In automotive testing, recovery testing can be used to evaluate the ability of vehicle systems and components to recover from failures or errors that could occur during regular operation.

Recovery testing in automotive testing involves intentionally inducing failures or errors in various vehicle systems and components, such as the engine, transmission, braking system, or electrical system, to evaluate the system's ability to detect and recover from the failure. This testing can help to identify potential issues and improve the reliability and safety of the vehicle.

For example, recovery testing might intentionally stall a vehicle's engine while it is running and then evaluate the ability of the engine control system to detect the failure and restart the engine. Similarly, recovery testing might intentionally cause a fault in a vehicle's braking system and then assess the ability of the system to detect the fault and continue to provide safe and reliable braking performance.

The ability of a vehicle system to recover from failures or errors is critical to the safety and reliability of the vehicle. By subjecting automotive systems and components to recovery testing, engineers can evaluate the effectives of their recovery mechanism and identify potential improvement areas.

Regression

Regression is a statistical technique that is used to analyze the relationship between one or more independent variables and a dependent variable. In regression analysis, the goal is to develop a mathematical model that can be used to predict the dependent variable's value based on the independent variables' values.

Regression analysis is used in various fields, including economics, finance, social sciences, and engineering. Regression analysis is used in these fields to model and understand the relationship between different variables and predict or forecast future outcomes.

Regression Testing

Regression testing is used to ensure that changes or updates to a software application do not have unintended effects on existing functionality.

Regression testing focuses on testing the software application as a whole rather than individual features or functions. This is because changes or updates to one feature or function can sometimes unintentionally affect other application parts. By testing the application as a whole, regression testing helps to ensure that these unintended effects are identified and resolved.

Regression testing is essential in the automotive industry because software is used extensively in the design and operation of modern vehicles. Any defects or errors in software can have serious consequences, including potential safety risks. By using regression testing, software developers can ensure that changes or updates to software applications do not introduce new defects or errors that could compromise the safety or reliability of a vehicle.

Regression testing can be automated using specialized software tools, which can help to improve the efficiency and effectiveness of the testing process. Automated regression testing tools can quickly and accurately retest large volumes of software code, allowing developers to identify and resolve defects or errors promptly and efficiently.

Regressive Bias

Regressive bias refers to the tendency for test results to become less extreme or variable over time as the testing process continues. This bias can occur for various reasons, such as fatigue or learning effects on the part of the test subject or changes in the testing environment that affect the results.

Regressive bias in testing is a significant concern when it leads to inaccurate or misleading test results. For example, if test subjects become fatigued or bored during a lengthy testing session, their performance may decline, producing artificially low test scores. Similarly, changes to the testing environment over time, such as changes in lighting or temperature, can affect the accuracy and reliability of test results.

It is essential to control testing conditions and procedures carefully and use appropriate statistical methods to analyze test results to reduce regressive bias. Identifying and correcting any bias or variability in the results ensures that they accurately reflect the true performance of test subjects or systems.

In addition, it may be useful to periodically retest subjects or systems to ensure that any changes in performance over time are accurately captured and accounted for. This can help to validate that the testing process remains accurate and reliable over time and that the results continue to provide useful and meaningful information for decision-making.

Release Candidate

A release candidate (RC) is a prerelease software application intended for testing purposes and is typically made available to a limited audience before the final release. An RC aims to identify and fix any remaining bugs or issues before the software's final release.

An RC is usually the last stage of the software development cycle, nearly representing the final version of the software. It is often preceded by alpha and beta releases, earlier software versions made available for testing to a broader audience.

During the RC phase, users and developers test the software extensively to identify any remaining bugs, issues, or compatibility problems. Feedback from users and developers is used to make final adjustments and fixes to the software in preparation for the final release.

The software's final release is made available to the public once the RC has been thoroughly tested and any remaining issues have been resolved. This last release is considered the stable, production-ready version of the software and is intended for general use by end users.

Release Notes

Release notes are a document or set of documents that describe the changes, new features, and bug fixes that are included in a software release. They are intended to inform users, customers, or stakeholders of the changes that have been made and how they will be affected. Release notes are commonly used in software development to communicate updates to a software product. Developers, testers, and other technical teams often use them to keep track of the changes and plan for future development. They also help customers and end users understand what has been changed, the new features, and how to use them. Release notes can be delivered in text files, PDFs, website pages, or release note platforms [57]. Some examples of what might be included in release notes include the following:

- A list of new features and improvements
- A list of bug fixes and known issues
- Any changes to the user interface or user experience
- Any API (Application Programming Interface) changes or breaking changes that developers should be aware of
- Any system requirements or dependencies that have changed
- Any relevant documentation or resources, such as user guides or tutorials
- Contact information for support or feedback

It's important to write release notes clearly and concisely and to use language that is easy to understand for the intended audience. They should be easy to read and scan through, so it's a good idea to use headings, bullet points, and other formatting techniques to make the information easy to digest.

Reliability

Reliability in product design refers to a product's ability to perform its intended function consistently and without failure over time. A highly reliable product can be expected to have a low failure rate, meaning it will not break down or malfunction frequently. Reliability affects customer satisfaction, brand reputation, and overall product performance [128].

There are several methods to improve the reliability of products; one of the most common is reliability testing, which tests a product under different conditions and over some time to determine its failure rate. Other methods include using robust materials and components, designing for ease of maintenance and repair, and incorporating redundancy and fail-safe mechanisms.

Failure rates are measured by the number of failures per unit of time or usage. One of the most commonly used units for λ is FIT, which stands for "failure in time" and represents the number of failures per billion (109) hours. Under the exponential distribution assumption, the mean time between failures is the inverse of the failure rate [129] (see also Mean Time between Failure).

To design a reliable product, it is essential to consider its intended use and environment and to thoroughly test and validate the product design throughout the development process. Additionally, it's essential to continually monitor and gather customer feedback to identify potential reliability issues and make improvements. For more information, see **J2958_202002**, **J2816_201804**, and **J3083_201703** [129, 130, 131].

Reliability and Confidence

Several key parameters and equations are relevant to reliability and confidence in product testing:

1. Reliability (R): Reliability represents the probability that a product will perform its intended function without failure over a specific period and under defined conditions. Product testing often determines reliability by subjecting products to various environmental conditions, stress levels, or usage scenarios. The testing results, such as the number of failures observed over time, are used to estimate the product's reliability.
2. Failure Rate (λ): The failure rate is a critical parameter in reliability analysis. It represents the rate at which failures occur in a product or system. The failure rate is often estimated by analyzing the observed failure data from product testing. The equation for the failure rate is λ = (Number of Failures) / (Total Test Time).
3. Mean Time between Failures (MTBF): MTBF is the average time between two consecutive product or system failures. It is a crucial reliability metric used to assess the expected reliability performance of a product. MTBF is calculated as the reciprocal of the failure rate (1/λ).
4. Confidence Level (C) and Confidence Interval (CI): Confidence level and CI are statistical measures used to quantify the level of certainty associated with reliability estimates. In product testing, confidence intervals provide a range of values within which the true reliability is estimated to lie with a certain level of confidence. The confidence level is often expressed as a percentage (e.g., 90% confidence level).
5. Statistical Distributions: Different statistical distributions, such as the exponential distribution, Weibull distribution, or log-normal distribution, are often used to model the failure data obtained from product testing. These distributions provide equations and parameters to estimate reliability and failure probabilities over time.
6. Accelerated Life Testing (ALT): ALT is used to assess reliability by subjecting a product to accelerated stress levels or conditions to simulate long-term usage in a shorter time frame. ALT involves applying stress factors such as temperature, humidity, or voltage to accelerate failure mechanisms. The test results are then used to

estimate product reliability and failure rates under normal operating conditions.

7. Reliability Growth Models: Reliability growth models, such as the Duane model or Crow-AMSAA model, are used to predict the improvement in reliability over time as product design or manufacturing issues are addressed. These models incorporate data from product testing and help estimate the future reliability performance of a product.

Organizations can assess and improve product reliability by conducting product testing and using the parameters and equations mentioned above. Testing provides valuable data for estimating reliability metrics, calculating failure rates, determining confidence intervals, and making informed decisions regarding product design, quality improvements, and warranty policies. Reliability and confidence analysis based on testing results enable organizations to deliver more reliable products to customers and enhance their overall satisfaction. For additional information, see **J2940_202002** [132].

Reliability Growth Model

In the automotive industry, the reliability growth model is used to analyze and predict the improvement of the reliability of automotive systems and components over time. The model is used to identify and correct defects and issues in automotive systems during the design, development, and testing phases to improve overall reliability and performance.

The automotive industry uses various techniques to improve the reliability of automotive systems and components, including testing, simulation, and data analysis. A reliability growth model is an essential tool in this process, as it provides a way to quantify and track the progress of testing and debugging efforts and to predict the remaining defects or issues that need to be addressed before an automotive system is released.

Automotive engineers and designers can use the model to predict the reliability of complex systems and components, such as engine control modules, transmissions, and powertrain systems. That information can then be used to quantify and analyze the reliability of automotive systems and components over time, and to make data-driven decisions about how to allocate resources to improve those systems.

Repeatable/Repeatability

The degree to which the findings of subsequent measurements of the same measure, when conducted under the same measurement conditions, coincide closely is known as repeatability or test-retest dependability. Repeatability is an important aspect of testing, which depends on the ability to produce consistent and reliable results when a test case is executed multiple times. In other words, a test case is considered repeatable if it produces the same results each time it is executed under the same conditions.

The ability to produce repeatable results is essential for ensuring the accuracy and reliability of testing. If a test case produces different results each time it is executed, it can be difficult to identify and isolate defects or issues in the system being tested.

To ensure repeatability in testing, it is important to carefully control the test environment and conditions, including any inputs or variables that could affect the results. This can involve creating standardized test procedures, using automated testing tools, and carefully documenting the test conditions and results.

Test cases are critical to testing, as they provide instructions or procedures for executing a specific test scenario. Test cases should be designed to be repeatable, so they can be executed multiple times to ensure the reliability and accuracy of the tested system.

Reproducibility

Reproducibility refers to the ability to replicate or reproduce the same results from a test or experiment when it is conducted by different individuals or under different conditions. In other words, if a test is reproducible, it means that other people can conduct the same test and obtain the same results.

Reproducibility is an important aspect of testing, as it helps to ensure that the test results are accurate, reliable, and unbiased. Reproducibility is particularly important in scientific research and other fields where test results are used to make important decisions or inform policy.

To ensure reproducibility in testing, it is important to carefully document all aspects of the test, including the procedures, equipment, materials, and environmental conditions. This can involve creating standardized test procedures, using automated testing tools, and carefully documenting the test conditions and results.

In addition, it is important to ensure that the test results are not affected by external factors not related to the system being tested. For example, suppose a test involves measuring the performance of a software application. In that case, the test environment should be standardized and controlled so that the results are not affected by other applications or processes running on the same computer.

Requirements

Requirements are the needs, expectations, and specifications that a product, system, or project must meet. They describe what the product or system must do, how it should behave, and what features and functions it should have. Requirements can be divided into two broad categories: functional and nonfunctional [50].

Functional requirements describe the specific behaviors and functions that the product or system must perform. For example, a functional requirement for a software application might be that it must allow users to create and save new documents.

Nonfunctional requirements, on the other hand, describe the characteristics and qualities of the product or system. These can include performance requirements (such as speed and reliability), usability requirements (such as ease of use and accessibility), and security requirements (such as data privacy and protection).

Requirements are typically identified and documented during the requirements gathering phase of a project, and they are then used as the basis for design, development, and testing [133].

Attributes of good requirements include the following:

- Cohesiveness
- Completeness
- Consistency
- Correctness
- Observability
- Feasibility
- Unambiguousness
- Necessity
- Verifiability
- Traceability

Requirements-Based Testing

Requirements-based testing is an essential approach in the development of automotive systems, as it helps to ensure that systems meet the needs and expectations of users and stakeholders, are safe and reliable, and perform as expected in various conditions.

In the automotive industry, requirements-based testing is used to verify that the system meets a wide range of requirements, including functional requirements (such as performance, safety, and reliability), and nonfunctional requirements (such as usability, maintainability, and scalability). This testing approach can be applied to various automotive systems, such as powertrain, braking, steering, and infotainment systems [133].

The process of requirements-based testing in the automotive industry typically involves several steps:

1. Requirements Analysis: The requirements documentation is carefully reviewed to identify the automotive system's key functional and nonfunctional requirements.

2. Test Planning: Based on the requirements analysis, a test plan is developed that outlines the testing approach, the test cases and scenarios that will be used, and the criteria for determining whether the tests have been successful.
3. Test Design: Test cases and scenarios are designed to cover all of the automotive system's key requirements and are relevant to the specific use cases and operating conditions of the system.
4. Test Execution: The test cases and scenarios are executed, and the results are carefully monitored and documented. This may involve testing the system under various conditions, such as different temperatures, speeds, loads, and environmental conditions.
5. Defect Reporting: Any defects or issues that are identified during testing are reported, and the testing process may be repeated until all defects have been resolved.

Requirements-based testing is essential in the automotive industry, as it helps ensure that the system meets the required safety (legal) and performance standards and can function effectively in various operating conditions. Testing against the requirements helps identify any potential issues or defects early on in the development process, saving time and costs in the long run.

Re-Simulation Testing

Re-simulation testing in the automotive industry involves running simulations of previously conducted tests or scenarios to validate and verify the performance of a system or component. This type of testing is beneficial when changes have been made to the design or components of a system and need to be validated before being implemented.

Re-simulation testing can be used for various purposes in the automotive industry:

1. Validating Design Changes: When changes are made to the design of a system or component, it is important to ensure that these changes do not negatively affect the performance of the system. By running simulations of the previously conducted tests or scenarios, it is possible to validate the changes and ensure that the system still performs as expected.
2. Verifying System Performance: Re-simulation testing can be used to verify the performance of a system under different operating conditions or scenarios. This can help to identify any potential issues or defects that may have been missed during the initial testing phase.

3. Improving Testing Efficiency: Reusing previously conducted tests or scenarios, re-simulation testing can help improve testing efficiency and reduce the time and cost associated with testing.
4. Validating System Safety: Re-simulation testing can also be used to validate the safety of a system under different conditions or scenarios, and to ensure that the system meets the required safety standards.

Re-simulation starts with recordings from actual drives, capturing that interaction within that drive, for playback in a simulation (e.g., hardware in the loop rigs):

- Vehicle sensor data
- Processing objects (POs)
- Virtual environment"

Resource Allocation Matrix

A resource allocation matrix is used in project management to assign resources to different tasks or activities. The matrix helps to ensure that resources are allocated efficiently and effectively, maximizing their use and minimizing waste.

The resource allocation matrix can be used in testing to allocate resources, such as people, equipment, and materials, to different testing activities. For example, the matrix can be used to allocate resources for unit testing, integration testing, system testing, and user acceptance testing. This ensures that the testing process is well-organized and that resources are used effectively.

Moreover, the resource allocation matrix can also be used to track the progress of testing activities and ensure that they are completed on time and within budget. The matrix can help to identify potential bottlenecks or issues that could affect the testing process, allowing project managers to take corrective action to keep the project on track.

Resources and projects are mapped out according to the timeline in a resource matrix. Due to its matrix view of resources in relation to projects, the matrix offers potent way to spot gaps. A resource manager can quickly identify which resources are not fully assigned with the help of this view.

Reviews

Reviews are a critical part of the software development process that involve examining and evaluating software code, documentation, and other project deliverables to ensure they meet specific quality standards. Reviews can help identify issues early in the development process, reducing the likelihood of defects and bugs in the final product.

Several types of reviews are commonly used in product development [12]:

1. Code Reviews: These involve examining the source code of a software application to ensure that it meets coding standards and best practices.
2. Design Reviews: These involve examining the overall design of a software application to ensure that it meets functional requirements and is scalable, maintainable, and extensible.
3. Test Plan Reviews: These involve examining the testing strategy and test plans for a software application to ensure that it covers all functional and non-functional requirements.
4. Documentation Reviews: These involve examining the project documentation, such as user manuals and system documentation, to ensure that it is accurate, complete, and well-written.
5. FMEA: Failure mode effects analysis is a specific type of review of the design or process artifacts. See DFMEA; Failure Mode Effects Analysis; PFMEA [134].

Reviews can be conducted using a variety of techniques, including informal walkthroughs, peer reviews, and formal inspections. These techniques involve different levels of participation from team members and can be tailored to suit the needs of a particular project and team.

Reviews are an important aspect of software development that can help ensure high-quality, reliable software products that meet the needs of the end users.

Risk Analysis

Risk analysis in project management involves identifying potential risks, assessing their likelihood and impact, and developing strategies to mitigate or manage them. Risk analysis is often performed in software development to identify risks that could impact a software application's quality, functionality, or usability [39, 122].

Testing is a critical component of risk management in software development because it helps identify defects and bugs before an application is released to users. Testing can help mitigate risks by identifying potential issues early in the development process, reducing the likelihood that they will affect the final product.

To effectively manage risks through testing, the testing strategy should be designed to address identified risks. This may involve specific types of testing, such as security, performance, or usability, depending on the nature of the risks. For example, if security risks are identified, specific security testing measures may be incorporated into the testing process to ensure that the software application is secure and protected against potential security threats.

Moreover, risk analysis can help inform the prioritization of testing activities. Risks that are identified as high-impact and high-likelihood can be given higher priority in the testing process, ensuring that they are thoroughly tested and addressed before the software is released to users.

Risk analysis and testing are complementary activities in software development. By performing risk analysis and developing a testing strategy that addresses identified risks, project teams can reduce the likelihood of defects and bugs in the software and deliver high-quality, reliable software products to users.

Risk-Based Testing

Risk-based testing focuses on testing the areas of a software application with the highest risk [135]. Risk-based testing aims to prioritize testing efforts based on the level of risk associated with different features, functions, or components of the software application; for example, testing the antilock brake system (ABS) is more important than testing the HVAC system.

Risk-based testing identifies, assesses, and ranks risks based on their probability and potential impact. The highest-ranked risks are then used to guide the testing process, ensuring that testing efforts are focused on areas of the software application that are most likely to be problematic or have the greatest impact on users.

Risk-based testing can be applied at different stages of software development, including requirements gathering, design, development, and testing. By incorporating risk-based testing into each stage of the software development process, project teams can proactively identify and address potential issues, reducing the likelihood of defects and bugs in the final product.

Benefits of risk-based testing include the following:

1. Efficient Use of Testing Resources: By focusing testing efforts on areas of the software application with the highest risk, testing resources can be used more efficiently.
2. Improved Software Quality: By identifying and addressing potential issues early in the development process, the overall quality of the software application can be improved.

3. Reduced Cost: Identifying and addressing potential issues early in the development process can reduce the cost of fixing defects and bugs in the software application.

Risk-based testing is a valuable approach to software testing that can help ensure high-quality, reliable software applications that meet the needs of users.

Risk Compensation (Peltzman Effect)

The Peltzman effect is an economic theory which suggests that people adjust their behavior in response to perceived levels of safety or risk. The theory suggests that when people feel safer, they may take more risks, which could increase accidents or negative outcomes.

The Peltzman effect can be applied to the behavior of software users. For example, if users perceive a software application as safe and reliable, they may be more likely to take risks or engage in behavior that could result in adverse outcomes. As a result, it is crucial that software developers and testers thoroughly test the software application to ensure that it is safe and reliable, even if users perceive it to be so.

Thorough testing can help mitigate the potential negative impact of the Peltzman effect by identifying and addressing potential risks and issues in the software application. Testing can help ensure that the software application is safe and reliable, even if users engage in behavior that may increase the risk of negative outcomes. Moreover, testing can help developers and testers identify areas of the software application that may be more prone to issues or vulnerabilities, allowing them to take proactive measures to mitigate potential risks.

Risk Control

Risk control is an important aspect of risk management in software development to mitigate or manage identified risks. In the context of testing, risk control measures can be implemented to reduce the likelihood or impact of potential issues in the software application.

Several risk control measures can be implemented during the testing process to mitigate potential risks:

1. Test Planning and Design: By developing a comprehensive testing plan that includes specific testing objectives and criteria, project teams can ensure that testing efforts are focused on areas of the software application that are most likely to be problematic or have the greatest impact on users.

2. Test Automation: Automated tests can help reduce the likelihood of human error and increase the efficiency of testing efforts. They can be run more frequently and consistently, allowing project teams to quickly identify and address potential issues.
3. Security Testing: Security tests can help identify potential vulnerabilities in the software application, allowing project teams to implement measures to mitigate the risks of security breaches or attacks.
4. Code Reviews: These reviews can reveal potential coding errors or issues before they are introduced into the software application. This helps reduce the likelihood of defects and bugs in the final product.
5. Continuous Testing: Ongoing testing of the software application throughout the development process, rather than only at the end of the process, can help identify potential issues early on, reducing the likelihood of defects and bugs in the final product.

By implementing risk control measures, project teams can reduce the likelihood or impact of potential issues in a software application, resulting in a higher-quality, more reliable product that meets the needs of end users.

Risk-Driven Testing

Risk-driven testing is an essential aspect of software development and testing in the automotive industry. Modern cars and other vehicles rely heavily on complex software systems to operate safely and efficiently. As a result, any errors or issues in these systems can have serious consequences, including accidents, injuries, and fatalities.

In the context of automotive software development, risk-driven testing identifies and assesses the potential risks associated with different parts of a software system. This includes identifying potential safety-critical functions, such as braking, acceleration, and steering, and ensuring that these functions are tested to mitigate potential risks [135].

Risk-driven testing in the automotive industry also means complying with regulatory standards and requirements, such as the ISO 26262 standard for functional safety. This standard provides guidelines for developing and testing safety-critical software systems in vehicles, with a focus on identifying and mitigating potential risks.

Some examples of risk-driven testing practices in the automotive industry include the following:

1. Failure Mode and Effects Analysis: FMEA is a systematic approach to identifying and assessing potential failure modes and their effects on the vehicle's software system. This helps project teams prioritize

testing efforts and ensure that safety-critical functions are thoroughly tested.

2. HIL Testing: Hardware-in-the-loop testing evaluates software systems in a simulated environment, using real hardware components. This allows project teams to thoroughly test the software system in a controlled environment, reducing the risk of errors or issues in the final product.
3. Safety Testing: Safety-critical functions, such as braking, acceleration, and steering, are thoroughly tested using a range of techniques, including manual testing, automated testing, and simulation testing.

Risk-driven testing is essential to software development and testing in the automotive industry. By identifying and mitigating potential risks early in the development process, project teams can ensure that vehicles are safe and reliable for consumers.

Risk Management

Risk management identifies, assesses, and prioritizes risks and takes action to minimize or mitigate their potential impact [125].

The risk management process in software development typically involves the following steps:

1. Risk Identification: This involves identifying potential risks to a software application, such as technical risks, schedule risks, or budget risks. This can be done through brainstorming sessions, reviews of past projects, or other methods.
2. Risk Assessment: This involves assessing the likelihood and potential impact of each identified risk. Risks can be prioritized based on their likelihood and impact, and a risk management plan can be developed to address each risk.
3. Risk Mitigation: This involves taking action to minimize or mitigate the potential impact of identified risks. This includes implementing risk control measures, such as testing, code reviews, or security measures, to reduce the likelihood of a risk occurring or to lessen its impact if it does occur.
4. Risk Monitoring and Control: This involves monitoring the software development process and implementing changes to the risk management plan as needed. This includes revisiting risk assessments, implementing additional risk control measures, or adjusting the project schedule or budget to account for identified risks.

Effective risk management can help ensure that software applications are developed in a timely and efficient manner, meet the needs of users, and are delivered on time and within budget. By identifying and mitigating potential risks early in the development process, project teams can avoid costly delays or setbacks and deliver high-quality software applications that meet end users' needs.

Risk Mitigation

Risk mitigation is taking action to reduce or eliminate the potential impact of identified risks. It is an essential part of risk management that involves developing and implementing strategies to minimize the likelihood of a risk occurring and minimize the impact should the risk come to pass.

Several strategies can be used for risk mitigation:

1. Risk Avoidance: This involves taking action to avoid the risk entirely, such as by not pursuing a project or by using a different approach or technology that reduces the likelihood of a risk occurring.
2. Risk Reduction: This involves taking action to reduce a risk's likelihood or potential impact. This can include implementing risk control measures, such as testing, code reviews, or security measures, to reduce the likelihood of a risk occurring or to lessen its impact if it does occur.
3. Risk Transfer: This involves transferring the risk to another party, such as through insurance or by outsourcing a particular function to a third-party provider.
4. Risk Acceptance: This involves accepting the risk and its potential impact, while taking action to minimize that impact on the project. This can include developing contingency plans or alternative strategies to address the risk if it occurs.

Effective risk mitigation involves identifying potential risks early in a project's life cycle, developing a risk management plan, and taking action to address identified risks throughout the project. By implementing risk mitigation strategies, project teams can reduce the likelihood of delays, setbacks, or other negative impacts on the project, and ensure that the project is delivered on time, within budget, and to the satisfaction of stakeholders.

ROI

Return on investment (ROI) and testing are closely related in software development, as testing is one of the most important investments that project teams can make to ensure that software applications are of high quality and meet users' needs. By investing in testing, project teams can identify and address potential issues and risks in the software application, which can lead to a positive ROI by reducing costs and increasing revenue.

$$ROI = \left(\frac{Net\ Profit}{Cost\ Of\ Investment} \right) \times 100\%$$

There are several ways that testing can impact the ROI of a software development project, including the following:

1. Reduced Development Costs: By identifying and addressing potential issues and risks early in the development process, testing can help reduce development costs by minimizing the need for rework and changes.
2. Improved Efficiency: By ensuring that the software application is of high quality and meets the needs of users, testing can help improve efficiency by reducing the time and effort required to support and maintain the application.
3. Increased Customer Satisfaction: By ensuring that the software application is of high quality and meets the needs of users, testing can help increase customer satisfaction, leading to increased revenue and customer loyalty.
4. Reduced Risk: By identifying and addressing potential issues and risks early in the development process, testing can help reduce the risk of costly errors or failures that could negatively impact the project's ROI.

Rollover Testing

There is no single type of rollover event, these events are linked to terrain, and vehicle shape and mass distribution. Figure R.1 shows diagrams of the metrics involved in various types of rollover events: (a) Static trip—the static trip limit is reached when $\vartheta i = \beta$, with the CoM (Center of Mass) directly over the contact point A. (b) Roll energy—the vehicle with roll rate ωi is rotating from ϑi toward β. The vehicle is unstable when the instantaneous roll rotational energy exceeds the remaining work to rotate the vehicle to β. (c) Net moment—when the instantaneous net force (Fi) acting at the vehicle CoM passes above the contact point A causing a net tipping moment, then the vehicle is unstable based on a moment balance, ignoring angular rate. (d) Roll energy + net moment—the vehicle with instantaneous roll rate ωi and instantaneous net force (Fi) passing below point A.

FIGURE R.1 Models of various rollover events [137].

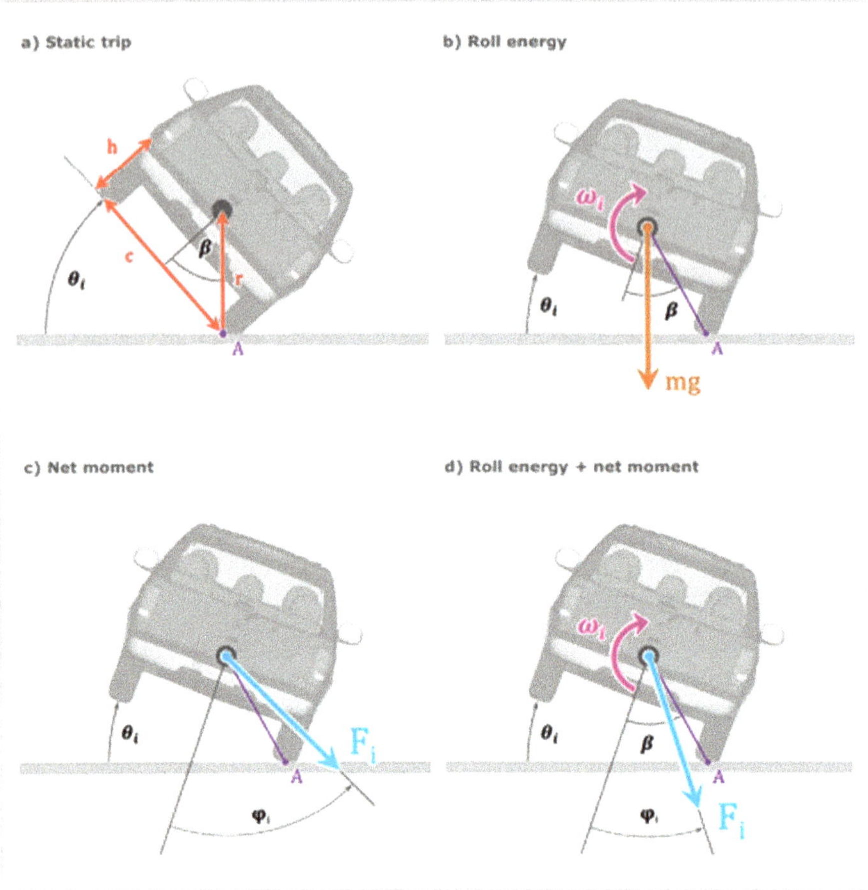

Vehicle rollover testing is a type of automotive testing that is used to evaluate the stability and safety of vehicles in the event of a rollover. Rollover accidents can be hazardous and can result in serious injuries or fatalities, so it is essential to ensure that vehicles are designed and built to withstand these accidents.

There are several types of vehicle rollover testing:

1. Dynamic Rollover Testing: This involves driving a vehicle on a track and making it tip over in a controlled manner. The objective of this test is to evaluate the stability and handling of the vehicle during a rollover.
2. Inverted Drop Testing: This involves dropping a vehicle upside down from a specified height onto a rigid surface. This test aims to evaluate the strength and integrity of the vehicle's roof structure and to ensure that it can withstand the impact of a rollover.

3. Sideways Impact Testing: This involves impacting a vehicle on its side with a moving barrier. This test aims to evaluate the effectiveness of the vehicle's side impact protection systems in the event of a rollover.
4. Rollover Resistance Testing: This involves measuring a vehicle's ability to resist tipping over during sudden steering maneuvers. The objective of this test is to evaluate the stability and handling of the vehicle during emergency maneuvers.

Vehicle rollover testing is an essential part of automotive safety testing and helps to ensure that vehicles are designed and built to withstand the impact of rollover accidents. By conducting these tests, automotive manufacturers can identify potential design flaws and improve the safety and stability of their vehicles, ultimately improving the safety of drivers and passengers on the road. For additional information, see **J2926_202110** and **J2114_201102** [138, 139].

Root Cause Analysis

Root cause analysis is a problem-solving method used to identify the underlying cause of a problem or incident. Root cause analysis aims to identify the underlying factors that contribute to the problem rather than merely address the symptoms. This allows for effective solutions that tackle the underlying cause and prevent the problem from recurring. Root cause analysis can be applied to various issues, including safety incidents, equipment failures, and process defects. Techniques used in root cause analysis include brainstorming, cause-and-effect diagrams, and statistical analysis. In the automotive world, there are generally two approaches to root cause analysis: the 8D and the A3.

8D (eight disciplines) is a problem-solving methodology that is used to identify and correct the root cause of a problem. It is a structured approach that involves eight steps:

1. Define the Problem: Clearly define the problem and its impact on the customer or the organization.
2. Develop a Team: Assemble a team of individuals who are knowledgeable about the problem and can work together to find a solution.
3. Describe the Problem: Gather and organize data about the problem, including its symptoms, causes, and effects.
4. Identify the Root Cause: Use the data collected in step 3 to identify the root cause of the problem.
5. Develop and Implement Corrective Actions: Develop a plan of action to address the root cause and implement it.

6. Verify the Corrective Actions: Verify that the corrective actions have effectively eliminated the problem.
7. Implement Permanent Corrective Actions: Implement permanent corrective actions to prevent the problem from recurring.
8. Recognize and Celebrate Success: Recognize and celebrate the team's and the organization's success solving the problem.

The 8D methodology is a powerful tool for identifying and addressing problems systematically and effectively. It is widely used in manufacturing, healthcare, and other industries to improve quality and performance.

A3 is a problem-solving method that uses a visual, structured approach to identify and address the underlying causes of a problem or issue. The A3 refers to the size of the paper used to document the process, which is typically 11x17 inches.

The A3 process typically follows these steps:

1. Identify and Define The Problem: Clearly state the problem and its impact on the organization or process.
2. Gather Data and Information: Collect observations, interviews, and measurements related to the problem.
3. Analyze the Data and Information: Use tools such as fishbone diagrams and Pareto charts to identify possible causes of the problem.
4. Identify the Root Cause: Determine the underlying cause of the problem, which is often referred to as the "root cause."
5. Develop and Implement a Solution: Create a plan to address the root cause and implement that solution.
6. Verify and Monitor: Verify that the solution is effective and monitor the problem to ensure it does not reoccur.

The A3 process is often used in manufacturing and healthcare settings but can be applied to any problem-solving situation. It is a useful tool for identifying the root cause of a problem and developing a plan to address it in a structured, data-driven way. For more information, see *Systemic Root Cause Early Failure Analysis during Accelerated Reliability Testing of Mass-Produced Mobility Electronics* [140].

Runtime

The final stage of a computer program's life cycle in which the code is executed on the computer's central processing unit as machine code is known as runtime, or execution time, in computer science. In other words, "runtime" refers to a program's execution phase.

S

"Men seek for vocabularies that are reflections of reality. To this end, they must develop vocabularies that are selections of reality. And any selection of reality must, in certain circumstances, function as a deflection of reality."

—Kenneth Burke

Safety-Critical Systems

Safety-critical systems are those for which a failure could seriously harm people, the environment, or property. Examples of safety-critical systems include medical devices, nuclear power plants, aviation systems, and automotive systems. Automotive safety-critical systems consist of active safety systems, such as blind spot warning, forward collision warning, lane departure, and so on [141]. Testing critical safety systems is essential to ensure they are designed and built to meet strict safety standards and prevent catastrophic failures.

Testing safety-critical systems involves a rigorous and systematic approach to identifying potential hazards, assessing the risk associated with those hazards, and verifying that the system is designed and built to mitigate those risks. Testing critical safety systems typically involves a combination of verification and validation activities:

1. Requirements Analysis: Analyzing the requirements to ensure they are complete, unambiguous, and consistent with safety standards.
2. Design Verification: Verifying that the system design meets the safety requirements and is free from defects.
3. Functional Testing: Evaluating the system's functions to ensure that they perform as intended and meet safety requirements.
4. Performance Testing: Evaluating the system's performance to ensure that it operates within safety limits and can handle various failure modes.
5. Environmental Testing: Examining the system's ability to withstand various environmental conditions, such as temperature and humidity, that may affect its safety.

6. Fault Tolerance Testing: Assessing the system's ability to handle and recover from various failures and faults without causing harm.
7. Certification Testing: Verifying that the system meets safety standards and regulatory requirements.

Testing safety-critical systems is a complex and challenging process that requires a high degree of expertise and attention to detail. It is essential to ensure that safety-critical systems are thoroughly tested before they are put into use to prevent accidents and protect people, the environment, and property.

Safety Monitoring Function

Safety monitoring functions are an essential aspect of safety-critical systems, designed to continuously monitor a system and detect any anomalies or failures that could lead to a safety hazard. Testing safety monitoring functions is essential to ensure that they are designed and implemented correctly and to verify that they can promptly detect and respond to safety hazards.

Testing safety monitoring functions typically involves a combination of verification and validation activities:

1. Requirements Analysis: Analyzing the requirements for the safety monitoring functions to ensure they are complete, unambiguous, and consistent with safety standards.
2. Design Verification: Verifying that the design of the safety monitoring functions meets the safety requirements and is free from defects.
3. Functional Testing: Evaluating the safety monitoring functions to ensure that they perform as intended and meet safety requirements. This can involve testing for specific failure modes and verifying that the system can detect and respond appropriately to these failures.
4. Performance Testing: Evaluating the performance of the safety monitoring functions to ensure that they operate within safety limits and can handle various failure modes.
5. Fault Tolerance Testing: Assessing the safety monitoring functions' ability to handle and recover from various failures and faults without causing harm.
6. Certification Testing: Verifying that the safety monitoring functions meet safety standards and regulatory requirements.

Testing safety monitoring functions is critical to the overall testing process for safety-critical systems. It is essential to ensure that these functions are thoroughly tested before the system is put into use to prevent accidents and protect people, the environment, and property.

Salt Fog (Spray)

Salt fog testing is a type of environmental testing commonly used in the automotive industry to assess the corrosion resistance of automotive components and materials.

During salt fog testing, the automotive components or materials are typically placed in a test chamber and exposed to a salt fog or mist. This mist contains a high concentration of salt that is intended to simulate the corrosive effects of saltwater and other harsh environmental conditions that the components or materials may encounter in the real world [20]. Tests for salt spray are also carried out in an enclosed testing chamber. A spray nozzle is used to apply a saltwater solution to a sample. The purpose of this thick saltwater fog/spray is to simulate a corrosive environment. The evaluation of oxide appearance depends on how long it takes for corrosion to occur and how corrosion-resistant a product is.

The duration and severity for both types of tests depends on the automotive manufacturer's specific requirements, the product's placement on the vehicle, and the materials being tested. Typically, test durations range from several hours to several weeks, and the severity of the test is determined by the concentration of salt and the temperature and humidity of the test environment.

Salt fog/spray testing is an important part of automotive testing to ensure that components and materials used in automotive applications can withstand harsh environmental conditions and maintain their performance and safety over time. Automotive manufacturers use the results of salt fog/spray tests to make design and material selection decisions and to ensure that their products meet safety and quality standards. For more information, see **SAE/USCAR-1** and **J1455_201703** [20, 142].

Saltwater Immersion

Saltwater immersion testing is commonly used in the automotive industry to assess the corrosion resistance of automotive components and materials. This testing involves immersing the components or materials in saltwater to simulate the corrosive effects of seawater and other harsh environmental conditions. The test evaluates the saltwater intrusion into sensitive parts of the product.

During saltwater immersion testing, the automotive components or materials are typically submerged in a saltwater solution for a specified period. The test duration and severity depends on the specific requirements of the automotive manufacturer and the materials being tested. Typically, test duration ranges from several hours to several weeks, and the severity of the test is determined by the concentration of salt in the solution and the temperature of the test environment.

Saltwater immersion testing is an integral part of automotive testing to ensure that components and materials used in automotive applications can withstand harsh environmental conditions and maintain their performance and safety over time. Automotive manufacturers use saltwater immersion test results to make design and material selection decisions and ensure that their products meet safety and quality standards.

Sample

Sample testing is essential to quality control and assurance processes in the automotive industry. Testing automotive components and materials is critical to ensure their reliability, durability, and safety.

In the manufacturing process of automotive components, sample testing is often used to evaluate the performance and durability of materials, such as metals, plastics, or composites. For example, a sample of a new material may be subjected to tensile, compression, or impact testing to determine its strength and stiffness. Similarly, a sample of a coating material may be tested for corrosion resistance, wear resistance, or adhesion strength.

Sample testing is also used to evaluate automotive systems, such as engines, transmissions, brakes, and suspension systems. In these cases, samples of individual components are tested to ensure their performance and compatibility within the larger system. For example, a sample of a brake pad may be tested for its stopping power and wear resistance, or a sample of a suspension spring may be tested for its load-carrying capacity and fatigue life.

In addition to evaluating the quality and performance of individual components, sample testing is also used to verify that automotive products comply with regulatory and safety standards. For example, samples of automotive components may be tested to ensure they meet emissions standards, crash test standards, or other safety regulations.

Sanity Testing

Sanity testing, also known as smoke testing, is performed to quickly assess whether a new software build or release is stable enough to proceed with more comprehensive testing. The purpose of sanity testing is to check the basic functionality of the software and ensure that it is ready for further testing.

Sanity testing is typically performed after a new build or release of software has been deployed, but before more thorough testing is performed. During sanity testing, a set of basic tests are executed on the software to verify that it is functioning as expected and meets the minimum requirements for further testing. This can include testing basic features and functionality, verifying that the software starts up and shuts down properly, and ensuring no major defects or errors.

The term "sanity testing" comes from the idea that testing is focused on verifying the sanity or basic functionality of the software rather than testing every feature and functionality in detail. Sanity testing is not intended to be comprehensive or exhaustive but rather to check the software's basic functionality and stability quickly.

Sanity testing is an important part of software testing because it helps to identify any major defects or issues early in the development process, before more comprehensive testing is performed. By catching these issues early, developers can save time and resources by addressing them before they become more difficult and expensive to fix.

Scalability Testing

Scalability testing in the automotive industry involves vehicle software systems, such as infotainment systems, driver assistance systems, and autonomous driving systems [143]. Automotive scalability testing aims to evaluate how well these systems handle increased or decreased workloads, such as the number of sensors or the amount of data being processed.

Scalability testing is typically performed by simulating different scenarios, such as driving in heavy traffic or extreme weather conditions, and measuring how the software systems respond. The tests may be conducted in a controlled environment, such as a testing facility, on a closed track, or on public roads during real-world driving conditions.

The performance of the software systems is measured and analyzed to identify any scalability issues, such as response time delays, system crashes, or data processing errors. The results of scalability testing can be used to improve the design and performance of the software systems, and to ensure that they can handle increased workloads without compromising safety or reliability.

Scalability testing in the automotive industry is important for ensuring the safety and reliability of software systems used in vehicles. As the automotive industry continues to incorporate more advanced software systems into vehicles, such as autonomous driving systems, scalability testing will become increasingly important to ensure that these systems can handle the demands of real-world driving conditions.

Scatter Plot

Scatter plots (also known as scatter diagrams) in testing are used to visualize the relationship between two variables and identify any data patterns or trends (see Figure S.1). For example, in software testing, scatter plots can be used to plot the relationship between the number of defects found and the time spent on testing. Each data point would represent a test cycle, with the number of defects found plotted on the y-axis and the time spent on testing plotted on the x-axis [40].

FIGURE S.1 An example of a range of scatter plots.

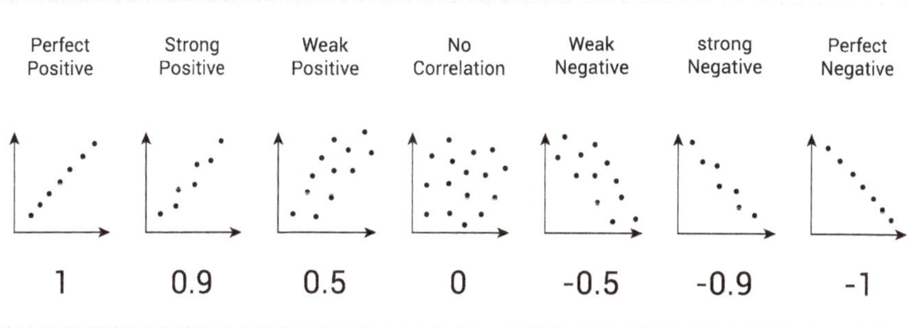

By analyzing the scatter plot, testers can identify whether there is a correlation between the number of defects found and the time spent on testing. They can also place any outliers or unusual data points that may indicate a problem with the system or the testing process.

Scenario

A scenario in testing refers to a specific situation or sequence of actions that a product might encounter. It is a narrative description of a use case or user story that helps to guide the testing process. In software (and hardware) testing, scenarios are used to design test cases that simulate real-world usage of the software. Each scenario typically includes a description of the user's goals, their actions to achieve those goals, and the expected results.

Scenarios can be used to guide both manual and automated testing. Manual testers can use scenarios to ensure that they cover all the major use cases and user stories, while automated tests can be designed to simulate specific scenarios and to validate that the software behaves as expected.

By using scenarios in testing, testers can ensure that they are testing the software in a way that is relevant to the user's needs and that covers all the major use cases and user stories. This can help improve the software's quality and usability and reduce the risk of defects and issues arising in production.

Scenario-Based Testing

Scenario-based testing involves designing and executing test cases based on various scenarios or use cases. It is a black-box testing technique, which means that testers focus on the external behavior of the software and do not have access to its internal workings.

Testers identify scenarios or use cases relevant to the software and design test cases based on those scenarios. For example, in a banking application, a scenario might involve a customer logging in, checking their account balance, transferring funds to another account, and then logging out. Testers would then design test cases to simulate that scenario and validate that the software behaves as expected.

The benefits of scenario-based testing include the following:

1. Improved Test Coverage: By designing test cases based on real-world scenarios, testers can ensure that they cover all the major use cases and user stories, reducing the risk of defects and issues arising in production.
2. Better Alignment with User Needs: Scenario-based testing ensures that testing focuses on the user's needs and goals rather than just the software's technical requirements.

3. Early Defect Identification: By testing the software in realistic scenarios, testers can identify defects and issues early in the development process, reducing the cost and time required to fix them.

Scenario-based testing can be conducted manually or using automated testing tools. Automated testing tools can be used to simulate many scenarios quickly and efficiently.

Scenario Database

A scenario database is a collection of predefined scenarios that can be used in software testing to guide the creation of test cases. The scenarios in the database represent typical usage patterns of the software and can ensure that the testing process covers all major use cases and user stories.

A scenario database are created by gathering input from various stakeholders, such as developers, business analysts, and end users. The scenarios are then categorized based on various factors, such as the functional area of the software they relate to, the user type, or the complexity of the scenario.

Using a scenario database provides several testing benefits:

1. Improved Test Coverage: A scenario database can ensure that all major use cases and user stories are covered by the testing process, reducing the risk of defects and issues arising in production.
2. Reduced Testing Time: Since the scenarios in the database are pr-defined, testers do not need to spend as much time creating new test cases, reducing the overall testing time.
3. Consistent Testing: By using the same scenarios for each testing cycle, testers can ensure that the testing is consistent and repeatable, making it easier to compare results across different test cycles.

A scenario database is a useful tool in testing to guide the creation of test cases and ensure that all major use cases and user stories are covered. However, it is important to ensure that the scenarios in the database are regularly updated and maintained to reflect changes in the software or end users' needs.

Scene

In the context of software testing, a scene refers to a specific scenario or use case that is being tested, for example, a particular sequence of user actions or interactions with the software.

Scenes are often used to guide the creation of test cases and ensure that all major use cases and user stories are covered in the testing process. By designing test cases based on specific scenes, testers can ensure that the software behaves as expected in different scenarios and use cases.

Scribe

In the context of formal reviews, a scribe refers to a person responsible for documenting the review results. A formal review is a structured process that involves a team of reviewers examining a software artifact (such as a design document, code, or test plan) to identify defects, errors, and areas for improvement. The review aims to improve the software artifact's quality and ensure that it meets the required standards and specifications [12].

During a formal review, the scribe typically notes the review team's discussions, decisions, and actions. The scribe may also document any defects or issues that are identified during the review, along with the recommended actions for resolving them. The scribe is responsible for creating a report summarizing the results of the review, which may include a list of defects, recommendations for improvements, and other feedback.

The role of the scribe is important for ensuring that the results of the review are documented and communicated effectively to the relevant stakeholders. Thus, the scribe helps ensure that the software artifact is improved and meets the required quality standards.

Scripting Language

Scripting languages can be used for various testing purposes, such as automated testing, load testing, and regression testing. Scripting languages are programming languages designed for scripting tasks, which typically involve automating repetitive tasks or performing tasks requiring minimal interaction with the user.

In the context of software testing, scripting languages are used to automate the execution of test cases. This can save time and effort compared to manually executing test cases, especially for repetitive or time-consuming tests. Popular scripting languages used for testing include Python, JavaScript, and Ruby.

Scripting languages are also used in load testing, which involves simulating a large number of users or transactions to test the performance and scalability of a system. By scripting user interactions and simulating large volumes of traffic, load testing can help identify potential bottlenecks and performance issues.

In addition, scripting languages are employed in regression testing, which involves retesting a software application after changes have been made to ensure that the changes have not introduced new defects or caused any regression issues. By automating regression tests using scripting languages, testers can confirm that the application behaves as expected after each change and catch any issues early in the development process.

There are many scripting languages that can be used for various tasks, including software testing. Here are a few examples of popular scripting languages:

1. Python is a versatile and easy-to-learn language that is commonly used in software testing, as well as data analysis, web development, and machine learning.
2. JavaScript is a scripting language that is commonly used for front-end web development, but it can also be used for automated testing using frameworks like Selenium.
3. Ruby is a dynamic scripting language that is often used in web development and automated testing using the popular framework RSpec.
4. Bash is a Unix shell scripting language commonly used to automate Unix/Linux systems tasks, including testing scripts.
5. PowerShell is a Microsoft-developed scripting language that is commonly used for automation tasks, including testing scripts on Windows systems.

Security Testing

Security testing in the automotive industry involves evaluating the security [144] of a vehicle's various electronic systems and components, such as the infotainment system, telematics systems, and connected car technologies. The goal of security testing is to identify potential vulnerabilities and weaknesses in these systems that hackers could exploit to gain unauthorized access, steal data, or cause damage.

Automotive security testing typically involves a combination of manual and automated testing techniques:

1. Vulnerability Scanning: This involves using automated tools to scan the various components of a vehicle's electronic systems to identify potential vulnerabilities, such as outdated software, weak passwords, or unsecured network connections.
2. Penetration Testing: This involves attempting to exploit identified vulnerabilities by simulating an attack on the system. This can include trying to gain unauthorized access, steal data, or manipulate the system.
3. Threat Modeling: This involves analyzing the potential threats and attack vectors that could be used against a vehicle's electronic systems and components and developing strategies to mitigate those risks.

4. Code Analysis: This involves reviewing the source code of the software used in a vehicle's electronic systems to identify potential security weaknesses or vulnerabilities.
5. Fuzz Testing: This involves testing the resilience of a system to unexpected or unusual inputs, such as malformed messages or data, in order to identify potential vulnerabilities or weaknesses.

Security Testing Tool

Various security testing tools are available for use in the software development and testing process. Some commonly used security testing tools include the following:

1. Burp Suite is a popular web application security testing tool that includes a wide range of tools for scanning, analyzing, and testing web applications for security vulnerabilities.
2. Metasploit is an open-source penetration testing tool that allows testers to simulate attacks against various systems and applications in order to identify potential vulnerabilities.
3. Nessus is a vulnerability scanner that can be used to scan networks, systems, and applications for known security vulnerabilities and potential risks.
4. Wireshark is a network protocol analyzer that can be used to monitor network traffic and identify potential security issues, such as unauthorized access or data breaches.
5. Nmap is a network exploration and security auditing tool that can be used to identify hosts and services on a network, as well as potential security vulnerabilities and attack vectors.
6. AppScan is an IBM tool that provides web application security testing and analysis, including scanning for vulnerabilities, analyzing code, and testing for various types of attacks.
7. Selenium is an open-source web application testing framework that can be used for automated functional testing as well as security testing.

These are just a few examples of the many security testing tools available. The choice of tool will depend on the specific needs and requirements of the testing project, as well as the expertise of the testing team.

Semmelweis Reflex

The Semmelweis reflex is a phenomenon in which a person or group rejects new information or evidence that contradicts their beliefs or practices. In the testing context, this is evident when testers have preconceived notions or biases that prevent them from considering certain scenarios or types of testing.

For example, if testers (or developer or management) believe that a particular software feature is not susceptible to security vulnerabilities, they may not perform security testing on that feature. However, if new information or evidence suggests the feature may indeed be vulnerable, the testers' reluctance to consider security testing on that feature would be an example of the Semmelweis reflex.

To avoid the Semmelweis reflex in testing, it is important for testers, developers, and management to approach each testing scenario with an open mind and be willing to consider all possibilities and potential risks, even if they may contradict existing beliefs or practices. Testers should also be helpful to adapt their testing strategies based on new information or evidence that emerges during the testing process and be willing to engage in constructive discussions with other testing team members to ensure that all potential risks and scenarios are considered.

Sensor Fusion Testing

Sensor fusion is combining data from multiple sensors to improve the overall quality and reliability of their data. This technique is used in many applications, such as robotics, autonomous vehicles, and surveillance systems. Sensor fusion testing evaluates the accuracy and performance of a sensor fusion system.

The testing process for sensor fusion involves several steps:

1. Sensor Calibration: This is the process of adjusting and verifying the accuracy of each sensor used in the fusion system.
2. Data Collection: Data is collected from each sensor and the quality of the data is evaluated.
3. Data Synchronization: The data from each sensor is synchronized to ensure that they are all recording data at the same time.
4. Data Fusion: The data is combined using various algorithms and techniques to produce an output that is more accurate than any individual sensor's output.
5. Validation: The accuracy and performance of the sensor fusion system are validated by comparing the output to a reference dataset.
6. Optimization: The system is optimized by adjusting the parameters and algorithms used in the fusion process to improve accuracy and performance.

Sensor Testing

Sensor testing in the automotive industry is a critical step in ensuring the safety and reliability of vehicles. There are several types of sensors used in automotive applications. Some of the sensors and their typical parameters are displayed in Figure S.2.

FIGURE S.2 Sensor types and typical parameters [145].

DESCRIPTION	DISPLAYED UNITS	SENSITIVITY	MAX/MIN READINGS	ENVIRONMENT
		ANALOG SENSORS		
Washer Level, Continuous	Volume	16 parts	MFG	E,U,F
Engine Oil Level, Continuous	Volume	0.1 Liter	MFG	U,F
Fuel Level	Volume	0.4 Liters	MFG	E,I,F
Battery Condition Volts	Volts	0.1 Volts	MFG	I,U
Alternator Output Amps	Amperes	0.5 Amps	MFG	U
Transmission Fluid Temperature	Deg. C	0.5 Degree	MFG	U,F
Coolant Temperature, Continuous	Deg. C	1.0 Degree	MFG	U,F
Air Blend Door Position	Position	1 Degree	MFG	I
		DIGITAL SENSORS		
Hood Latch	N/A	ON/OFF	O/C	E
Trunk Latched	N/A	ON/OFF	O/C	E,I
Door Locked, per door	N/A	ON/OFF	O/C	I
Washer Level, Low Limit	N/A	ON/OFF	L/H	E,U,F,
Brake Fluid, Low Limit	N/A	ON/OFF	L/H	U,F
Engine Oil Level, Low Limit	N/A	ON/OFF	L/H	U,F
Transmission Oil Level, Low Limit	N/A	ON/OFF	L/H	U,F
Coolant Temperature Limit	N/A	ON/OFF	L/H	U,F
		OPERATOR SWITCHES		
Power Door Lock Switch	N/A	On/Off	O/C	I
Power Window, Up Switch	N/A	On/Off	O/C	I
Power Mirror, In Switch	N/A	On/Off	O/C	I
Power Seat, Backward Switch	N/A	On/Off	O/C	I
Power Seat, Recline Switch	N/A	On/Off	O/C	I
HVAC Fan, Low Speed Switch	N/A	On/Off	O/C	I
HVAC Fan, High Speed Switch	N/A	On/Off	O/C	I
Rear Window Defogger Switch	N/A	On/Off	O/C	I
NOTE: Metric Units Assumed	LEGEND:	O/C - Open/Closed L/H - Low/High MFG - Manufacturer Dependent	E - Exterior I - Interior U - Under Hood F - Fluid Immersion	

Modern vehicles include the following sensors among others:

1. Lidar (Light Detection and Ranging): These sensors use laser beams to detect and measure the distance to objects, providing accurate 3D maps of the environment around a vehicle.
2. Radar (Radio Detection and Ranging): These sensors use radio waves to detect the presence of objects and measure their distance and speed.
3. Cameras: Cameras are used for various vehicle purposes, including lane departure warnings, blind spot detection, and parking assistance.
4. Ultrasonic Sensors: These sensors use sound waves to detect the distance to objects, often used for parking assistance and obstacle detection.

During the sensor testing process, several tests are performed to evaluate the accuracy and reliability of each sensor. Some of the standard tests include the following:

1. Environmental Testing: Sensors are tested in extreme weather conditions, such as extreme temperatures and high humidity, to ensure they operate correctly in different environments.
2. Durability Testing: Sensors are tested for durability and reliability, simulating the wear and tear they experience over time.
3. Performance Testing: Sensors are tested for accuracy and precision in detecting and measuring distance, speed, and other parameters.
4. Compatibility Testing: Sensors are tested to ensure they can work with other systems in the vehicle, such as the navigation and control systems.

Severity Levels

Severity levels represent the degree of impact or severity of a defect or issue on a software system under test. Defect severity is typically assigned a level or priority during software testing, indicating the urgency of addressing the issue [39].

There are different methods of categorizing defect severity levels, but a commonly used model includes the following:

1. Critical: Defects that cause the system to crash or severely malfunction, making it impossible to use the software.
2. High: Defects that cause the system to malfunction, resulting in significant issues that affect the software's functionality, performance, or usability.
3. Medium: Defects that cause minor issues or inconveniences that do not impact the software's overall functionality.
4. Low: Defects that cause minor cosmetic issues or other noncritical problems that do not impact the software's functionality or usability.

Assigning severity levels to defects during testing helps the development team prioritize which issues need to be addressed first. Critical and high severity defects are typically resolved as a top priority because they can significantly impact the system's performance, while medium and low severity defects may be addressed later in the development cycle or in a future release.

It is essential to establish clear guidelines for assigning severity levels during testing to ensure that all team members consistently evaluate and prioritize defects. The goal is to ensure the product (software) is functional, reliable, and meets the end users' needs.

Shared Information Bias

Shared information bias, or the "common knowledge effect," occurs when groups spend more time discussing and considering information already known to all group members rather than new or unique information. This bias can impact the testing process in several ways.

Shared information bias can affect testing by limiting the range of test cases considered. Team members may focus on test cases that are already familiar to everyone, overlooking more unique or complex test cases that could potentially reveal critical defects or issues.

Another way shared information bias can impact testing is by limiting the scope of testing. For example, if a group is focused on a particular feature or function of the software, they may overlook potential issues in other areas of the system.

It's important to establish clear testing objectives and guidelines to mitigate shared information bias during testing. Testers should be encouraged to consider a range of test cases and scenarios, including those that may not be immediately obvious or familiar to all team members. It may also be helpful to rotate team members or bring in outside perspectives to avoid groupthink and encourage diverse perspectives.

In addition, employing a structured testing approach, such as risk-based testing, can help ensure that testing efforts are focused on the most critical areas of the system and that all potential issues are considered. Awareness of shared information bias and its potential impact on testing is essential to ensure a thorough and effective testing process.

Simulation

Simulation is a valuable tool in the automotive industry for designing, testing, and evaluating vehicle systems and components. Simulation can be used at various stages of the product development process, from initial concept design to final validation and testing [146].

One important use of simulation in the automotive industry is vehicle crash testing. Crash simulation helps engineers design safer vehicles by modeling a vehicle's and its occupants' behavior in various crash scenarios. Using simulations, engineers identify potential safety issues and design improvements to reduce the risk of injury in the event of a crash.

Another area where simulation is used in automotive engineering is designing and testing powertrain systems, such as engines, transmissions, and drivetrains. Simulations can model fluid dynamics (cooling packages), electrical systems, and communications networks, as well as the performance of these systems under various conditions. The information gleaned from simulations is used to optimize their efficiency, power output, and fuel economy.

Simulation are also used to test and evaluate the performance of advanced driver assistance systems (ADAS), such as collision avoidance systems, lane departure warning systems, and adaptive cruise control. Simulations can identify potential issues with these systems and optimize their performance to improve safety and performance [146].

Simulation is an essential early tool used to economically and virtually explore a component or system, allowing engineers to design safer, more efficient, and higher-performing vehicles while reducing the cost and time required for physical testing and validation.

Simulation-Based Fault Injection Testing

Simulation-based fault injection testing (SBFIT) is a technique used in the automotive industry to evaluate the safety and reliability of a vehicle's electronic systems. With the increasing complexity of modern vehicles and the integration of electronic systems, SBFIT has become an essential tool for ensuring the safety and reliability of automotive systems [146].

In SBFIT for automotive systems, a simulation model of a vehicle's electronic system is created, and simulated faults are injected into the model to simulate various failure scenarios. The faults can be virtually injected at different system levels, such as the sensors, control units, or communication networks. The system's behavior is then observed under these fault conditions to determine how it responds and whether it can recover from the faults.

SBFIT is used to test the safety and reliability of various vehicle systems, such as the braking system, steering system, engine control system, and advanced driver assistance systems (ADAS). By simulating faults, SBFIT can help identify potential issues before the vehicle is deployed, reducing the risk of failures in the field.

One advantage of SBFIT in the automotive industry is that it is often more cost-effective and efficient than physical fault injection testing, which can be time-consuming and expensive. SBFIT can also to test a wider range of fault scenarios and is easily repeated and modified to evaluate different aspects of the vehicle's behavior.

However, as with any testing technique, SBFIT has some limitations, including the need for an accurate simulation model of a vehicle's electronic systems and the potential for the model to not fully capture all the complexities and interactions of real-world vehicle performance. Therefore, SBFIT should be used with other testing techniques, such as physical testing and validation, to ensure a comprehensive evaluation of the vehicle's safety and reliability.

Simulation Model

A simulation model is a computer program that represents a real-world system or process. Simulation models are used to simulate the system's behavior and predict its outcomes. They are widely used in engineering, science, economics, and other fields to study complex systems and test new ideas or scenarios [39].

Simulation models can be created using different techniques and tools, depending on the system being modeled and the objectives of the simulation. Some common techniques for creating simulation models include the following:

1. Mathematical Models: These models use equations to describe the behavior of a system. They are often used in physics, engineering, and other quantitative fields.
2. Agent-Based Models: These models simulate the behavior of individual agents, such as people or animals, and their interactions with each other. They are often used in social sciences and economics.
3. Discrete Event Models: These models simulate the behavior of a system based on discrete events, such as arrivals, departures, or failures. They are often used in manufacturing and logistics.
4. System Dynamics Models: These models simulate the behavior of a system over time and are often used to study complex systems, such as ecosystems or economies.

Once a simulation model is created, it can be run using different scenarios or inputs to predict the behavior of the system under different conditions. The simulation results can then be analyzed to evaluate the system's performance, identify potential issues or improvements, and make decisions about how to optimize the system.

Simulator

A device used for training that has controls similar to genuine vehicles, aircraft, or other complex systems to provide a realistic replica of how they operate.

Situation

In testing, the situation refers to the environment, conditions, and context in which a test is conducted. The situation can significantly impact the testing process and the results obtained from testing.

Some factors that can affect the situation in testing include the following:

1. Testing Objectives: These can influence the testing situation, such as the type of tests that are performed, the level of testing, and the resources required.
2. Test Environment: This can affect the situation by providing or limiting access to resources, hardware, and software required for testing.
3. Test Data: The quality and quantity of test data can affect the situation by influencing the accuracy and validity of the test results.
4. Time Constraints: These can affect the situation by limiting the amount of time available for testing, which can impact the depth and scope of the testing.
5. Stakeholder Expectations: The expectations of customers, management, and others can influence the situation by setting expectations for the testing process and the results.

It is essential to consider the situation when planning and conducting testing to ensure that the testing is appropriate and effective. By understanding the situation, testers can identify potential challenges and limitations and develop strategies to overcome them. Additionally, considering the situation can help ensure that the testing results are relevant and useful for the intended audience.

Six Sigma

Six Sigma is a data-driven methodology to improve business processes and reduce defects. Motorola developed it in 1986, and organizations in various industries have widely adopted it. The name refers to the statistical concept of having a process that is six standard deviations from the mean, equating to a defect rate of only 3.4 defects per million opportunities. Six Sigma aims to achieve near-perfect quality of products and services. The methodology uses statistical tools and techniques to identify and eliminate the root cause of defects and improve processes. The Six Sigma approach (DMAIC) involves *defining* the problem, *measuring* current performance, *analyzing* data to *identify* causes of defects, improving the process, and *controlling* future performance (see Figure S.3) [147].

FIGURE S.3 The Six Sigma acronym DMAIC is a key to understanding the process.

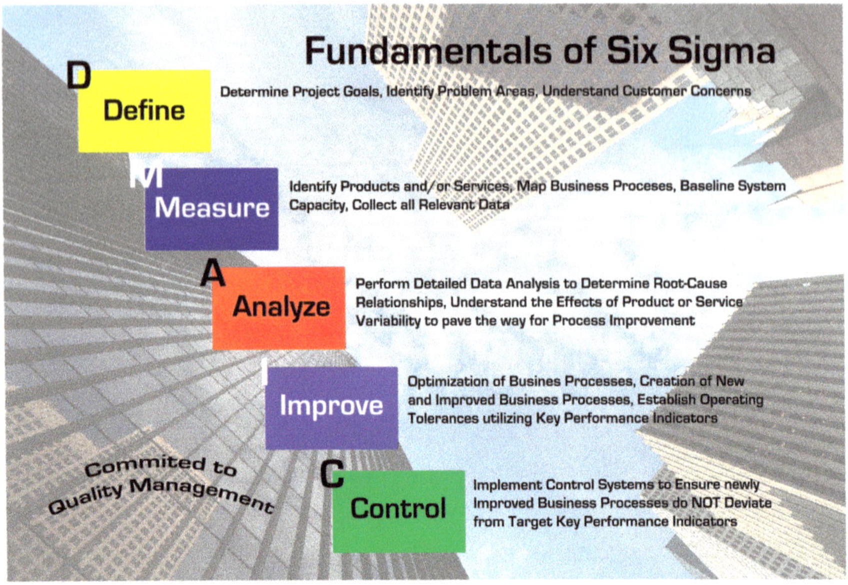

Fotoluminate LLC/Shutterstock.com.

Skewness

In testing, skewness refers to the distribution of data points in a sample or population. Skewness is a measure of the asymmetry of the data around the mean, with positive skewness indicating that the data is skewed to the right and negative skewness indicating that the data is skewed to the left.

Skewness can significantly impact testing because it can affect the validity and reliability of the results obtained from testing. For example, if the data is skewed, it may not represent the population being tested, leading to inaccurate conclusions.

Skewness can also impact testing in other ways:

1. Sample Size: Skewed data can require larger sample sizes to obtain statistically significant results.
2. Test Selection: The selection of tests used for testing can be affected by skewness, with certain tests being more appropriate for skewed data.
3. Statistical Analysis: Skewed data may require different types of statistical analysis to account for the skewness and ensure accurate results.
4. Outlier Detection: Skewed data can make detecting outliers more challenging because they may be more difficult to identify.

To account for skewness in testing, it is essential to carefully analyze the data and understand its distribution. Depending on the extent of the skewness, testers may need to adjust the testing methods or analysis techniques to account for the skewness and ensure accurate results. Additionally, when reporting test results, it is important to include information about the skewness of the data to help stakeholders interpret the results correctly.

Smoke Testing

See Sanity Test.

Soak Testing

Soak testing subjects a system to a sustained load or stress over an extended period to evaluate its stability and performance under normal and heavy usage conditions. Soak testing is designed to identify potential performance issues that may occur over time, such as memory leaks or other types of degradation.

During soak testing, a system is subjected to realistic conditions that simulate typical usage patterns to see how it performs over an extended period. This can involve simulating a large number of users accessing the system, performing various actions, and generating a high volume of data. Soak testing may affect other factors, such as changes in system configurations or stress testing of network components [20].

Some benefits of soak testing include the following:

- Identifying issues that may only occur over an extended period
- Evaluating the stability and performance of a system under normal and heavy usage conditions
- Providing insights into system behavior and performance trends over time
- Improving system reliability and robustness by identifying and addressing potential performance issues before they impact users

Soak testing is challenging because it can be time-consuming and resource-intensive. It also sometimes requires specialized tools and expertise to simulate realistic usage patterns and monitor system performance over an extended period. However, despite these challenges, soak testing is an essential component of performance testing and helps to ensure that systems are stable, reliable, and perform well under various usage scenarios.

Software Applications

Software applications are an increasingly important part of automotive systems, enabling a wide range of functionalities and features that enhance the driving experience and improve safety. here are just some examples of software applications in the automotive industry:

1. Infotainment Systems: These software applications provide drivers and passengers with access to entertainment, communication, and navigation features, such as music streaming, hands-free calling, and GPS navigation.
2. Advanced Driver Assistance Systems (ADAS): These software applications use sensors and cameras to provide drivers with real-time information about their surroundings, such as lane departure warnings, blind spot detection, and automatic emergency braking.
3. Engine Management Systems: These software applications control and optimize the performance of the vehicle's engine, including fuel injection, ignition timing, and emissions control.
4. Telematics: These software applications enable vehicle-to-vehicle and vehicle-to-infrastructure communication, providing drivers with real-time traffic updates, weather alerts, and other safety and convenience features.

Developing software applications for the automotive industry requires specialized expertise in software engineering, embedded systems, and cybersecurity. Additionally, automotive software applications must be tested rigorously to ensure their safety, reliability, and performance under various driving conditions.

As vehicles become increasingly connected and automated, software applications are expected to play an even larger role in the automotive industry, enabling new functionalities such as autonomous driving and vehicle-to-everything (V2X) communication.

Software-Based Fault Injection Testing

Software-based fault injection testing (SW-FIT) is used in the automotive industry to evaluate the robustness and safety of vehicle software systems. SW-FIT involves injecting faults or errors into the software code to simulate real-world scenarios and evaluate how the system responds.

In the automotive industry, SW-FIT evaluates safety-critical software systems such as those used in advanced driver assistance systems (ADAS), engine management systems, and brake control systems. Fault injection testing can identify potential errors or vulnerabilities in the software that could lead to safety hazards or system failures.

Several types of fault injection techniques are used in SW-FIT:

1. Random Fault Injection: In this technique, faults are randomly injected into the software code to evaluate the system's response.
2. Time-Based Fault Injection: This technique involves injecting faults into the software code at specific times or intervals to simulate real-world scenarios.
3. State-Based Fault Injection: This technique involves injecting faults into the software code based on the system's current state.

SW-FIT can be a complex and time-consuming process requiring specialized tools and expertise to perform effectively. However, it is an essential component of software testing in the automotive industry to ensure the safety and reliability of vehicles.

The results of SW-FIT can be used to identify potential issues in the software code and improve the system's robustness and safety. By detecting and addressing potential faults and errors in the software, SW-FIT can help reduce the risk of accidents and improve the overall performance of automotive systems.

Software Build

A software build is a process of compiling and linking source code files into a standalone executable or library that can be executed or used by other software applications. A software build typically includes several steps, such as compiling, linking, and packaging the code into a format that can be distributed and deployed [148].

The software build process may include other tasks like code analysis, testing, and quality assurance. Depending on the complexity of the software project, the build process may be automated using build tools such as Make, Gradle, or Jenkins [7].

The build process typically starts with compiling the source code files into object files, which are then linked together to create an executable or library file. The linker resolves dependencies between the object files and links in any required external libraries [149].

Once the software build is complete, the resulting executable or library can be tested and deployed to production. The build process may also include packaging the software into a distribution format, such as a .zip or .tar.gz file, for easy distribution and installation.

The software build process is an important component of the software development life cycle (SDLC), as it ensures that the software is compiled correctly and that any errors or issues are caught and addressed before deployment. By automating the build process and using tools and best practices, software development teams can streamline the build process and improve the quality and reliability of their software applications.

Software Design Document

A software design document (SDD) is a comprehensive outline of a software project's architecture, design, and implementation details. The SDD provides a detailed description of the software system's components, functions, and how they interact [150].

The software design document typically includes the following sections:

1. Introduction: This section provides an overview of the software project, including its purpose, scope, and objectives.
2. Architecture: This section describes the overall architecture of the software system, including its high-level components, their relationships, and the interfaces between them.
3. Design: This section details the creation of each component of the software system, including its functionality, data structures, algorithms, and interfaces.
4. Implementation: This section explains how the software system will be implemented, including the programming language, development environment, and tools that will be used.
5. Testing: This section outlines the testing strategy for the software system, including the types of testing that will be performed, the testing tools and techniques that will be used, and the testing environments.
6. Deployment: This section describes how the software system will be deployed, including the hardware and software requirements, the installation process, and the system configuration.
7. Maintenance: This section outlines the maintenance plan for the software system, including the procedures for bug fixing, updates, and upgrades.

The software design document serves as a road map for the development team, providing a detailed blueprint of the software system that they are building. It also serves as a reference for future maintenance and updates to the software system. The SDD is an important component of the software development life cycle (SDLC) and helps ensure that the software system is developed according to the specifications and requirements of the stakeholders.

Software Fault

A software fault—a software defect or bug—is an error or flaw in a computer program that causes it to behave unexpectedly or incorrectly. Various factors, such as programming errors, design flaws, incorrect assumptions, or hardware failures, can cause software faults [7, 151].

Software faults can manifest in different ways, such as system crashes, data corruption, incorrect outputs, security vulnerabilities, or performance issues. Software faults can be difficult to detect and fix, as they can result from complex interactions between different software system components.

Detecting and fixing software faults is an important software development process. Techniques such as debugging, testing, and code reviews can help identify and correct software faults before they become serious issues.

There are several types of software faults:

1. Syntax Errors: These are errors in the programming language syntax, such as missing or extra punctuation, that prevent the code from compiling or executing.
2. Logic Errors: These are errors in the program logic that cause it to produce incorrect or unexpected results, even though it may compile and execute without errors.
3. Runtime Errors: These are errors that occur during program execution, such as memory access violations, divide-by-zero errors, or invalid function calls.
4. Integration Errors: These are errors that occur when different components of the software system interact with each other in unexpected ways, such as communication failures or data mismatches.
5. Security Vulnerabilities: These are weaknesses in the software system that can be exploited by attackers to gain unauthorized access or perform malicious actions.

It's important for software developers to be aware of these different types of software faults and to take appropriate measures to detect and fix them, in order to ensure that the software system is reliable, secure, and performs as expected.

Software-in-the-Loop Simulation

Software-in-the-loop (SIL) testing involves integrating a software component or subsystem into a simulated or emulated environment to test its functionality and performance. The simulation models the behavior of the system or hardware that the component will interact with in the actual deployment environment.

SIL testing can be used to validate and verify software behavior early in the development process before the system is integrated with hardware or deployed in a real-world environment. It can also be used to test software changes or updates in a safe and controlled environment without physical hardware or equipment.

SIL testing typically involves the following steps:

1. Design a simulation or emulation environment that models the behavior of the system or hardware that the software will interact with.
2. Integrate the software component or subsystem into the simulation environment.
3. Run tests on the software to verify its functionality and performance in the simulated environment.
4. Analyze the test results to identify any issues or defects that need to be fixed.

SIL testing can be used in a variety of software development contexts, including embedded systems, control systems, and autonomous vehicles. It can help reduce development time and costs, improve software quality and reliability, and mitigate risks associated with hardware testing.

SIL testing is often used in conjunction with other types of testing, such as unit testing, integration testing, and hardware-in-the-loop testing, to provide a comprehensive testing strategy that covers all aspects of the software system.

Software Integration

Software integration is the process of combining different software components or subsystems into a larger software system that functions as a single unit. Integration is an important phase of the software development life cycle (SDLC), as it ensures that the different software components work together properly and achieve the desired functionality and performance [149].

Software integration typically involves the following steps:

1. Component Testing: Each software component or subsystem is tested independently to ensure that it functions correctly and meets its specifications.
2. Integration Testing: The software components are combined and tested together to ensure that they interact correctly and perform their intended functions.
3. System Testing: The entire software system is tested as a whole to ensure that it meets the system requirements and specifications.

There are several approaches to software integration:

1. Top-Down Integration: This approach involves first integrating the higher-level software components, followed by the lower-level components. This method is useful when the overall system behavior is well understood and can be modeled accurately.

2. Bottom-up Integration: This approach involves first integrating the lower-level software components, followed by the higher-level components. This method is useful when the individual components have complex behaviors and interactions that need to be thoroughly tested.
3. Hybrid Integration: This approach combines top-down and bottom-up integration, starting with the most critical or complex components and gradually integrating the others.

Software integration can be a complex and time-consuming process, especially in large-scale software systems. To ensure successful integration, it's important to have a clear understanding of the system requirements and specifications and a comprehensive testing strategy that includes unit testing, integration testing, and system testing.

Software Life Cycle Models

Software life cycle models are systematic approaches to software development that outline the phases, activities, and deliverables involved in the software development process. Different software life cycle models emphasize different aspects of the development process, such as requirements gathering, design, testing, or maintenance. Some common software lifecycle models include the following [150]:

1. Waterfall Model: This is a sequential software development process in which each development phase is completed before the next phase begins. This model is well-suited for projects with well-defined requirements and a fixed scope.
2. Agile Model: This model emphasizes flexibility and collaboration between development teams and stakeholders. It focuses on delivering working software in small, incremental iterations, with continuous feedback and adaptation.
3. Spiral Model: This is an iterative software development process combining waterfall and iterative models elements. It emphasizes risk management and incorporates multiple development, prototyping, and testing iterations.
4. V-Model: This is a variation of the waterfall model that emphasizes the relationship between testing and development. It involves defining test plans and cases at each stage of development, and testing each component before moving on to the next stage.
5. Iterative Model: This is an incremental software development process involving multiple development, testing, and evaluation iterations.

Each iteration builds upon the previous one, with continuous feedback and adaptation.

6. DevOps Model: This is a software development model that emphasizes collaboration between development teams and operations teams. It involves continuous integration, delivery, and testing, focusing on delivering high-quality software quickly and efficiently.

Each software life cycle model has advantages and disadvantages, and the choice of model depends on factors such as the project requirements, scope, and constraints. It's important to choose a project-appropriate software life cycle model and adapt it as necessary throughout development.

Software Maintenance Activity

Software maintenance is the process of modifying, updating, and repairing software after it has been deployed. Software maintenance activities can be classified into four categories: corrective maintenance, adaptive maintenance, perfective maintenance, and preventive maintenance.

1. Corrective Maintenance: These activities fix defects or errors in the software that are discovered after it has been deployed. This includes debugging, troubleshooting, and problem resolution.
2. Adaptive Maintenance: These activities modify the software to adapt to environmental or requirements changes. This includes adjusting the software to work with new hardware or operating systems or adding new functionality to meet changing user needs.
3. Perfective Maintenance: These activities improve the software's performance, reliability, or maintainability. This includes activities such as optimizing code, improving documentation, or enhancing the user interface.
4. Preventive Maintenance: These activities proactively identify and address potential issues before they become problems. This includes activities such as analyzing usage data to identify areas for improvement or conducting code reviews to identify potential defects.

In addition to these four categories, software maintenance activities may include software configuration management, software documentation management, and software quality assurance.

Software Module

A software module is a self-contained, reusable software program component that performs a specific function or set of related functions. It can be thought of as a building block or a piece of the larger software system.

Software modules are typically designed to be modular and independent, so they can be easily integrated into a larger software system or replaced with alternative modules. This makes software development more efficient and allows developers to reuse code across multiple projects, reducing development time and cost.

Some examples of software modules include the following:

1. Libraries: These are collections of prewritten code that perform specific functions. They can be linked to a software program to provide additional functionality.
2. Plugins: These can be added to a software program to extend its functionality. Plugins are often used in applications such as web browsers, where users can install plugins to add new features.
3. Drivers: These are modules that allow the software to communicate with hardware devices. Drivers translate commands from the software into signals that the hardware can understand.
4. Components: These are self-contained units of functionality that can be integrated into a larger software system. Components can be developed independently and then combined to create a larger system.

Software modules are important for software development because they promote reusability, maintainability, and scalability. By breaking down complex software systems into smaller, more manageable modules, developers can create more flexible and adaptable software that can be easily modified or updated over time.

Software Quality Assurance

Software quality assurance (SQA) is a set of activities and processes that are performed to ensure that a software product meets its requirements and satisfies its intended purpose. SQA aims to identify defects and potential problems early in the software development life cycle and prevent them from becoming bigger issues later on [153].

SQA involves a range of activities:

1. Planning: This involves defining the objectives, scope, and approach of the SQA process, and identifying the resources and timelines required.
2. Requirements Analysis: This involves reviewing the software requirements to ensure they are complete, consistent, and testable.
3. Design Review: This involves assessing the software design to ensure it is adequate, complete, and meets the requirements.

4. Code Review: This involves examining the software code to ensure it is well written, efficient, and maintainable.
5. Testing: This involves conducting various types of evaluation, including functional testing, performance testing, and security testing, to ensure that the software meets its requirements and is of high quality.
6. Reporting and Tracking: This involves documenting the results of the SQA activities and tracking any defects or issues that are identified.
7. Continuous Improvement: This involves using the results of the SQA activities to identify areas for improvement in the software development process and implementing changes to improve the quality of the software.

SQA is crucial in software development because it helps to identify defects and potential problems early in the software development life cycle, which reduces the cost and time required to fix them later on. It also helps to ensure that the software meets its requirements and is of high quality, which improves customer satisfaction and reduces the risk of costly software failures.

Software Quality Metrics

Software quality metrics are a set of quantitative measurements used to evaluate the quality of software products. These metrics are used to identify potential issues, monitor progress, and improve the quality of software during the development process. There are many different types of software quality metrics, but some of the most commonly used ones include the following:

1. Code Coverage: This metric measures the percentage of code that is executed during testing. A higher code coverage indicates that more code has been tested, and therefore the software is more likely to be of higher quality.
2. Defect Density: This metric measures the number of defects found in the software per unit of code. A lower defect density indicates that the software is of higher quality.
3. Cyclomatic Complexity: This metric measures the complexity of the software's control flow. A higher cyclomatic complexity indicates that the software is more complex and therefore more difficult to test and maintain.
4. Maintainability Index: This metric measures the ease with which the software can be maintained and modified. A higher maintainability index indicates that the software is easier to maintain and modify.

5. Performance: This metric measures how well the software performs under different loads and conditions. Higher performance indicates that the software is more efficient and effective.
6. Usability: This metric measures how easy the software is to use and how well it meets user needs. A higher usability score indicates that the software is more user-friendly.

Software quality metrics are important because they provide a way to objectively measure the quality of the software and identify potential issues. By using these metrics, developers can identify areas for improvement and make changes to improve the quality of the software.

Software Requirements Document

A software requirements document (SRD) is a formal document that outlines the requirements for a software product. It provides a detailed description of the software's functionality, performance, design, and other aspects. The SRD serves as a communication tool between the development team and the stakeholders to ensure that everyone understands what is expected from the software [150].

The contents of an SRD typically include the following elements:

1. Introduction: This section provides an overview of the software and its purpose.
2. Scope: This section outlines the boundaries and limitations of the software, as well as any assumptions that have been made during the planning phase.
3. Functional Requirements: This section describes the specific functions and features that the software must perform. It includes detailed descriptions of the inputs, outputs, and processing logic required for each function.
4. Nonfunctional Requirements: This section describes the nonfunctional requirements of the software, such as performance, reliability, usability, and security.
5. Design Requirements: This section describes the design specifications for the software, including any technical constraints, hardware and software platforms, and design standards.
6. Acceptance Criteria: This section outlines the criteria that must be met in order for the software to be considered acceptable to the stakeholders.
7. Glossary: This section provides definitions of key terms and concepts used throughout the document.

The SRD is a critical document in the software development process as it provides a clear understanding of the software requirements and serves as a reference for the entire project team. It helps to ensure that everyone is working toward the same goals and that the software is delivered on time, within budget, and meets the expectations of the stakeholders.

Software Testing

Software testing is the process of evaluating a software product or system to ensure that it meets all specified requirements and performs as expected. Software testing aims to identify defects or errors in the software and ensure that they are corrected before the product is released to end users.

There are several types of software testing:

1. Unit Testing: This type of testing evaluates individual components or units of the software in isolation to ensure that they are functioning correctly.
2. Integration Testing: This type of testing looks at how individual units of the software interact with each other to ensure that they work together correctly.
3. System Testing: This type of testing evaluates the entire system as a whole to ensure that it meets the specified requirements and performs as expected.
4. Acceptance Testing: This type of testing looks at how end users employ the software to ensure that it meets their needs and expectations.
5. Regression Testing: This type of testing evaluates the software after changes have been made to ensure that no new defects have been introduced.
6. Performance Testing: This type of testing examines the software's performance under various conditions to ensure that it meets performance requirements.
7. Security Testing: This type of testing assesses the software's security features to ensure that it is secure from potential threats.

Software testing is an essential part of the software development process and is critical to ensuring that the software is of high quality and meets the needs of the users. It is important to perform testing at every stage of the software development process to identify and correct defects as early as possible to minimize the project timeline and cost impact.

Solder Bridge

A solder bridge is an unintended electrical connection that occurs when a small amount of solder connects two or more adjacent solder joints. This can happen when excess solder is applied to a circuit board or when solder wicks between the pins or pads during reflow.

Solder bridges can cause several problems in electronic circuits, including short circuits and high resistance connections, which can affect the performance and reliability of the circuit. In some cases, solder bridges can also cause electrical damage to the components.

It is important to use proper soldering techniques and tools to prevent solder bridges from occurring. The solder should be applied in the precise amount and at the correct temperature to prevent excess solder from flowing into unwanted areas. Proper placement of components can also help prevent solder bridges by ensuring that the pins or pads are adequately spaced apart.

If a solder bridge does occur, it can be removed using a soldering iron or a desoldering tool to carefully separate the excess solder. However, it is important to avoid damaging the surrounding components or the circuit board during this process.

Solder Whiskers

Solder whiskers are tiny, hairlike metal protrusions that can grow on the surface of soldered connections over time. They are typically made of tin or lead and can cause short circuits or other electrical problems in electronic circuits.

Solder whiskers can form when there is residual stress in the solder or substrate material after the soldering process. As the solder cools and solidifies, this stress can cause the formation of whiskers that grow over time. Other factors that can contribute to the formation of solder whiskers include temperature changes, humidity, and the presence of impurities in the solder or substrate material.

Solder whiskers can be difficult to detect, as they are often too small to be seen with the naked eye. However, they can cause serious problems in electronic circuits, such as short circuits or component failure.

To prevent solder whiskers from forming, it is important to use high-quality solder and substrate materials and to minimize the amount of residual stress in the soldered connection. In addition, proper storage and handling of electronic components can help prevent the formation of whiskers over time.

If solder whiskers are detected in a circuit, they can be removed using a soldering iron or a desoldering tool to carefully remove the affected area of solder. However, it is important to take care not to damage the surrounding components or the circuit board during this process.

Specification

In the context of software engineering, a specification refers to a detailed description of the requirements and functionality of a software system or application. A specification serves as a formal document that defines the scope of the software project, including its purpose, goals, and user requirements.

A software specification typically includes information about the system architecture, data models, algorithms, user interfaces, and other technical details. It may also describe the software's performance requirements, security features, and other quality attributes.

The purpose of a software specification is to provide a clear and concise description of the software requirements and functionality, which can be used as a reference by developers, testers, and other stakeholders throughout the software development life cycle. A well-written specification can help ensure that the software meets the needs of the users and performs as intended.

There are several different types of software specifications, including functional specifications, nonfunctional specifications, and design specifications. Functional specifications describe the features and behavior of the software system, while nonfunctional specifications describe the performance, security, and other quality attributes of the system. Design specifications describe the technical details of the software architecture and implementation.

Spectrum Analyzer

A spectrum analyzer is used to analyze the frequency spectrum of a signal. It is an electronic test instrument that measures and displays the frequency spectrum of an input signal. The frequency spectrum is a graphical representation of the frequency content of a signal, showing the amplitude of each frequency component.

A spectrum analyzer typically consists of a display screen and a control panel, with various knobs and buttons to adjust the frequency range, resolution bandwidth, and other settings. The input signal is usually connected to the spectrum analyzer via a coaxial cable or other type of input connector.

The spectrum analyzer works by taking an input signal and converting it into a digital signal that can be analyzed and displayed on the analyzer's screen. The frequency spectrum of the input signal is then displayed, with the amplitude of each frequency component shown as a function of frequency.

Spectrum analyzers are commonly used in the field of electronics and telecommunications to measure and analyze signals in the frequency domain. They can be used to troubleshoot problems in radio frequency (RF) systems, such as identifying interference sources or measuring the power and frequency of a signal. They are also used in the design and testing of electronic circuits and components, such as filters and amplifiers.

Stability

Product stability refers to the ability of a software product or application to remain functional and perform as expected over time and under various conditions. In other words, a stable product should continue to work reliably and consistently without crashing or producing unexpected results.

Testing plays a critical role in ensuring product stability. By testing the software thoroughly, developers can identify and fix any bugs, errors, or other issues that could affect the stability of the product. This includes testing the software under different conditions, such as varying workloads, different operating systems or devices, and varying network conditions.

One of the key types of testing for product stability is regression testing. Regression testing involves retesting the software after making changes or updates to ensure that previously working features and functionality continue to work as expected. This helps prevent any unintended consequences or side effects that could affect the stability of the product.

In addition to testing, other factors can also affect product stability, such as the quality of the code and the software architecture. Developers can help ensure product stability by following best practices for software development, such as writing clean, well-structured code and designing software that is easy to maintain and update.

Stakeholder

A stakeholder is any individual, group, or organization that has an interest in a project or product or that is affected by it. Stakeholders can be internal or external to the organization, and can include a wide range of individuals and groups, including customers, employees, shareholders, partners, suppliers, regulatory agencies, and the general public.

Stakeholders may have different levels of interest and influence on a project or product, and it is important to identify and manage stakeholders effectively in order to achieve project or product success. Stakeholder management involves identifying and prioritizing stakeholders, understanding their needs and expectations, and engaging with them throughout the project or product development process to ensure their concerns are addressed and their feedback is incorporated.

Effective stakeholder management can lead to a variety of benefits, such as increased support and buy-in for the project or product, improved communication and collaboration, and a better understanding of customer needs and preferences. It can also help identify potential risks or issues early on in the project, allowing for timely mitigation or resolution.

Standard Deviation

Standard deviation is a statistical measure that shows how much variation or dispersion there is in a dataset. It measures the average distance of data points from the mean (average) of the dataset.

In other words, standard deviation measures how spread out the data is from the average. If the standard deviation is low, the data points are close to the mean and the data is tightly clustered around the average. On the other hand, if the standard deviation is high, the data points are further away from the mean and the data is more spread out.

Standard deviation is calculated by taking the square root of the variance of the dataset. The formula for calculating standard deviation is as follows:

$$\sigma = \frac{\sqrt{\Sigma}\left(x_i - \mu\right)^2}{N}$$

Where:

σ = population standard deviation
N = the size of the population
x_i = each value from the population
μ = population mean

Standard deviation is commonly used in science and engineering fields to analyze data and make informed decisions. It can help identify outliers in a dataset and assess the reliability and consistency of the data.

Standards

Standards are established guidelines or criteria that define a set of best practices, technical specifications, or quality requirements for a particular industry or field. Standards help to ensure that products, services, and processes are safe, reliable, and consistent across different organizations and locations.

The various types of standards include the following:

1. Technical: These standards define the technical requirements and specifications for products, processes, and systems. Examples of technical standards are ISO standards, IEEE standards, and ASTM standards.
2. Quality: These standards define the requirements for quality management systems and practices. Examples of quality standards are ISO 9001, Six Sigma, and total quality management (TQM).

3. Safety: These standards define the safety requirements for products, processes, and systems. Examples of safety standards are OSHA standards and ANSI standards.
4. Environmental: These standards define the environmental requirements and guidelines for products, processes, and systems. Examples of environmental standards are the J standards from SAE International.

Various organizations develop and maintain standards, including national and international standards organizations, industry associations, and regulatory bodies. Compliance with standards can help organizations improve the quality of their products and services, increase efficiency and productivity, and enhance their reputation and credibility.

State

In engineering and computer science, system state refers to the condition or configuration of a system at a particular moment in time. A system can be any physical or abstract entity comprising multiple components and operating according to specific rules or principles.

The state of a system can include information about the values of its internal variables, the status of its components or subsystems, and any other relevant parameters that describe its behavior or performance. For example, the state of a computer system might include the amount of memory used, the status of its peripherals, and the current processes running on it.

This information will enable the testers to develop equipment and procedures to test the conditions of the states and transitions between states [154]. In software testing, system state is an essential factor to consider when designing and executing tests. Testing a system in different states can help identify defects or issues that may not be apparent in other states.

One approach to testing system state is to use test cases that cover different combinations of inputs and expected outputs as defined by the various states of the system and the mechanisms which will move the system state from one to another. For example, a test case might involve evaluating how the system responds when a specific input is provided or behaves when a particular subsystem is in a specific state.

Another approach is to use tools or techniques that simulate different system states, such as fault injection or stress testing. These methods can help identify the system's behavior under extreme conditions or when specific components malfunction.

State Diagram

A state diagram is used to represent the behavior of a system or process over time. It is a graphical representation of a system or process's different states and the transitions between them.

State diagrams are commonly used in software engineering, control systems, and other fields where the behavior of a system or process needs to be understood and modeled. They are also known as state machines or state-transition diagrams [154].

A state diagram typically includes the following elements:

1. States: Represented by circles or rectangles, states represent the different conditions or situations that a system or process can be in.
2. Transitions: Represented by arrows, transitions show the movement of a system or process from one state to another.
3. Events: Represented by labels on the transitions, events are the triggers that cause the system or process to move from one state to another.
4. Actions: Represented by labels on the transitions or states, actions are the activities or processes that occur as a result of a transition or state.
5. Initial State: Represented by a filled-in circle or a label, the initial state is the starting point for the system or process.
6. Final State: Represented by a filled-in circle or a label, the final state is the ending point for the system or process.

State diagrams are useful for modeling and understanding the behavior of complex systems, identifying potential issues or problems, and developing testing conditions. They can also be used to design and implement software systems, control systems, and other systems that respond to different conditions and events.

State Table

A state table is a tool used to help design and analyze digital systems and circuits. The table lists all the possible states that a system can be in, along with the inputs and outputs that cause transitions between the states.

Current State	Input (Key Action)	Next State
Off	Insert key	Acc
Off	Press start	Ignition
Acc	Press start	Ignition
Ignition	Turn key left	Off
Ignition	Turn key right	Start
Start	Release key	Ignition
Ignition	Press stop	Off

In this table, we have four possible states: Off, Acc (Accessory), Ignition, and Start. The key actions or inputs are represented in the second column. The third column indicates the resulting state after the given input.

For example, if the current state is Off and the user inserts the key, the next state would be Acc (Accessory). Similarly, if the current state is Ignition and the user turns the key left, the next state would be Off.

The state table helps to identify the possible states of the system, the conditions that cause transitions between states, and the outputs generated in each state. This enables test case development and identification of what constitutes pass/fail behavior.

State Transition Diagram

See State Diagram.

State Transition Testing

A black-box test technique in which test cases are designed to exercise elements of a state transition model [154]. In computer science and software engineering, a state transition refers to the change in the behavior or state of a system as a result of an event or input.

	Step1	Step 2	Step 3
Current State	OFF	ON	OFF
Input	Switch ON	Switch OFF	Switch ON
Output	Light ON	Light OFF	Light ON
Finish State	ON	OFF	ON

A system can have multiple states, each with its own behaviors and rules. When an event or input occurs, the system transitions from one state to another. A set of rules, called a state transition diagram, describes the mechanisms that shift the system from one state to another and defines the possible transitions between states and the conditions under which they occur.

For example, in a simple vending machine, the initial state is "Waiting for Input." When a coin is inserted, the state transitions to "Coin Accepted"; and when a button is pressed, the state transitions to "Dispensing Item." Finally, when the item is dispensed, the state transitions to "Waiting for Input" again.

State transitions are also used in areas such as robotics, control systems, and manufacturing processes. In these systems, state transitions control the behavior of the system in response to different inputs or events.

In conclusion, state transition refers to the change in the state of a system or process as a result of an event or input. It is a fundamental concept in computer science, software engineering, and other fields that involve complex systems and processes.

Statement Coverage

Statement coverage measures the degree to which the statements in a program have been executed during software testing.

In statement coverage, each statement in the code is evaluated to determine whether it has been executed by at least one test case. The coverage is expressed as a percentage of the total number of statements in the code. For example, if a program has 100 statements and 80 of them have been executed during testing, the statement coverage would be 80%.

Statement coverage can help identify parts of the code that have not been tested, which can indicate potential defects or areas of the code that may require additional testing. However, statement coverage alone does not guarantee that all possible paths through the code have been tested or that all possible defects have been identified.

Some limitations of statement coverage include the following:

1. It only measures whether or not a statement has been executed. It does not consider the inputs or conditions that may affect the behavior of the statement.
2. It does not consider the flow of control in the program, so it may miss code paths that are not executed under certain conditions.
3. It does not guarantee that all possible defects have been identified, as there may be defects that are not related to specific statements but instead to the interaction between different parts of the code.

Despite its limitations, statement coverage can be a useful metric for assessing the thoroughness of testing and identifying areas of the code that may require additional attention.

$$Statement\ Coverage = \frac{Tested\ Executable\ Statements}{Total\ Executable\ Statements}$$

Statement Testing

Statement testing is a white-box test technique in which test cases are designed to execute statements. Statement testing is closely related to statement coverage testing, a software testing technique that involves creating test cases to ensure that statements in a program have been executed at least once during testing. Statement testing aims to verify that the program's code has been completely executed and that no statements have been missed or left untested.

To perform statement testing, testers typically use a coverage tool to track the execution of each statement during testing. This tool can provide a report indicating which statements have been executed and which have not, allowing testers to identify any gaps in coverage.

Testers can also use boundary-value analysis and equivalence partitioning techniques to design test cases that exercise different input and output conditions and ensure that all statements are executed. Additionally, techniques such as code inspection and peer review can be used to verify that each statement is necessary and correct.

Static Analysis

Static analysis is the process of evaluating a component or system without executing it based on its form, structure, content, or documentation (ISO24765) [162]. In software testing, it involves analyzing a program's source code or other artifacts without actually executing them. Static analysis aims to identify potential defects, security vulnerabilities, or performance issues before the code is compiled or executed. Static analysis tools can detect issues such as syntax errors, coding standards violations, memory leaks, and potential security vulnerabilities.

Static analysis is often combined with other testing techniques, such as unit and integration testing, to ensure that a program is thoroughly tested and meets all requirements. Static analysis can help reduce the cost and time required for testing and debugging by identifying potential issues early in the development process.

Static analysis can be automated using specialized tools that analyze a program's source code or other artifacts. These tools can provide reports highlighting potential issues and suggest ways to address them. Some popular static analysis tools include SonarQube, FindBugs, and PMD.

While static analysis can be a useful testing technique, it has some limitations. For example, it cannot detect issues that arise during runtime or when a program is executed in a specific environment. Additionally, static analysis tools can produce false positives, identifying potential issues that do not exist or are irrelevant to the program.

Static Analysis Tool

This tool performs static analysis on a program's source code or other artifacts to detect potential defects, security vulnerabilities, or performance issues. Static analysis tools use automated techniques to analyze a program's source code or other artifacts without actually executing the program.

Static analysis tools can be integrated into the software development process to inspect code quality and security continuously. Static analysis tools can help developers save time and resources and deliver more reliable and secure software by detecting potential issues early in the development process.

Static Testing

Static testing does not involve the execution of a test item [159]. In software testing, it involves reviewing and evaluating the software without executing the code. It is a way to identify defects and issues in the early stages of the software development process, before the code is actually run. Here is a general walkthrough of the static testing process:

1. Identify the SCOPE of the Testing: Determine what components of the software will be tested, including specific modules, functions, and code sections.
2. Gather Necessary Materials: Obtain the necessary documents, such as the requirements specification, design documents, and user manuals, to understand the expected behavior of the software.
3. Perform Code Reviews: Review the code line by line to identify potential issues, such as syntax errors, logical errors, and security vulnerabilities. This can be done manually or with the use of tools such as static analysis tools.
4. Review Design Documents: Evaluate the design documents to ensure that the software is being developed according to the specified requirements and design.
5. Conduct Inspections: Conduct formal inspections, also known as "walkthroughs," to review the code and design documents with a team of developers and testers.
6. Document Findings: Document any defects or issues found during the static testing process, including the location and severity of the issue.
7. Fix Defects: Work with the development team to fix any defects identified during the static testing process.
8. Retest: Reevaluate the code and design documents to ensure that the defects have been properly addressed and that the software is ready for further testing.

Statistical Testing

Statistical testing involves analyzing data collected during testing to make inferences about the behavior or performance of a software program. Statistical testing determines whether the differences observed in the data are statistically significant or could have occurred by chance [112].

Statistical testing is often used in performance testing to compare the performance of two or more software applications or different versions of the same application. For example, statistical testing can be used to compare the response times of different applications under different loads.

Statistical testing typically involves collecting data and performing hypothesis testing. Hypothesis testing involves formulating a null hypothesis and an alternative hypothesis, collecting data, and then using statistical methods to determine whether to reject or fail to reject the null hypothesis.

Statistical testing is a powerful tool for evaluating the performance and behavior of software applications. By analyzing data collected during testing, developers can identify performance bottlenecks and areas for improvement. However, statistical testing can be complex and it requires a solid understanding of statistics and data analysis.

Statistics

Statistics is a branch of mathematics that deals with data collection, analysis, interpretation, presentation, and organization [40]. In the context of product testing, statistics can be used to analyze the data collected during testing and to make inferences about the performance or behavior of a program.

There are two main branches of statistics: descriptive statistics and inferential statistics. Descriptive statistics involve summarizing and presenting data using mean, median, and standard deviation measures. Inferential statistics include using sample data to make inferences about a population.

Some common statistical measures used in software testing include the following:

1. Mean: The mean is the average value of a set of data.
2. Median: The median is the middle value in a set of data.
3. Standard Deviation: The standard deviation measures the amount of variability or dispersion in a set of data.
4. Confidence Interval: The confidence interval (CI) is a range of values within which a population parameter is estimated to lie.
5. Hypothesis Testing: Hypothesis testing involves formulating a null hypothesis and an alternative hypothesis, collecting data, and then using statistical methods to determine whether to reject or fail to reject the null hypothesis.

Statistics can be used in software testing for a variety of purposes, including performance testing, usability testing, and acceptance testing. By analyzing the data collected during testing, developers can identify areas for improvement and make data-driven decisions about how to optimize their software.

Steam Cleaning and Pressure Washing

Steam cleaning is a method of cleaning that uses steam to remove dirt and grime from surfaces. In the automotive industry, steam cleaning is often used to clean engines, wheels, and other vehicle parts. Steam cleaning is effective at removing stubborn dirt and grime and is also environmentally friendly since it does not require harsh chemicals [163].

Pressure washing is a method of cleaning that uses high-pressure water to remove dirt and grime from surfaces. In the automotive industry, pressure washing is often used to clean vehicles, particularly trucks and buses. Pressure washing is effective at removing dirt, grease, and other debris from surfaces, but it can also damage paint and different finishes if used improperly.

Both steam cleaning and pressure washing, collectively referred to as "blast cleaning," are opportunities for destructive water intrusion and chemical agents on vehicle components. For more information, see **J1455**.

Stochastic Testing

Stochastic testing randomly generates test cases and inputs to evaluate the behavior and performance of a software application. In contrast to deterministic testing, which involves predefined test cases, stochastic testing uses randomness to simulate real-world scenarios and explore the behavior of a system under unpredictable conditions [39].

Stochastic testing is beneficial for identifying edge cases and unexpected behavior in a system. By randomly generating test cases, developers can uncover scenarios that might not have been considered in a traditional test plan. Stochastic testing can also be used to evaluate the performance and scalability of a system by simulating real-world usage patterns.

Stochastic testing can be performed using various techniques:

1. Random Input Generation: This technique involves randomly generating inputs to a system and evaluating its behavior and output.
2. Fuzz Testing: This technique involves feeding unexpected or malformed data into a system to identify vulnerabilities and unexpected behavior.
3. Monte Carlo Simulation: This technique involves using random inputs to simulate a system's behavior under different conditions.

Stochastic testing can be a powerful software testing tool but can also be resource-intensive and time-consuming. As with any testing technique, it's important to balance the benefits of stochastic testing with a realistic examination of the time and resources required to perform it effectively.

Storage Testing

Storage testing and memory testing are important aspects of automotive testing, as they ensure that the electronic systems in vehicles are reliable and perform as intended.

Storage testing involves evaluating the nonvolatile storage media used in automotive systems, such as flash memory, hard drives, and SD cards. The goal of storage testing is to ensure that the storage media are able to store and retrieve data accurately and reliably, even under adverse conditions such as temperature and vibration. This is particularly important in automotive systems, where reliable storage is critical for safety-critical applications such as airbag deployment and engine control.

Memory testing examines the volatile memory used in automotive systems, such as DRAM and SRAM. Memory testing aims to ensure that the memory can store and retrieve data accurately and reliably, without errors or corruption. This is important in automotive systems, where reliable memory is critical for maintaining the performance and stability of the system.

Both storage and memory testing can be performed using various techniques, including functional testing, stress testing, and environmental testing. Functional testing involves testing the system under normal operating conditions to ensure that it performs as intended. Stress testing involves subjecting the system to extreme conditions, such as high temperatures or heavy loads, to evaluate its performance and reliability. Environmental testing involves subjecting the system to various environmental conditions, such as temperature and humidity, to assess its performance and reliability under different conditions.

Stress Testing

See HALT and HASS.

Stress Testing Tool

Stress testing tools for automotive systems are specifically designed to test the performance, reliability, and safety of electronic systems used in automobiles.

Structural Analysis

Structural analysis is a technique used in software engineering to analyze and verify the structure of a software system. The structural analysis aims to identify potential defects and design flaws in the software system and to ensure that the software meets the desired quality attributes such as reliability, maintainability, and scalability.

Structural analysis techniques include code reviews, static, and dynamic analysis:

1. Code reviews involve manual inspection of the code by one or more developers to identify potential defects and design flaws. Code reviews can be done formally or informally and conducted at different stages of the software development life cycle.

2. Static code analysis involves using automated tools to analyze the source code of a software system without actually executing it. These tools can identify potential defects and design flaws in the code, such as coding errors, security vulnerabilities, and performance issues.
3. Dynamic analysis involves testing the software system by executing it with various inputs and analyzing its behavior during runtime. Dynamic analysis can be used to identify defects and design flaws that are difficult to detect through static analysis.

Structural analysis is an important part of software testing and quality assurance, as it helps ensure that the software system is reliable, maintainable, and scalable.

Structural Testing

Structural testing, also known as white-box testing or clear box testing, is used in the automotive industry to assess the internal structure, design, and implementation of hardware and software components. It involves analyzing the underlying structure and code of automotive systems to ensure their integrity, reliability, and compliance with predefined specifications.

In automotive hardware, structural testing focuses on evaluating the physical components, such as engine systems, electrical circuits, sensors, and actuators. It aims to detect and rectify potential weaknesses, defects, or malfunctions that may affect the vehicle's performance, safety, or durability.

Structural testing of automotive software involves examining the source code, modules, and interfaces of software applications and embedded systems used in automobiles. It helps identify programming errors, logic flaws, or vulnerabilities that could lead to functional failures, security breaches, or system crashes. This testing approach ensures that the software components adhere to coding standards, follow industry best practices, and meet the specific requirements of the automotive domain.

Structural testing techniques commonly employed in the automotive industry include code coverage analysis, unit testing, integration testing, and static code analysis. These methods aim to thoroughly assess the internal structure, data flows, control flows, and interactions within automotive hardware and software, enabling engineers to detect and rectify potential issues early in the development process.

By conducting structural testing, automotive manufacturers and developers can enhance their products' overall quality, reliability, and safety, ensuring compliance with regulatory standards and delivering optimal performance to end users.

Structure-Based Testing

Structure-based testing is a type of white-box testing that focuses on the internal structure of software, and it is used to ensure that the software performs as intended by verifying its internal structure. Structure-based testing aims to detect defects and errors in a software system by examining its internal structure and then designing test cases to assess that structure. This approach aims to identify defects and errors in the software code, such as missing or incorrect logic or improper implementation of algorithms, which other testing methods may not detect.

There are different types of structure-based testing techniques:

1. Control Flow Testing: This technique tests the paths and control structures in the software code by examining the flow of control in the program. This method detects errors such as infinite loops, unreachable code, and missing code blocks.
2. Data Flow Testing: This technique is used to test the data structures and data flows in the software code. It involves analyzing how data is used and manipulated within the program, which can help identify issues such as uninitialized variables, incorrect data types, and data dependencies.
3. Branch Testing: This technique involves testing all possible outcomes of conditional statements (if/else statements) in the software code, which can help detect errors such as missing conditions, incorrect logic, and unexpected results.
4. Path Testing: This technique is used to test all possible paths through the software code, which can help detect errors such as incorrect logic, unexpected results, and missing code blocks.

Stub

A stub is a skeletal or special-purpose software component implementation used to develop or test a component that calls or is otherwise dependent on it. It replaces an interface component [164]. Also known as a method stub, a stub can temporarily replace unfinished code or emulate the behavior of code that has already been written.

In automotive software testing, stubbing involves replacing a software component with a simplified version, called a stub, that simulates the behavior of the actual component. This is done in order to isolate the component being tested and ensure that it is functioning correctly on its own, without interference from other components.

Stubbing is particularly useful in testing complex systems, where it may be difficult or impractical to test all components simultaneously. By using stubs to simulate the behavior of other components, testers can focus on testing one component at a time, without worrying about the behavior of other components. For example, a stub could be used to simulate the behavior of a sensor that provides input to an automotive control system. By using a stub to simulate the sensor, testers can verify that the control system responds correctly to the input, without having to connect and test the sensor itself physically.

Subadditivity Effect

The subadditivity effect has implications for testing, particularly in situations where multiple events or factors can affect the outcome of a test. For example, multiple factors can impact the performance of a software system, such as the number of users, the complexity of the data, and the hardware specifications. If each of these factors is tested individually, the test results may not accurately reflect the actual performance of the system in real-world scenarios where multiple factors are present simultaneously. The subadditivity effect may cause testers to underestimate the impact of the combined factors on the system's performance.

To address this issue, testers can use techniques such as factorial testing, which involves testing all possible combinations of factors, to ensure that the system performs well under various conditions. Additionally, testers can use statistical methods such as analysis of variance (ANOVA) to analyze the results of the tests and determine the impact of each factor and their interactions on the system's performance.

Testers need to be aware of the subadditivity effect and its implications for testing. By considering the joint probabilities of multiple factors and using appropriate testing techniques and statistical methods, testers can ensure that their tests accurately reflect the real-world performance of the system under various conditions.

Subjectivity Validation

Subjectivity validation is important in product testing because subjective factors, such as user preferences, opinions, and experiences, can significantly influence a product's performance and usability.

For example, in a user experience (UX) test, the subjective experience and feedback of the test participants can provide valuable insights into the product's usability, satisfaction, and overall appeal. However, the interpretation and analysis of these subjective factors can also be influenced by personal biases, expectations, and contextual factors.

To address the issue of subjectivity in product testing, it is important to use a variety of methods and metrics to evaluate the product's performance and the user's experience. This can include objective measures, such as task completion time, error rates, system performance, and subjective criteria, such as user feedback, surveys, and interviews.

It is also important to consider the context and diversity of the test participants to ensure that the test results reflect the needs and preferences of the target audience. This can involve recruiting a diverse range of participants, conducting tests in realistic scenarios, and analyzing the data in a way that accounts for individual differences and contextual factors.

Subjectivity validation is an essential aspect of product testing, as it helps to ensure that the test results are accurate, reliable, and representative of the target audience's needs and preferences. By using various methods and metrics to evaluate the product's performance and user experience and by considering the context and diversity of the test participants, testers can increase the validity and reliability of the test results.

Sunk Cost Fallacy (Irrational Escalation)

The sunk cost fallacy refers to the tendency to continue investing resources in a project or activity, even when it is no longer rational or beneficial, simply because of previous investments. This can lead to poor decision-making and wasted resources, as the decision to continue is based on past investments rather than on future benefits. An example of business case evaluation is return on investment, this compares costs with potential profit of the endeavor.

$$Return\,on\,Investment = \left(\frac{Profit - Cost\ of\ Investment}{Cost\ of\ Investment}\right) x\,100\%$$

In testing, the sunk cost fallacy is evident when a significant amount of time and resources have already been invested in a particular testing approach or tool, and there is resistance to changing course or adopting new approaches, even though they may be more effective or efficient. Another business evaluation approach, arguably more detailed, is the net present value. This equation accounts for interest impacts over the years.

$$Net\ Present\ Value = \Sigma R_t / (1+i)^t$$

For example, if a testing team has invested significant time and resources in developing custom testing equipment, rigs, and test software, they may be hesitant to switch to a new commercial tool, even if it is more efficient and effective. The decision to continue using the custom framework is based on the sunk costs of the previous investment rather than on the potential benefits of the new tool.

To avoid the sunk cost fallacy in testing, it is important to regularly evaluate the effectiveness and efficiency of the testing approaches and tools being used and to be willing to change course if necessary. This may involve conducting pilot studies, gathering data on the performance and effectiveness of different testing approaches, and analyzing the costs and benefits of each approach.

It is important to avoid framing decisions in terms of sunk costs and instead to focus on the potential benefits and costs of future investments. By taking a forward-looking approach and considering the potential benefits of new approaches, testers can avoid the sunk cost fallacy and make more rational and effective decisions.

Symptom

Automated software testing techniques try to identify symptoms because they are observable system behaviors. A cause cannot be observed by itself.

System

A computer or other technological device, as well as all of its dependencies, are collectively referred to as a system in computing.

Systems of Systems

A system of systems is made up of several task-oriented or dedicated systems that combine their resources and talents to produce a larger, more sophisticated system that provides more functionality and performance than the sum of its parts.

T

"Concern for man and his fate must always form the chief interest of all technical endeavors. Never forget this in the midst of your diagrams and equations."

—Albert Einstein

Temperature Cycle

Cycling between two temperature extremes, usually at relatively rapid rates of change, is known as temperature cycling (see Figure T.1). It is a type of environmental stress test used in manufacturing to identify latent, long-term faults by producing failure through thermal fatigue. For more information, see **J1455**.

FIGURE T.1 Examples of thermal cycle [20].

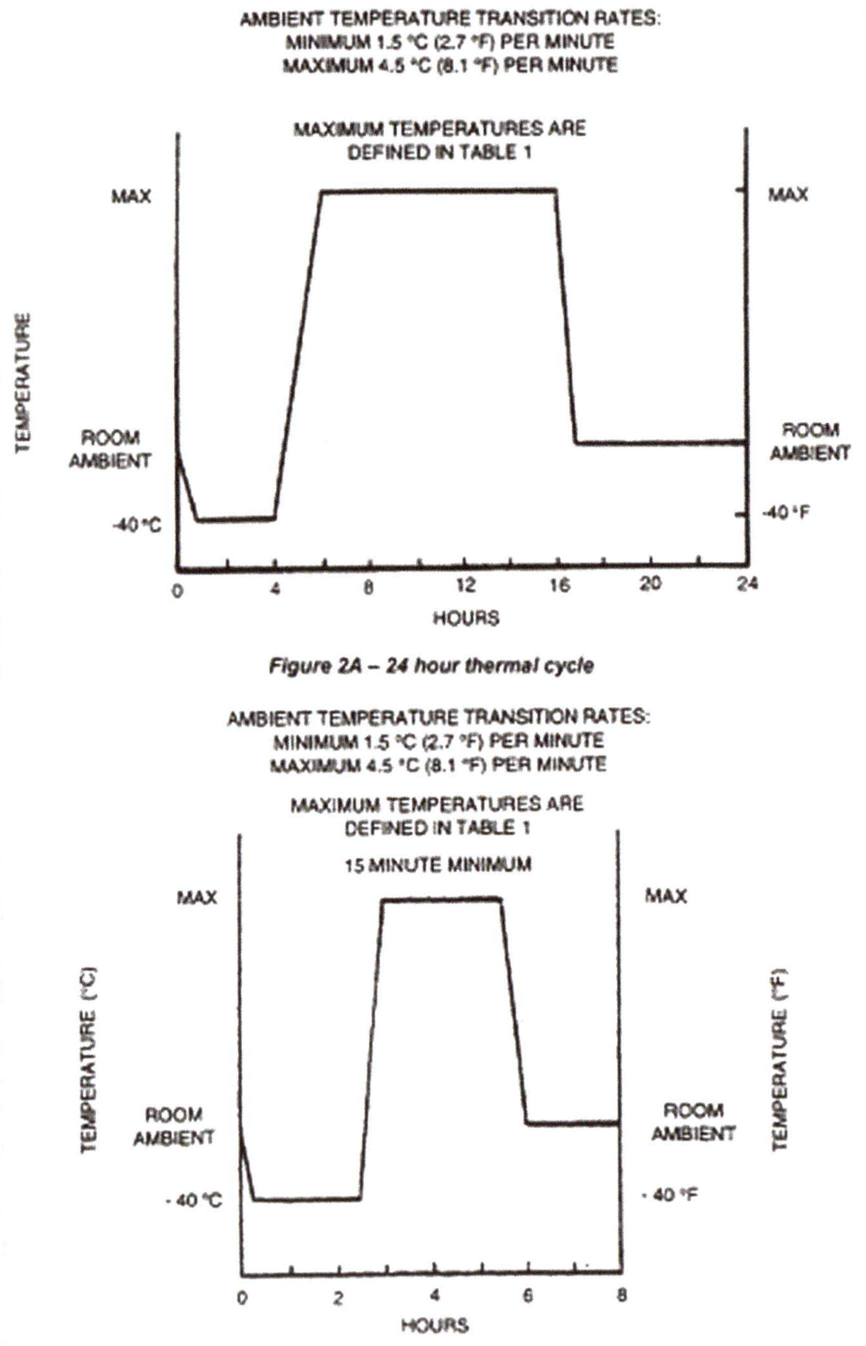

Test and Evaluation Master Plan

A test evaluation master plan (TEMP) is a document that outlines the strategy and approach for evaluating the effectiveness and efficiency of a testing effort. It typically includes the following elements: [39] [122]

1. Objectives: The TEMP should clearly state the objectives and goals of the testing effort, including the scope, timelines, and resources required.
2. Test Methods: The TEMP should describe the methods and techniques that will be used to evaluate the effectiveness and efficiency of the testing effort. This may include test automation, exploratory testing, and performance testing, among others.
3. Test Deliverables: The TEMP should identify the deliverables that will be produced, including test plans, test cases, test reports, and other documentation.
4. Test Environment: The TEMP should describe the environment, including the hardware, software, and network configurations, required for testing, as well as any tools and technologies that will be used.
5. Test Data: The TEMP should outline the approach for generating and managing test data, including the types of data required and how it will be stored, managed, and analyzed.
6. Test Schedule: The TEMP should provide a detailed schedule for the testing effort, including milestones, timelines, and dependencies.
7. Test Team: The TEMP should outline the roles and responsibilities of the testing team, including the skills and expertise required for each role.
8. Test Risks: The TEMP should identify and assess the risks associated with the testing effort, including technical risks, resource risks, and schedule risks, and outline the approach for mitigating and managing those risks.

The TEMP is a critical document for ensuring that a testing effort is well planned, well executed, and well evaluated, and that the testing outcomes meet the objectives and goals of the project. It helps to ensure that the testing effort is aligned with the overall project goals and objectives, and that the testing outcomes are effective, efficient, and of high quality.

Test Bed

A test bed is a platform or environment specifically designed and set up to evaluate a particular system, product, or service. It is typically used for developmental or experimental purposes and can include various components or resources, such as hardware, software, data, and other necessary tools and resources. Test beds are often used in the development of new technologies or products, as they provide a controlled and isolated environment in which to test and evaluate the performance and functionality of a particular system or component.

Test Campaign

A test campaign is a series of testing activities designed to evaluate the performance, functionality, or quality of a product, system, or application. It typically involves a range of testing techniques and methods, such as functional testing, integration testing, performance testing, security testing, and user acceptance testing, among others.

The primary goal of a test campaign is to identify and resolve defects and issues in the product or system being tested, and to ensure that it meets the requirements and expectations of the stakeholders. A test campaign may be conducted at various stages of the development process, such as during the initial development phase, after each sprint in an Agile development process, or before a product is released to the market.

A test campaign typically involves the following steps:

1. Planning: This involves defining the scope, objectives, and goals of the testing effort, as well as the resources, timelines, and methodologies that will be used.
2. Test Design: This involves creating test plans, test cases, and test scenarios that will be used to evaluate the product or system being tested.
3. Test Execution: This involves carrying out the testing activities as per the test design and collecting data and results for analysis.
4. Test Analysis: This involves analyzing the data and results collected during the testing process and identifying any defects or issues that need to be resolved.
5. Reporting: This involves documenting the results of the testing effort and presenting them in a format that can be easily understood by the stakeholders.
6. Retesting: This involves carrying out additional testing activities to verify that any defects or issues identified during the initial testing effort have been resolved.

Test Case

A set of preconditions, inputs, actions (where applicable), expected results, and postconditions developed based on test conditions [161]. A test case is a set of instructions or steps describing how to verify whether a particular aspect of a software application, system, or product is working correctly. It is a detailed description of a specific test scenario, including the expected results, inputs, and testing conditions.

Software testers create test cases as part of the testing process to ensure that the software meets the expected functional and nonfunctional requirements. They are used to verify that the software works as intended and that all the features and functionalities of the software are properly implemented.

A typical test case includes the following elements:

1. Test Case ID: A unique identifier assigned to the test case to keep track of it during testing.
2. Test Case Name/Description: A brief description of the test case that explains the purpose and objective of the test case.
3. Preconditions: The conditions that must be met before the test case can be executed.
4. Test Steps: The detailed instructions that describe the sequence of steps to be executed to carry out the test case.
5. Expected Results: The expected output or behavior of the software after the test case has been executed.
6. Actual Results: The actual output or behavior of the software after the test case has been executed.
7. Pass/Fail Status: Whether the test case has passed or failed.

Test cases are an essential component of software testing as they provide a structured and systematic approach to verify the software's functionality and behavior. Test cases help testers identify defects and issues early in the development process, which reduces the cost and time required for development and improves the quality of the software.

Test Cell

A test cell is a specialized facility or laboratory used to test and evaluate various equipment, systems, or products. In the context of automotive engineering, a test cell is typically used for evaluating internal combustion engines, hybrid and electric powertrains, and other vehicle components.

A typical automotive test cell is equipped with a range of instruments and equipment for measuring various parameters such as temperature, pressure, torque, speed, and emissions. These instruments are connected to the equipment or system being tested, and data is collected and analyzed to evaluate its performance and behavior.

Automotive test cells can be used for a variety of testing activities:

1. Durability Testing: This involves running the equipment or system under different loads and conditions to evaluate its reliability and durability over time.
2. Performance Testing: This involves measuring the performance and efficiency of the equipment or system under different operating conditions.
3. Emissions Testing: This involves measuring the emissions of the equipment or system to ensure compliance with environmental regulations.
4. Calibration Testing: This involves calibrating the equipment or system to ensure that it is operating within its specified range.
5. Development Testing: This involves testing and evaluating new equipment or systems during the development process to identify and resolve issues before the product is released.

Test cells are an essential tool for automotive engineers and manufacturers, as they provide a controlled environment for testing and evaluating equipment and systems under a range of operating conditions. By using test cells, automotive engineers can improve the quality and reliability of their products, reduce development time and costs, and ensure compliance with regulatory requirements.

Test Charter

A test charter outlines the objectives, scope, and approach of a testing session. It is a high-level test plan that directs testers toward what to test and how to test it.

The test charter is usually created by the test manager or lead and is based on the test strategy and requirements. It defines the test objectives, the testing techniques to be used, and the expected outcomes. The document also includes any risks associated with the testing and any dependencies or constraints that may affect the testing process.

A typical test charter contains the following elements:

1. Test Objective: A statement that describes the purpose of the testing session and what is to be achieved.
2. Scope: A description of the areas and functionalities to be tested, as well as any areas that are out of scope.

3. Test Approach: A description of the testing techniques and methodologies to be used, such as manual or automated testing, exploratory testing, regression testing, and so on.
4. Test Environment: A description of the hardware and software environment required for the testing, including any tools or resources needed.
5. Test Data: A description of the data needed to execute the tests, including any sample data or test cases that may be required.
6. Risks: A description of any potential risks associated with the testing, such as security or performance issues, and how these risks will be mitigated.
7. Dependencies: A description of any dependencies or constraints that may affect the testing process, such as the availability of resources or access to specific data.

A test charter provides a clear road map for the testing process and helps ensure that the test is focused and efficient. By using a test charter, testers can prioritize their efforts and ensure that all relevant areas are covered, while minimizing the risk of overlooking critical issues.

Test Coverage

Test coverage is a measure of the extent to which a software application or system has been tested. It is a metric used to evaluate the effectiveness and completeness of the testing process, and to identify areas that require further testing.

Test coverage can be defined in different ways, depending on the specific objectives of the testing process. The most common types of test coverage are as follows:

1. Statement Coverage: This measures the percentage of statements in the source code that have been executed during the testing process. This type of coverage focuses on the individual statements in the code, ensuring that each statement has been tested at least once.
2. Branch Coverage: This measures the percentage of branches in the source code that have been executed during the testing process. A branch is a decision point in the code that can lead to different outcomes, and branch coverage ensures that all possible outcomes have been tested.
3. Path Coverage: This measures the percentage of all possible paths through the code that have been executed during the testing process.

This type of coverage ensures that all possible combinations of branches and conditions have been tested.

4. Function Coverage: This measures the percentage of functions or modules in the software application that have been tested. This type of coverage ensures that all functions and modules have been tested, and that they are working as intended.

Test coverage can be evaluated manually or using automated tools. Automated testing tools can generate reports that show the coverage achieved during the testing process and can identify areas of the code that have not been tested.

Test coverage is an important metric for evaluating the effectiveness of the testing process and ensuring that the software application or system is thoroughly tested. By achieving high test coverage, software developers and testers can improve the quality and reliability of the software and reduce the risk of defects and issues in the production environment.

Test Data

This is the data needed for test execution [135]. Data is prepared or chosen to fulfill the prerequisites for execution and the input information necessary to run one or more test cases. The purpose of test data is to ensure that the software application or system is functioning correctly under various input conditions and scenarios.

Test data can take different forms, depending on the type of testing being performed:

1. Unit Testing: Test data for unit testing involves providing inputs to individual software units, such as functions or methods, and checking their outputs against expected results.
2. Integration Testing: Test data for integration testing involves testing how different software modules or components interact with each other and checking that the system as a whole is functioning correctly.
3. System Testing: Test data for system testing involves testing the software application or system as a whole, including its interfaces with other systems, and checking that it meets the specified requirements and functional specifications.

Test data can be created manually or generated using automated tools. Manual test data creation involves identifying input values that are likely to exercise different aspects of the software functionality, such as edge cases or negative scenarios. Automated test data generation tools can generate large volumes of test data, using algorithms to ensure that the data covers a wide range of possible input values and scenarios.

In addition to creating test data, it is important to manage test data effectively to ensure that it is up-to-date, accurate, and relevant to the testing being performed. This involves storing test data in a centralized repository, tracking changes to the data over time, and ensuring that the data is protected and secured.

Test-Driven Development

Test-driven development (TDD) is a software development process that emphasizes writing automated tests before writing the actual code [165]. The process involves three main steps:

1. Write a Failing Test: The first step in TDD is to write a test that checks a specific functionality or behavior of the software. This test should fail, indicating that the functionality does not yet exist or is not working correctly.
2. Write the Minimum Code to Pass the Test: The next step is to write the minimum amount of code necessary to pass the test. This code should only implement the specific functionality being tested.
3. Refactor the Code: Once the test is passed, the code can be refactored to improve its quality and readability without changing its functionality. This step helps to ensure that the code is maintainable and scalable over time.

By following the TDD process, software developers can ensure that the code is thoroughly tested and meets specified requirements before it is integrated with the rest of the system. TDD can also help to identify potential issues and defects early in the development process, when they are easier and less expensive to fix.

In addition to improving software quality and reducing the risk of defects, TDD can also help to improve the efficiency and productivity of the development process. By focusing on writing tests first, developers can avoid the need for manual testing and reduce the time and effort required for debugging and troubleshooting.

Test Driver (Software Testing)

In software testing, a test driver is a software component or tool that is used to execute test cases and collect test results. The test driver acts as an interface between the test cases and the software application or system being tested, and it is responsible for controlling the execution of the tests and capturing the results.

Test drivers can take various forms depending on the type of testing being performed. For example, in unit testing, the test driver may be a simple script or program that calls the unit being tested with various inputs and checks the outputs. In integration testing, the test driver may be a more complex software component that orchestrates the execution of multiple software modules or components.

In addition to executing test cases and collecting test results, test drivers may also perform other functions:

- Creating and managing test data
- Monitoring system resources such as central processing unit (CPU) and memory usage during testing
- Logging and reporting test results
- Coordinating the execution of tests across multiple machines or environments

Test Driver (Vehicle Testing or Live Fire)

In automotive testing, a test driver refers to a person responsible for operating a vehicle during testing. The test driver is typically a professional driver trained to operate the vehicle under different testing conditions (on a track and over the road) and scenarios and can provide feedback on the vehicle's performance and handling.

Test drivers are often used during various stages of automotive testing:

- Development Testing: In this stage, test drivers are used to evaluate the performance and handling of prototype vehicles in controlled testing environments. This may include testing a vehicle's acceleration, braking, handling, and stability on a test track or other closed course.
- Validation Testing: Once the vehicle has been developed, test drivers are used to validate the vehicle's performance and handling under real-world driving conditions. This may include testing the vehicle on public roads, highways, and other driving environments.
- Compliance Testing: In some cases, test drivers may be used to perform compliance testing to ensure that the vehicle meets regulatory and safety standards.

During testing, test drivers use various tools and technologies to collect data on the vehicle's performance and handling, including telemetry systems, data loggers, and sensors. Engineers and designers then analyze this data to identify areas for improvement and optimization.

Test drivers play a critical role in automotive testing, providing valuable feedback on the performance and handling of vehicles under different conditions and scenarios. Their expertise and insights help to ensure that vehicles are safe, reliable, and perform to the highest standards.

Test Environment

A test environment is a software and hardware setup designed to support software testing activities. The test environment is separate from the production environment. It is used to verify that software works as expected before being deployed to production.

A test environment typically includes the following components:

1. Hardware: This includes the physical servers, workstations, and other hardware components necessary to run the software being tested.
2. Software: This includes the operating system, middleware, database management systems, and other software components that are required to run the software being tested.
3. Test Tools: This includes tools and software applications that are used to manage the testing process, such as test management tools, defect tracking systems, and automated testing tools.
4. Test Data: This includes sample datasets, test scenarios, and other data that is used to validate the software under different conditions and scenarios.
5. Network: This includes the local area network (LAN) and wide area network (WAN) connections that are necessary to support testing activities.

The test environment is typically configured to simulate the production environment as closely as possible, including hardware specifications, network bandwidth, and system configurations. This helps to ensure that the software is tested under conditions that are representative of the production environment.

A dedicated test environment is essential for ensuring that software is thoroughly tested and validated before being deployed to production. It helps to identify and resolve issues and defects early in the development process, reducing the risk of system failures or other issues in the production environment.

Test Estimation

Estimation of the effort, time, and resources required to complete software testing activities involves breaking down testing tasks into smaller units, evaluating the time and resources required for each task, and then combining these estimates to develop an overall estimate for the testing effort. The aim is to develop an accurate and realistic estimate for the testing effort that can help project managers plan and allocate resources, identify potential risks and issues, and make informed decisions about the project schedule and budget.

Test estimation typically involves the following steps:

1. Identify the Testing Scope and Objectives: This involves defining the testing requirements, such as the types of tests to be performed, the testing environment, and the expected outcomes.
2. Break Down the Testing Tasks: This involves grouping the testing tasks into smaller units, such as test cases, test scripts, and test scenarios.
3. Estimate the Time and Effort Required for Each Task: This involves assessing how much time and effort are needed for each testing task based on factors such as the complexity of the task, the testing environment, and the skill level of the testing team.
4. Combine the Estimates to Develop An Overall Estimate: This involves totaling the estimates for each testing task to develop an overall cost for the testing effort.

Test estimation can be a complex process. Various techniques and methodologies can be used to improve the accuracy of estimates, including historical data analysis, expert judgment, and statistical modeling techniques such as Monte Carlo simulation.

Test First Design

See Test-Driven Design.

Test Framework

In the context of automotive software testing, a test framework is a software tool or system that provides a structure and set of guidelines for creating and executing tests. It automates the testing process and allows testers to standardize and execute tests efficiently.

A test framework for automotive software testing typically includes the following components:

1. Test Libraries: This includes a set of reusable code modules that can be used to create and execute tests.
2. Test Runner: This component executes the tests and provides feedback on the results.
3. Test Data Management: This includes tools and systems for managing data and test cases.

4. Reporting: This includes tools and systems for generating reports on the results of testing.
5. Integration: This includes tools and systems for integrating the test framework with other tools and systems used in the development process.

Automotive test frameworks may also include features such as traceability, which allows testers to track the requirements that have been tested and the test results, and support for compliance with industry standards and regulations.

Some examples of automotive test frameworks include the VectorCAST/C++ framework, the TESSY test framework, and the dSPACE SystemDesk framework. These frameworks provide a structure and set of guidelines for creating and executing tests in the automotive software development process, helping to ensure that software is thoroughly tested and validated before it is deployed to production.

Test Harness

A test harness is a collection of stubs and drivers needed to execute a test [166]. It includes various components, such as test runners, test frameworks, test libraries, and test data. Test harnesses are often used in software development to automate the testing process and ensure that tests run consistently and efficiently.

Test harnesses are used in various contexts, including unit testing, integration testing, and acceptance testing. They are typically designed to be flexible and customizable, allowing developers to create and run tests that are tailored to their specific needs.

Test harnesses help improve the efficiency and reliability of the testing process by automating repetitive tasks and providing a consistent and repeatable way to run tests. They can also help to identify and debug issues more quickly by providing detailed test results and error messages.

Test Inspection Evaluation Master Plan

An extension of the test evaluation master plan (TEMP) that includes inspection activities.

Test Item

In software testing, a part of a test object that is used in the test process [167]. A test item could be a single function, module, subsystem, or an entire system.

Before evaluating a test item, it's essential to understand the requirements, design specifications, and acceptance criteria for the test item. The testing team should also develop a plan, including the test cases, test data, and expected results, to ensure the test item is thoroughly tested.

Assessing test items aims to uncover defects or issues in the software that could affect its performance, reliability, or usability. Test items can be evaluated using different testing techniques, such as functional, regression, integration, performance, security, and usability.

Test Management

Test management refers to planning, organizing, and coordinating the testing efforts of a software development project. It involves developing and implementing a testing strategy, selecting and managing the testing tools and resources, and overseeing the testing process to ensure that it is conducted effectively and efficiently [39, 122].

Test management is an important aspect of the software development process, as it helps to ensure the quality and reliability of the software being developed. Effective test management involves identifying the appropriate testing methods and tools for the project, allocating resources and establishing schedules for testing activities, and coordinating the work of the testing team.

Test management also involves tracking and reporting on the progress and results of the testing efforts and identifying and addressing any issues or problems that may arise during the testing process. It is typically the responsibility of a test manager or leader's responsibility to manage a project's testing efforts.

Test Management Tool

Test management tools are essential for ensuring the quality of software in the automotive industry.

Test Manager

A test manager is responsible for planning, organizing, and coordinating the testing efforts for a software development project. The test manager is responsible for developing and implementing a testing strategy, selecting and managing the testing tools and resources, and overseeing the testing process to ensure that it is conducted effectively and efficiently.

Test managers typically works closely with the development team and other stakeholders to ensure that the testing efforts align with the overall project goals and objectives. They may also be responsible for communicating the results of the testing efforts to management and other stakeholders, as well as making recommendations for improving the testing process.

In addition to their technical skills and knowledge, test managers should also have strong leadership, communication, and problem-solving skills, as they will be responsible for managing and coordinating the testing team's work and working with various stakeholders.

Test Metrics

Test metrics are measurements or statistics collected during the testing process that can provide valuable insights into the quality and reliability of the software tested and the effectiveness of the testing process itself [39]. There are many different types of test metrics:

- Test coverage measures the percentage of the codebase or functionality that has been tested.
- Defect density measures the number of defects (errors or issues) found in the software per unit of code or functionality.
- Defect severity measures the impact that defects have on a system.
- Mean time to repair measures the average time it takes to fix a defect once it has been discovered.
- Test case pass/fail rate measures the percentage of cases that pass or fail.
- Test execution time measures the total time it takes to run a test or a suite of tests.
- Test case execution efficiency measures the efficiency of the process, considering factors such as the number of test cases that are run, the time it takes to run them, and the number of defects found.

Collecting and analyzing test metrics can help organizations identify areas for improvement in the testing process and make informed decisions about the quality and reliability of their software.

Test Monitoring

Test monitoring refers to observing and tracking the progress and results of a test or series of tests. It involves collecting data about the test environment, the test object (the software entity being tested), and the test itself, as well as recording the results and any issues that may arise during the testing process [168]. Test monitoring can be performed manually, by a human tester observing the test and recording the results, or it can be automated, using tools and software designed to monitor the test and collect data automatically.

Test monitoring provides visibility into the testing process, ensuring that tests are being conducted correctly and are producing accurate and reliable results. It can also help identify any issues or problems arising during the testing process, allowing them to be addressed promptly and resolved before the software is released.

Test Object

A test object is the software or hardware entity that is being tested [156]. It can be a single function, a class, a module, or an entire system. The test object is the test's target, and the test's purpose is to determine whether the test object behaves as expected and/or meets certain requirements.

The test object is often called the "system under test" (SUT). The test object can interact with other software entities, such as external dependencies or supporting components, but the focus of the test is on the behavior of the test object itself. See Device Under Test.

Test Oracles

A test Oracle is a reference point or set of criteria against which the results of a test can be compared. It is used to determine whether a test has passed or failed. The term "oracle" comes from the idea that the reference point should be as reliable and accurate as an Oracle in ancient Greek culture, which was believed to be able to provide divine knowledge or guidance.

There are several different types of test oracles:

- Human Oracles: These are individuals who use their knowledge and expertise to determine whether the results of a test are correct.
- Expected Output Oracles: These are predetermined sets of output that a test is expected to produce, and the test results are compared against these expected outputs.
- Code Oracles: These are pieces of code used to compare a test's results with the expected results.
- Statistical Oracles: These are statistical models used to determine the probability that a test result is correct based on data from previous tests.

A precise and reliable test Oracle allows testers to accurately determine whether a test has been successful.

Test Plan

A test plan is a document that outlines the approach, objectives, and scope of testing activities for a software project. It comprehensively describes the testing process, including the test strategy, test environment, test deliverables, test schedule, and test execution criteria.

The primary purpose of a test plan is to ensure that testing is conducted efficiently and effectively and that the software meets all requirements and specifications. A well-designed test plan provides clear guidance to the testing team, ensures that testing is focused on critical areas of the software, and provides a basis for measuring progress and evaluating results.

A typical test plan includes the following components:

1. Introduction: An overview of the software project and its objectives.
2. Test Strategy: A high-level description of the testing approach, including the testing techniques, tools, and resources to be used.
3. Test Environment: A description of the hardware and software environments in which testing will be conducted.
4. Test Deliverables: A list of the test artifacts that will be produced during testing, such as test cases, test scripts, and test reports.
5. Test Schedule: A timeline for the testing activities, including milestones and dependencies.
6. Test Execution Criteria: The criteria for starting and ending testing, such as the completion of specific test cases or the achievement of certain performance goals.
7. Risks and Issues: A list of potential risks and issues that could affect the testing process and their associated mitigation strategies.
8. Sign-Off: The process for obtaining approval and sign-off from stakeholders.

Test Point Analysis

Test point analysis is a technique used to identify the specific points in a system or component that need to be tested to ensure that it is functioning correctly. It involves analyzing the system or component in detail to determine which specific points or elements are critical to its operation and then designing tests to validate the behavior of those points or elements. Black-box testing estimates are derived using test point analysis. It is comparable to function point analysis, which estimates results from white-box testing [169].

A variety of factors can be considered when performing test point analysis, such as the design and architecture of the system or component, the requirements and specifications that it is intended to meet, and the potential failure modes or vulnerabilities that it may have.

Test point analysis is an important part of the testing process because it helps ensure that tests are focused on the most critical elements of the system or component and are designed to validate the behavior of those elements comprehensively and reliably. By performing test point analysis, testers can more quickly and effectively identify and address potential issues, improving the system's or component's overall quality and reliability.

Test Procedure

A test procedure is a detailed set of instructions that describes how to execute a specific test or series of tests (ISO 29119-1) [160]. Test procedures are often used in software testing and can automate testing tasks or guide manual testing efforts. They can be written in various formats, such as natural language, pseudo-code, or a specific programming language.

Test procedures typically include the following types of information:

- Preconditions: Any setup or preparation that is required before the test can be run.
- Test Steps: The specific actions that must be performed to execute the test.
- Expected Results: The expected outcomes of the test, including any expected performance or functionality improvements.
- Pass/Fail Criteria: The specific criteria that will be used to determine whether the test has passed or failed.
- Cleanup: Any actions that need to be taken to reset the system or environment after the test has been run.

Test procedures are an important part of testing because they provide a clear and detailed set of instructions for executing tests. They also help to ensure that tests are run consistently and efficiently.

Test Progress Tracking

Test progress tracking is the process of monitoring and recording a test or series of tests as they are being executed [39, 122]. It is an important aspect of the testing process, as it allows developers and testers to track their progress and identify any issues that may arise.

There are a variety of ways to track test progress:

- Test Planning and Tracking Tools: These tools allow testers to create and track test plans, assign tasks to team members, and monitor progress.
- Test Management Software: This software provides a centralized repository for managing and tracking tests, including test cases, test results, and defects.
- Spreadsheets: Simple spreadsheets can be used to track test progress by recording the status of each test case (e.g., not started, in progress, passed, failed).
- Manual Tracking: Testers can manually track progress by keeping notes or using a whiteboard to record the status of each test case.

In addition to ensuring that tests are run efficiently and that any issues are identified and addressed promptly, tracking test progress can also help improve team members' communication by providing a clear and up-to-date view of the testing process.

Test Resumption

Test resumption is continuing a test or series of tests that has been interrupted or stopped. This might be necessary for various reasons, such as hardware or software failures, power outages, or other issues that prevent the test from being completed.

There are various ways to handle test resumption, depending on the specific circumstances and the available tools and resources. Some common strategies include the following:

- Saving Test Progress: If possible, tests can be designed to save progress at regular intervals, allowing them to be resumed from where they left off if they are interrupted.
- Restarting from a Checkpoint: Tests can be designed to create checkpoints at specific points in the test process, allowing them to be restarted from the checkpoint if they are interrupted.
- Restarting the Entire Test: In some cases, it may be necessary to restart the entire test if it is interrupted. This can be time-consuming and may require additional setup or preparation.

Test resumption can be important when designing and executing tests, especially for long or complex tests that may be more prone to interruption. By planning for potential disruption and developing strategies for handling test resumption, it is possible to ensure that tests are completed as efficiently and reliably as possible.

Test Script

In software testing, a test script is a set of instructions or code that is used to automate the testing process. iThe script comprises a series of steps written to simulate user actions on the software being tested and verify that the software behaves as expected.

Test scripts can be written in different programming languages, such as Python, Java, Ruby, and JavaScript. The scripts can be created manually or generated using a test automation tool.

The main benefits of using test scripts are as follows:

1. Efficiency: Test scripts automate repetitive and time-consuming testing tasks, enabling the testing team to focus on more critical testing activities.
2. Consistency: Test scripts ensure that the same testing procedures are followed each time, reducing the likelihood of human error.
3. Reusability: Test scripts can be reused for regression testing and can be modified to accommodate changes in the software.

4. Accuracy: Test scripts ensure that tests are performed consistently, increasing the accuracy and reliability of the testing results.
5. Traceability: Test scripts can be linked to specific requirements, providing traceability from the requirements to the test results.

Test scripts can be used for different types of testing, such as functional, integration, performance, and security. Developing and maintaining a comprehensive library of test scripts is essential for ensuring that testing is efficient and effective.

Test Sequence

A test sequence is the order in which a series of tests will run or perform. It is typically defined in a test plan or test specification and outlines the specific tests that will be run and the order in which they will be executed. The test sequence is often designed to ensure that tests are run logically and efficiently and to provide a clear understanding of the dependencies between different tests.

For example, in a software development context, a test sequence might include unit tests, integration tests, and acceptance tests, with the unit tests being run first, followed by the integration tests, and finally, the acceptance tests. In a product testing context, a test sequence might involve running different tests on various product components, such as testing the battery life of a mobile device before testing its display quality.

Here is an example of a test sequence that might be used to test an automotive component:

1. Prepare the test environment by setting up any necessary equipment, such as a dynamometer or test rig.
2. Install the component to be tested on the test vehicle or equipment.
3. Perform any necessary calibrations or setup procedures to ensure the test equipment is configured correctly.
4. Run a series of predetermined test cases, applying various inputs to the component and measuring the outputs.
5. Record the test results and relevant data, such as performance metrics or diagnostic information.
6. Evaluate the test results using the predetermined evaluation criteria.
7. If the component passes the test, move on to the next test in the sequence. If the component fails the test, troubleshoot the issue and determine the root cause.
8. Repeat the test sequence until all necessary tests have been completed.
9. Document the test results and any issues that were encountered.

10. Remove the component from the test vehicle or equipment and clean up the test environment.

Test Specification

A test specification is a document that outlines the details of a test or series of tests. It typically includes information about the test's purpose, the conditions under which it will be performed, the criteria used to evaluate the test, and the expected results. Test specifications can be used in various contexts, such as software development, product testing, and quality assurance. They are often used to ensure that tests are designed and executed in a consistent and repeatable manner, and to provide a clear understanding of what is being tested and how the test results will be evaluated.

Here is an example of the contents that might be included in a test specification:

- Introduction: This section provides an overview of the purpose and scope of the test, as well as any relevant background information.
- Objectives: This section outlines the specific goals or objectives of the test, along with what is expected to be learned or achieved as a result of the test.
- Test Items: This section lists the specific items or components that will be tested, such as a software application, a hardware device, or a process.
- Test Environment: This section describes the conditions under which the test will be performed, including any hardware, software, or other resources that are required.
- Test Design: This section outlines the approach that will be taken to test items listed in the Test Items section. It might include details about the specific test cases that will be run, the expected inputs and outputs, and the criteria that will be used to evaluate the test results.
- Test Procedure: This section provides step-by-step instructions for how the test should be conducted, including any required setup or preparation.
- Evaluation Criteria: This section defines the specific criteria that will be used to evaluate the test results, such as pass/fail criteria, performance metrics, or other measures of success.
- Expected Results: This section outlines the expected outcomes of the test, including any expected performance or functionality improvements.
- Pass/Fail Criteria: This section defines the specific criteria that will be used to determine whether the test has passed or failed.

- Reporting: This section outlines the process for documenting and reporting the test results, including any relevant templates or forms that should be used.
- Test Closure: This section describes any actions that need to be taken to close out the test, such as cleaning up the test environment, archiving test artifacts, or updating relevant documentation.

Test Specimen

In materials testing, a test specimen is a sample of a material or component that is used to perform physical or mechanical tests to evaluate its properties and performance. Test specimens are typically prepared in accordance with specific standards or procedures, depending on the type of test being conducted.

For example, in analyzing the tensile strength of metals, a test specimen is usually a thin, flat strip of metal with a specific size and shape that is loaded in tension until it breaks. In compression testing, a specimen may be a cube or cylinder of a certain size that is compressed until it deforms or fails.

In addition to tensile and compression testing, other types of material testing, such as flexure, torsion, or impact tests, will require different types of specimens. The properties that are evaluated using test specimens include strength, stiffness, ductility, toughness, and fatigue resistance.

The preparation of test specimens is a critical step in materials testing, as the material's properties and behavior may be affected by factors such as the specimen's size, shape, orientation, and surface finish. Careful attention must be given to specimen preparation to ensure accurate and reliable test results.

Test Strategy

A test strategy is a high-level plan that outlines a software project's testing approach and objectives. It provides an overview of how testing will be conducted, the resources required, and the types of testing that will be performed. The main goal of a test strategy is to ensure that testing is undertaken efficiently and effectively and that the software meets the requirements and specifications.

A test strategy typically includes the following elements:

1. Testing Objectives: The goals and objectives of the testing effort, such as verifying functionality, performance, security, or usability.
2. Testing Scope: The boundaries of the testing effort, such as the features or modules of the software that will be tested.
3. Testing Approach: The testing techniques, methodologies, and tools that will be used to achieve the testing objectives.

4. Test Environment: The hardware, software, and network configurations required for testing, including the test lab setup and data requirements.
5. Test Deliverables: The artifacts that will be produced during testing, such as test plans, test cases, test scripts, and test reports.
6. Test Schedules and Timelines: The time frame for the testing activities, including milestones and dependencies.
7. Testing Resources: The human resources, hardware, and software resources required for testing, such as testers, test automation tools, and testing infrastructure.
8. Risk and Issue Management: Identifying, mitigating, and managing risks and issues related to the testing effort.

The test strategy is an essential document that guides the testing effort and ensures that testing is focused, efficient, and effective. It provides a framework for planning, executing, and evaluating the testing process and serves as a reference point for all stakeholders involved in the software development life cycle.

Test Stub

See Stub

Test Suite

In software testing, a test suite is a set of scripts or procedures to be executed in a specific test run designed to evaluate a software application or system's functionality, performance, and quality [158]. A test suite may include a single test case or multiple test cases that are organized into a logical grouping based on the type of testing or the functionality being tested.

A test suite typically includes the following elements:

1. Test Case Name and Description: A brief summary of the test case and its purpose.
2. Test Case Steps: A step-by-step description of the actions that testers should perform to execute the test case.
3. Expected Results: A description of the expected outcome or behavior of the software when the test case is executed.
4. Test Data: The input data or conditions required to execute the test case.
5. Preconditions: The conditions that must be met before the test case can be executed.
6. Postconditions: The conditions that must be true after the test case is executed.

Test suites can be created for different types of testing, such as functional testing, integration testing, system testing, and acceptance testing. Test suites are usually generated using a test management tool, which allows testers to organize, execute, and track test cases and their results.

The main advantages of using a test suite are as follows:

1. Efficiency: A test suite allows the tester to execute multiple test cases efficiently and effectively.
2. Consistency: A test suite ensures that the same testing procedures are followed each time, reducing the likelihood of human error.
3. Reusability: Test cases in a test suite can be reused for regression testing and modified to accommodate software changes.
4. Scalability: A test suite can be expanded to include new test cases as the software evolves or new requirements are added.

Test Suspension

In software testing, a test suspension is a temporary or permanent halt or pause in the testing process due to some circumstances or issues that must be addressed before testing can resume. A test suspension aims to ensure that the testing is conducted effectively and efficiently and that the problems are resolved before further testing is performed.

Testers or the test manager may initiate a test suspension, based on a variety of factors:

1. A defect that requires further investigation or fixing before testing can continue.
2. An issue with the test environment, such as hardware or software failures.
3. Unavailability of test resources, such as testers or test equipment.
4. Changes to the software requirements or design that require modifications to the test cases or test plan.
5. Security or safety concerns that need to be addressed before testing can proceed.
6. The project or product is terminated due to closure or cancellation or budget exhaustion.

When a test suspension occurs, it is essential to document both the reason for the pause and its expected duration. The testing team should work together to address the issues that caused the suspension and to develop a plan for resuming testing as soon as possible.

The main benefits of a test suspension are as follows:

1. Ensuring the quality of the testing process and results
2. Reducing the risk of missing defects or issues that could impact the software
3. Enabling the testing team to address issues more effectively and efficiently
4. Improving the overall testing process by identifying and resolving issues in a timely manner

Test Tools

Test tools are software applications that are designed to assist with various aspects of the software testing process. These tools can be used to automate testing tasks, improve test efficiency, and enhance test accuracy. Many different types of test tools are available, each with unique features and capabilities.

Here are some of the most common types of test tools used in software testing:

1. Test Management Tools: These tools help manage the entire testing process, including test planning, test case creation, test execution, and defect tracking.
2. Test Automation Tools: These tools automate the execution of test cases, which can help to speed up testing and reduce the risk of human error.
3. Load Testing Tools: These tools are used to simulate a high volume of user traffic on the system being tested to evaluate its performance and scalability.
4. Security Testing Tools: These tools are used to identify vulnerabilities and security weaknesses in the software being tested.
5. Code Coverage Tools: These tools are used to determine the amount of code that is covered by the test cases to ensure that all code is thoroughly tested.
6. Test Data Management Tools: These tools are used to generate and manage test data to ensure that the software is tested with a variety of realistic data scenarios.
7. Test Reporting Tools: These tools generate reports and metrics to help assess the testing process's quality and identify improvement areas.
8. Exploratory Testing Tools: These tools support manual techniques such as ad hoc, risk-based, and exploratory testing.

Using test tools has a number of benefits:

1. Improved test accuracy and consistency
2. Increased test efficiency and productivity
3. Better test coverage and identification of defects
4. Enhanced collaboration and communication among testing team members
5. Improved testing process and reporting capabilities

Test Track

An automotive test track is a facility used to test vehicles' performance and reliability under a wide range of conditions. These tracks are typically designed to simulate real-world driving conditions and include straightaways, curves, hills, and rough surfaces.

When it comes to testing vehicles on a test track, various vibration elements are utilized to assess the vehicle's performance, durability, and comfort. Here are some common vehicle test track vibration elements:

Road Profile: These profiles are artificially created surfaces that simulate different road conditions. They include various types of bumps, undulations, potholes, and irregularities, which help evaluate the vehicle's suspension system, ride comfort, and handling characteristics.

Rough Road Simulation: This simulation involves subjecting the vehicle to continuous and repetitive vibrations resembling the vibrations experienced on rough or uneven roads. This test evaluates the durability of vehicle components, such as the chassis, suspension, and body structure, by replicating real-world road conditions.

Sinusoidal Vibration: These tests involve subjecting the vehicle to vibrations that oscillate at a single frequency and amplitude. They assess the vehicle components' structural integrity, fatigue life, and dynamic response under controlled vibrational conditions.

Random Vibration: This type of testing simultaneously introduces a wide range of frequencies and amplitudes, resembling the random vibration encountered during normal driving conditions. This test is used to evaluate the durability, reliability, and performance of various vehicle systems, including electronics, interior components, and fasteners.

Braking Vibration: These tests simulate the vibrations experienced during braking maneuvers. These tests assess the braking system's effectiveness, evaluate the vehicle's stability, and detect any abnormalities or issues related to braking performance.

Acceleration/Deceleration Vibration: These tests involve subjecting the vehicle to vibrations that mimic the forces experienced during rapid acceleration or deceleration. They help assess the vehicle's powertrain, drivetrain, and overall stability under different driving conditions.

Vertical and Lateral Excitation: These tests simulate vehicle movements in vertical and lateral directions, evaluating suspension systems, ride quality, and handling characteristics. They analyze the vehicle's response to uneven road surfaces, cornering forces, and dynamic maneuvers.

Test tracks are equipped with various monitoring and measuring equipment, such as sensors, cameras, and data loggers, to collect data on the performance of the vehicles. This data can be used to identify any issues that need to be addressed and to make improvements to the design and engineering of the vehicles.

Testability

Testability refers to the ease with which a product, system, or component can be tested. A testable product is one that can be easily and thoroughly examined to determine whether it meets specified requirements and/or quality standards.

Several factors can impact the testability of a product:

- The Design of the Product: A well-designed and modularized product is often easier to test than one that is complex and monolithic.
- The Availability of Documentation: Clear and comprehensive documentation can make it easier for testers to understand the product and how it should be tested.
- The Presence of Debugging Tools: Debugging tools, such as logging and tracing, can help testers identify and troubleshoot issues more easily.
- The Use of Standard Technologies: Products that use standard technologies are often easier to test, as there are more resources available for testing these technologies.

Ensuring that a product is testable can help reduce the time and resources required for testing, ultimately leading to a higher-quality product.

Testable Requirements

Testable requirements are specific, measurable, and verifiable statements that define the expected behavior or characteristics of a product, service, or system. Testable requirements are important because they provide a clear and objective basis for creating test conditions and evaluating the performance or functionality of the product. They can be used to design and execute tests and to compare the results of the tests to the expected behavior or characteristics specified in the requirements.

To be testable, requirements should be specific enough to be unambiguous and include clear criteria for determining whether they have been satisfied.

They should also be measurable, meaning there should be a way to quantify or observe the requirement. Testable requirements can help to ensure that a product meets the needs and expectations of its users and stakeholders and can help to identify and address any issues or defects that may affect the quality or usability of the product.

Here are some attributes of testable requirements:

1. Specific: Testable requirements should be clear and specific, without any ambiguity or room for interpretation.
2. Measurable: Testable requirements should be measurable, meaning that there should be a way to quantify or observe the requirement in some way.
3. Verifiable: Testable requirements should be verifiable, meaning that there should be a way to confirm whether the requirement has been satisfied.
4. Unambiguous: Testable requirements should be written in a way that is free of ambiguity or confusion.
5. Complete: Testable requirements should be complete, meaning that they should include all necessary information and details.
6. Traceable: Testable requirements should be traceable, meaning that there should be a clear link between the requirement and the product or system it pertains to.
7. Prioritized: Testable requirements should be prioritized, with the most important or critical requirements being addressed first.

By ensuring that requirements are testable, it is possible to design and execute tests that effectively evaluate the performance or functionality of a product, service, or system.

Testing

Testing is evaluating a product, system, or component to determine whether it meets specified requirements and/or quality standards. Testing is an important part of the software development process because it helps identify defects, bugs, and other issues that must be addressed before a product is released.

There are many different types of testing, including unit testing, integration testing, system testing, and acceptance testing, among others. The specific type of testing chosen will depend on the specific goals and the stage of development the product is in.

Testing can be performed manually by a human tester or it can be automated using specialized software. In either case, testing aims to ensure that the product is of high quality and will function as intended for end users.

Testing Artifacts

Testing artifacts are any documents, tools, or other materials used to test a product, service, or system. Testing artifacts include test plans, test cases, test data, test results, and other documents that describe or record the testing process. They can also include test automation tools, test management software, test environments, and other tools or resources that are used to design, execute, or analyze tests. Testing artifacts serve a variety of purposes: helping to define the scope and objectives of the testing, documenting the test approach and results, and facilitating the management and reporting of the testing process.

Testing Dependencies

Testing dependencies refer to the relationships between different tests or test cases, such that the results of one test may affect the performance or interpretation of another test. These dependencies can arise for various reasons, such as the shared use of resources, the mutual dependence of different features or functions, or the cascading effects of a failure in one part of a system on other parts.

Managing testing dependencies is an important aspect of testing, as it can help to ensure that tests are conducted in an orderly and efficient manner and that the results of the tests are reliable and meaningful. To manage testing dependencies, it may be necessary to carefully plan the sequence of tests, coordinate the use of shared resources, isolate different tests from one another, or design tests independent of one another.

Testing dependencies can be complex and require specialized tools or techniques to identify and resolve. It is important for those involved in testing to be aware of the potential for dependencies between tests and to take steps to manage them to ensure the integrity and reliability of the testing process.

Automotive testing can have several dependencies, which are factors that can affect the accuracy, reliability, or validity of the test results. Some potential dependencies in automotive testing include the following:

1. Environmental Conditions: Temperature, humidity, wind, and other weather conditions can all affect the performance of a vehicle, and so these factors should be controlled for or accounted for in testing.
2. Vehicle Condition: The age, mileage, and maintenance history of the vehicle can all affect its performance, and so these factors should be considered when interpreting test results.
3. Test Procedure: The specific steps followed during the test, as well as any equipment or tools used, can affect the results of the test. It is important to ensure that the test procedure is well defined and followed consistently.

4. Test Subjects: The characteristics of the people operating the vehicle, such as their age, experience, and physical condition, can influence the test results. It may be necessary to control for these factors or to use a representative sample of test subjects.
5. Data Collection and Analysis: The accuracy and completeness of the data collected during the test and the methods used to analyze the data can affect the reliability and validity of the test results. It is important to use appropriate methods for data collection and analysis.

Testing Ethics

Testing ethics refers to the principles and standards that guide the behavior and decision-making of software testers. Ethics are essential in software testing because they ensure that the testing process is conducted fairly, responsibly, and transparently, respecting all stakeholders' rights and interests.

Here are some of the key ethical principles and standards in software testing:

1. Professional Competence: Testers should possess the necessary skills, knowledge, and experience to perform their job effectively and continuously develop their skills through training and education.
2. Confidentiality: Testers should maintain the confidentiality of sensitive information, such as personal data or trade secrets, and use it only for legitimate testing.
3. Objectivity: Testers should remain impartial and objective in their testing activities, avoiding bias or conflicts of interest that could compromise the testing results.
4. Transparency: Testers should communicate openly and honestly with all stakeholders about the testing process, including its objectives, methods, and outcomes.
5. Respect for Stakeholders: Testers should respect the rights and interests of all stakeholders, including end users, customers, and other stakeholders, and ensure that testing activities do not harm them.
6. Quality: Testers should prioritize quality in their testing activities, aiming to identify and report all defects, issues, and vulnerabilities that could impact the software.
7. Compliance: Testers should comply with all relevant laws, regulations, and industry standards that apply to their testing activities.
8. Continuous Improvement: Testers should continuously improve their testing processes and practices to ensure they are effective, efficient, and up-to-date.

By adhering to these ethical principles and standards, software testers can ensure that their activities are fair, responsible, and transparent and contribute to developing high-quality software that meets the needs and expectations of all stakeholders.

Testing Object-Oriented Systems

Testing object-oriented systems involves evaluating the functionality and performance of systems based on object-oriented design and programming principles. Object-oriented systems are characterized by their use of objects, self-contained data units, and behavior that interact through well-defined interfaces.

There are several approaches to testing object-oriented systems:

1. Unit Testing: This involves testing individual objects or small groups of objects to ensure that they function correctly.
2. Integration Testing: This involves testing the interactions between objects to ensure that they work together correctly.
3. System Testing: This involves testing the entire system to ensure it meets the required functional and nonfunctional requirements.
4. Acceptance Testing: This involves testing the system from the end users' perspective to ensure that it is usable and meets the users' needs.

In addition to these approaches, testing object-oriented systems may involve performance testing to ensure that systems perform optimally under different load conditions and security testing to ensure that systems are secure against external threats.

Thermal Cycling

Thermal cycling testing is a type of environmental testing that is used to evaluate the effects of temperature changes on products and materials. It involves subjecting the test sample to a series of temperature changes in a controlled laboratory environment.

Thermal cycling testing simulates the temperature changes a product or material may experience in real-world use. It is commonly used to test the durability and reliability of products and materials in various industries, including aerospace, automotive, and electronics [20].

During thermal cycling testing, the test sample is typically mounted on a fixture and placed in a chamber that can control the temperature and humidity of the environment. The chamber temperature is then cycled between the high and low temperature limits, and the test sample is monitored for any changes or failures (see Figure T.2).

FIGURE T.2 An example of a thermal chamber for testing components.

Thermal cycling testing is an important part of the product development and quality assurance process, as it helps manufacturers ensure that their products will perform reliably and consistently in various temperature conditions.

Thermal Stress

In the automotive industry, thermal stress is caused by temperature changes on various components and systems of a vehicle. Thermal stress can occur in a variety of automotive applications, including engines, transmissions, exhaust systems, and braking systems.

Thermal stress is caused by a number of factors, including the temperature of the surrounding environment, the temperature of the components themselves, and the rate at which the temperature changes. It can also be caused by the expansion and contraction of different materials as they are subjected to different temperatures.

Thermal stress can have a number of negative effects on automotive components and systems:

- Decreased strength and durability
- Increased wear and tear
- Decreased performance and efficiency
- Increased risk of failure

To prevent thermal stress, automotive manufacturers and engineers must carefully design and test components and systems to ensure they can withstand the thermal stresses they will encounter. This may involve using materials that are resistant to thermal stress, designing components with adequate thermal expansion margins, and implementing cooling and insulation systems to mitigate the effects of thermal stress.

TIEMPO

Short for test inspection evaluation master plan organized. The organization explains the movement of the testing from low level of product competency to higher levels of systems performance. See Test Inspection Evaluation Master Plan [39].

Time-Saving Bias

Time-saving bias occurs when decision-makers prioritize speed and efficiency over accuracy and thoroughness in their decision-making process. This bias can lead to flawed decisions that are based on incomplete information or inadequate analysis.

In the testing context, time-saving bias can occur when testers are pressured to complete testing quickly, due to project timelines or resource constraints. This can lead to testers skipping certain steps or tests, or not spending enough time on analysis and verification, which can result in undetected defects or bugs.

Establishing clear testing objectives and priorities is important to address time-saving bias in testing and allocating sufficient time and resources to testing activities. This may involve prioritizing critical tests and focusing on the most important areas of the system, or using automated testing tools to help reduce testing time without sacrificing accuracy.

It is important to ensure that testers have the necessary skills and training to perform their testing duties effectively and efficiently. This may involve providing training in testing methodologies, tools, and techniques, as well as promoting a culture of quality and continuous improvement within the testing team.

Timing Synchronization

Timing synchronization, which selects the right time intervals to sample the incoming signal, is a technique used by receiver nodes. Carrier synchronization is the process by which a receiver modifies the frequency and phase of its local carrier oscillator to match those of the received signal.

TMMi

Test Maturity Model integration is a framework for assessing and improving an organization's testing processes and capabilities. TMMi was developed by the TMMi Foundation, a nonprofit organization that promotes and supports its use.

TMMi consists of five levels, each representing a different level of testing maturity within an organization: [170]

Level 1—Initial: Testing activities are ad hoc and unstructured, with no formal processes or procedures.

Level 2—Managed: Testing activities are planned, documented, and monitored, with basic processes in place to ensure consistency and repeatability.

Level 3—Defined: Testing activities are well defined, with formal processes and procedures in place that are consistently followed across the organization.

Level 4—Quantitatively Managed: Testing activities are measured, monitored, and controlled, focusing on continuous improvement and optimization of testing processes.

Level 5—Optimizing: Testing activities are continuously improved and optimized, focusing on innovation, learning, and excellence in testing.

The TMMi framework provides a road map for organizations to improve their testing processes and capabilities and achieve higher testing maturity levels. It includes a set of best practices, guidelines, and metrics that used to assess and improve testing processes. The model can be applied to any organization, regardless of its size or industry. TMMi is particularly useful for organizations that rely heavily on software systems and applications, as it can help ensure the quality and reliability of these systems.

Tool Chain

A tool chain is a set of software tools that are used in a particular sequence to perform a specific task or set of tasks. In software development, a tool chain is a set of tools used to develop, build, test, and deploy software.

A typical software development tool chain include a variety of tools:

1. Code editors or integrated development environments (IDEs) for writing and editing code
2. Version control systems for managing and tracking changes to code
3. Build tools for compiling code into executable files or libraries
4. Test automation frameworks for automating testing and ensuring code quality
5. Continuous integration/continuous delivery (CI/CD) tools for automating the build, testing, and deployment processes.

Other tools that can be added to a tool chain include static code analysis tools, profiling tools, debuggers, and more. The specific tools used in a tool chain will depend on the needs of the development team and the particular software development project.

Tool Confidence Level

Tool confidence level measures the reliability and accuracy of a software testing tool. It indicates the level of confidence that can be placed in the tool's ability to detect defects or issues in the product being tested correctly.

Tool confidence level is typically assessed using a predefined set of criteria considering such factors as a tool's functionality, ease of use, documentation, and support. An organization can establish its own requirements for a tool or use industry standards such as the Test Maturity Model integration (TMMi) [170].

The tool confidence level is usually expressed as a percentage or a rating system:

1. Level 1—Low Confidence: The tool is unreliable and should not be used for testing purposes.
2. Level 2—Medium Confidence: The tool has some reliability but may require additional testing and validation before it can be fully trusted.
3. Level 3—High Confidence: The tool is reliable and can be used for testing with a high level of confidence.

Achieving a high tool confidence level is essential for ensuring the accuracy and effectiveness of software testing. It helps minimize the risk of false positives or negatives, leading to costly defect leaks or delays in the product development process.

To achieve a high tool confidence level, it is important to evaluate testing tools before they are used in testing or manufacturing. This may involve conducting a series of tests, comparing the results to manual testing, and ensuring that the tool is regularly updated and maintained. It may also involve training and support to testers to ensure they use the tool correctly and effectively.

Tool Error Detection

Tool error detection refers to the ability of a (software) testing tool to detect errors or defects in the software being tested. This includes detecting coding, logic, and functional errors affecting the software's performance, security, or usability.

To detect errors effectively, software testing tools use a variety of techniques:

1. Static analysis involves analyzing the software's source code or design to identify potential issues or errors before the software is executed.
2. Dynamic analysis involves analyzing the software's behavior during runtime to identify errors that may not be detectable through static analysis.

3. Test automation involves automating the process of executing test cases to ensure consistent and thorough testing and to detect errors that may be missed through manual testing.
4. Code coverage analysis involves analyzing the code to ensure that all possible paths and scenarios have been tested and to identify areas of the code that may be more prone to errors.

To ensure that testing tools can detect errors effectively, it is important to select the right tool for the job, based on the needs of the project and the types of errors that need to be detected. It is also important to ensure that the tool is configured correctly and that the testing team is trained to use the tool effectively.

Regular testing and analysis of the tool's performance and error detection capabilities can also help to identify and address any issues or limitations in the tool, and to improve the testing process continuously.

Tool Impact

Tool impact and testing are closely related concepts in software testing. The impact of a testing tool is its effect on the testing process and, ultimately, the quality of the software being tested.

Testing tools can have a significant impact on the testing process in several ways:

1. Increased Efficiency: Testing tools can help automate repetitive testing tasks, saving time and increasing the testing process's efficiency.
2. Improved Accuracy: Testing tools can help to detect defects and issues more accurately than manual testing, reducing the risk of errors and improving the quality of the software.
3. Consistency and Repeatability: Testing tools can help to ensure that tests are repeatable and performed consistently, reducing the risk of variations in testing results.
4. Coverage and Completeness: Testing tools can help to ensure that all relevant test cases are executed, improving the coverage and completeness of the testing process.
5. Cost Savings: Testing tools can help to reduce the cost of testing by automating tasks and reducing the need for manual testing.

However, a testing tool can have a negative impact if it is not used effectively. Poorly designed or implemented testing tools can lead to false positives, false negatives, or other errors that can compromise the accuracy and effectiveness of the testing process.

To maximize the impact of testing tools, it is important to carefully select the right tool for the job, based on the needs of the project and the types of tests that need to be performed. Proper configuring and customizing the tool to fit the project's specific needs is also important. Regular testing and analysis of the tool's impact on the testing process can help to identify any issues or limitations in the tool, and to improve the testing process continuously.

Top-Down Testing

Top-down software testing starts from the highest level of a system and works its way down to the lower levels. This approach tests the system's overall functionality before testing the smaller components that make up the system.

One advantage of top-down testing is that it allows testers to first focus on the high-level functionality of the system, which can help them identify any significant issues early on in the testing process. This can save time and resources by identifying problems when they are easier to fix. The disadvantage is that sorting out the root cause of a system's perceived poor performance can be difficult.

Top-down testing is often used in conjunction with bottom-up testing, which involves first testing the lowest-level components of the system and working up to the higher levels. This combination of top-down and bottom-up testing can provide a more comprehensive view of systems and help ensure they function correctly.

Torque Testing

Torque testing is important in automotive testing, particularly in developing and quality assurance of engines, transmissions, and other mechanical systems. Torque is the measure of rotational force applied to a component, and it is critical for ensuring that automotive systems operate safely and efficiently. An ISO power plot is often used to depict engine power maps or power curves. It shows the power output of an engine at different operating points, typically represented by engine speed (revolutions per minute, rpm) and engine load (typically measured in terms of throttle position or manifold pressure) [171].

Torque testing in the automotive industry typically involves measuring the torque required to rotate a component or system, such as an engine or a transmission (see Figure T.3). This can be done using specialized equipment such as dynamometers, torque sensors, or torque wrenches.

FIGURE T.3 The list of parameters for torque testing.

	Steady-State		*Transient*	*Reference*
Applicability	non-road engines (ex. stationary, lawn equipment, generator, industrial)	on-road vehicle applications only	Light duty passenger vehicle applications only	section 9 (introduction)
Charge Air Cooler (CAC) Outlet Conditions	Use production CAC.	Use production CAC or lab CAC with equivalent pressure restriction. Measure CAC outlet temperature and pressure during steady-state testing in the application environment at each engine speed and match on dyno test stand.	Use production CAC or lab CAC with equivalent pressure restriction. Measure CAC outlet temperature and pressure during transient fixed gear vehicle testing. Match temperature and pressure on dyno test stand to vehicle values at each engine speed. All vehicle data must come from same fixed gear run.	section 7.5.5 section 6.2.3 section 9.1.4
Exhaust Back Pressure	Use full exhaust system in dynamometer test room.	Use full exhaust system in dynamometer test room or match back pressure after first reflective point based on data from steady-state vehicle testing with vehicle held at constant rpm for at least 2 min. Vehicle gear can vary by engine speed.	Match backpressure in dynamometer test room to value measured at peak power speed during transient vehicle acceleration maneuver. Any gear may be used, but all data must come from this same transient maneuver.	section 7.2.7 section 7.5.15 section 9.1.3
ECU Control Settings (fuel inj. timing, ignition timing, valve timing, EGR, tuning valve position, etc.)	Engine control unit (ECU) fully enabled and automatically controlling engine electronics (if applicable).	(1) Engine control unit (ECU) fully enabled and automatically controlling engine electronics OR (2) match ECU parameter settings according to values measured from vehicle testing with vehicle held at constant rpm for at least 2 min.	Measure parameter during transient fixed gear vehicle testing. Match value on dyno test stand to vehicle values at each engine speed. All vehicle data must come from same fixed gear run. Use same method of measurement in vehicle and on dyno stand.	section 7.5.4 section 9.3
Boost	Allow boost to stabilize according to engine control settings.	Allow boost to stabilize according to engine control settings.	Match manifold absolute pressure measured during transient vehicle acceleration maneuver.	section 7.5.6
Data Acquisition	Torque (±1%), engine speed (±1%), and all temperatures (±2°C) stabilized for at least 1 minute.	Torque (±1%), engine speed (±1%), and all temperatures (±2°C) stabilized for at least 1 minute.	(1) Torque (±1%), engine speed (±1%), and all temperatures (±2°C) stabilized for at least 1 minute, OR (2) torque and engine speed stabilized for 5 sec. then take 10 Hz data for 10 seconds.	section 7.5.3

Several different types of torque tests are commonly used in automotive testing:

1. Breakaway Torque: This measures the torque required to start a component rotating from a stationary position.
2. Running Torque: This measures the torque required to maintain the rotation of a component while it is operating under load.

3. Peak Torque: This measures the maximum torque a component can withstand before it fails or becomes damaged.
4. Torque Distribution: This measures torque distribution across multiple components or systems, such as the torque distribution between a four-wheel drive vehicle's front and rear wheels.

Total Quality Management

Total quality management (TQM) is an approach to continuously improve the quality of a company's products and processes. It is a customer-focused approach that involves the whole organization, from top management to front-line employees, in the continuous improvement of processes, products, services, and the culture in which they are provided (see Figure T.4).

FIGURE T.4 TQM touches many areas of an organization.

ski14/Shutterstock.com.

TQM is based on the idea that quality is not just the responsibility of the quality control department but the responsibility of every employee in the organization. It is an ongoing effort to improve all aspects of the business, including product and service design, development, production, and delivery.

TQM can be applied to any organization but is often used in manufacturing, healthcare, and service industries. It is a holistic approach that seeks to continuously improve all aspects of an organization, aiming to increase customer satisfaction and loyalty.

TQM involves several key principles:

1. Customer Focus: Meeting the needs and expectations of customers is the top priority.
2. Continuous Improvement: TQM involves ongoing efforts to identify and eliminate waste and improve efficiency.
3. Employee Involvement: All employees are encouraged to participate in the continuous improvement process.

4. Process-Oriented Approach: TQM focuses on the processes used to produce products and services, rather than individual employees or departments.
5. Data-Driven Decision-Making: TQM relies on data and factual analysis to make decisions rather than relying on personal opinions or subjective judgments.

Various TQM tools and techniques are employed to achieve these objectives: [40] [112]

1. Cause and Effect Diagram (also known as fishbone diagram or Ishikawa diagram): This tool helps identify and visualize the possible causes of a problem or quality issue by categorizing them according to different factors, such as people, processes, equipment, materials, and environment.
2. Check Sheets: These simple data collection forms are used to record and track data related to specific quality characteristics or defects. They help in collecting accurate and consistent data for further analysis.
3. Pareto Chart: This bar graph displays the frequency or occurrence of different categories or factors in descending order. It helps identify and prioritize the most significant problems or issues based on their frequency or impact.
4. Control Charts: These statistical tools are used to monitor and control process performance over time. They help identify any variations or trends that may indicate a process is out of control and in need of corrective action.
5. Histogram: This graphical representation of data distribution displays the frequency or count of data points falling within specific intervals or bins. Histograms help visualize the shape, central tendency, and variability of data.
6. Scatter Diagram: This diagram shows the relationship between two variables on a Cartesian plane. It helps identify any correlation or patterns between variables, which can be useful in identifying potential causes of quality issues.
7. Flowcharts: These charts visually illustrate the sequence of steps and decision points in a process. They help identify bottlenecks, inefficiencies, and opportunities for improvement in a circle.
8. 5 Whys: This technique is used to identify the root cause of a problem by repeatedly asking Why? until the underlying cause is identified. It helps uncover deeper causes rather than focus only on symptoms.

9. Quality Function Deployment (QFD): QFD is a structured approach to capture customer requirements and translate them into specific design and process characteristics. It helps ensure that customer needs and expectations are effectively incorporated into product development and process improvement.
10. Benchmarking: This technique compares an organization's processes, products, or services against the best practices of industry leaders or competitors. It helps identify areas for improvement and establish performance targets.

These are just a few examples of the tools used in TQM. Each tool serves a specific purpose and can be applied at different stages of the quality improvement process. The selection and effective use of these tools depend on an organization's particular needs and context.

The TQM process follows a plan–do–check–act (PDCA) cycle, a four-step process that involves planning, implementing, checking the results, and taking corrective action as needed. The PDCA cycle can be applied to any aspect of an organization, from processes and products to services and systems. It is a continuous process that allows organizations to make incremental improvements on an ongoing basis.

Here is a brief overview of each step in the PDCA cycle:

1. Plan: Identify an opportunity for improvement and develop a plan to address it.
2. Do: Implement the plan and collect data on the results.
3. Check: Analyze the data and assess the effectiveness of the changes made.
4. Act: Take corrective action as needed and make any necessary adjustments to the plan.

By following the PDCA cycle, organizations can continuously improve their processes, products, and services, thus increasing customer satisfaction and loyalty.

Traceability

Testing and traceability are closely related in the software development process. Testing is evaluating a system or its components to find defects. Traceability is the ability to trace the relationship between different elements in the software development process, such as requirements, design, and implementation.

Traceability is essential in testing because it helps ensure that tests are comprehensive and cover all relevant requirements. It also helps identify the source of any defects found so that they can be addressed appropriately.

Traceability is typically established through the use of a traceability matrix, which maps requirements to the design and implementation elements that fulfill them. Testers can then use the matrix to verify that all requirements are properly implemented and that all design and implementation elements can be traced back to a requirement.

By establishing traceability between requirements, design, and implementation, testers can ensure that the software being developed meets the end users' needs and is free of defects. It also helps to identify any gaps or inconsistencies in the software's requirements, design, or implementation, which can be addressed before the final product is released.

Traceability Matrix

Software traceability testing focuses on tracing requirements through the different stages of the development process. It is used to ensure that all requirements are properly captured and traceable to the design and implementation of the software, and that the final product satisfies all of those requirements.

Testers create a traceability matrix that maps each requirement to the design elements and code that implement it. They then verify that all requirements are traceable to the appropriate design and implementation elements and that all design and implementation elements can be traced back to a requirement (see Figure T.5).

FIGURE T.5 An example of a test to requirement link and traceability matrix.

	Test_Case1	Test_Case2	Test_Case3	Test_Case4	Test_Case5	Test_Case6	Total Test Cases
Req_#001	X			X			2
Req_#002	X		X	X			3
Req_#003		X					1
Req_#004			X				1
Req_#005		X			X	X	3
Req_#006				X		X	2

Traceability testing is another important way to ensure that the software being developed is free of defects and meets the needs of end users. It also helps identify gaps or inconsistencies in the software's requirements, design, or implementation, which can be addressed before the final product is released.

Tread Testing

Tread testing refers to the process of evaluating the performance characteristics of tire treads. The tread is the part of the tire that comes into contact with the road surface, and it plays a critical role in providing traction, handling, and stability to the vehicle.

Tread testing is performed to assess various properties of tire treads:

1. Tread Wear: This test evaluates the ability of the tire tread to resist wear and tear over time. The test is typically performed by measuring the depth of the tread grooves before and after a specified distance of driving.
2. Wet Traction: This test evaluates the ability of the tire tread to maintain grip on wet or slippery surfaces. The test is typically performed by measuring the stopping distance of a vehicle on a wet surface at a specified speed.
3. Dry traction: This test evaluates the ability of the tire tread to maintain grip on dry surfaces. The test is typically performed by measuring the stopping distance of a vehicle on a dry surface at a specified speed.
4. Rolling Resistance: This test evaluates the energy efficiency of the tire by measuring the amount of energy required to keep the tire rolling at a specified speed.
5. Noise: This test evaluates the amount of noise generated by the tire tread when it comes into contact with the road surface. The test is typically performed by measuring the sound level produced by the tire at various speeds.

Trim Testing

Trim testing is integral to automotive testing, particularly in the development and quality assurance of interior components [173]. Trim refers to a vehicle's interior's decorative and functional elements, such as seats, dashboards, door panels, and other interior surfaces. Automotive trim testing typically involves evaluating these components' performance, durability, and safety under various conditions, such as exposure to extreme temperatures, humidity, UV radiation, and resistance to wear and tear and abrasion (see Figure T.6).

FIGURE T.6 An example of a trim scuff fixture.

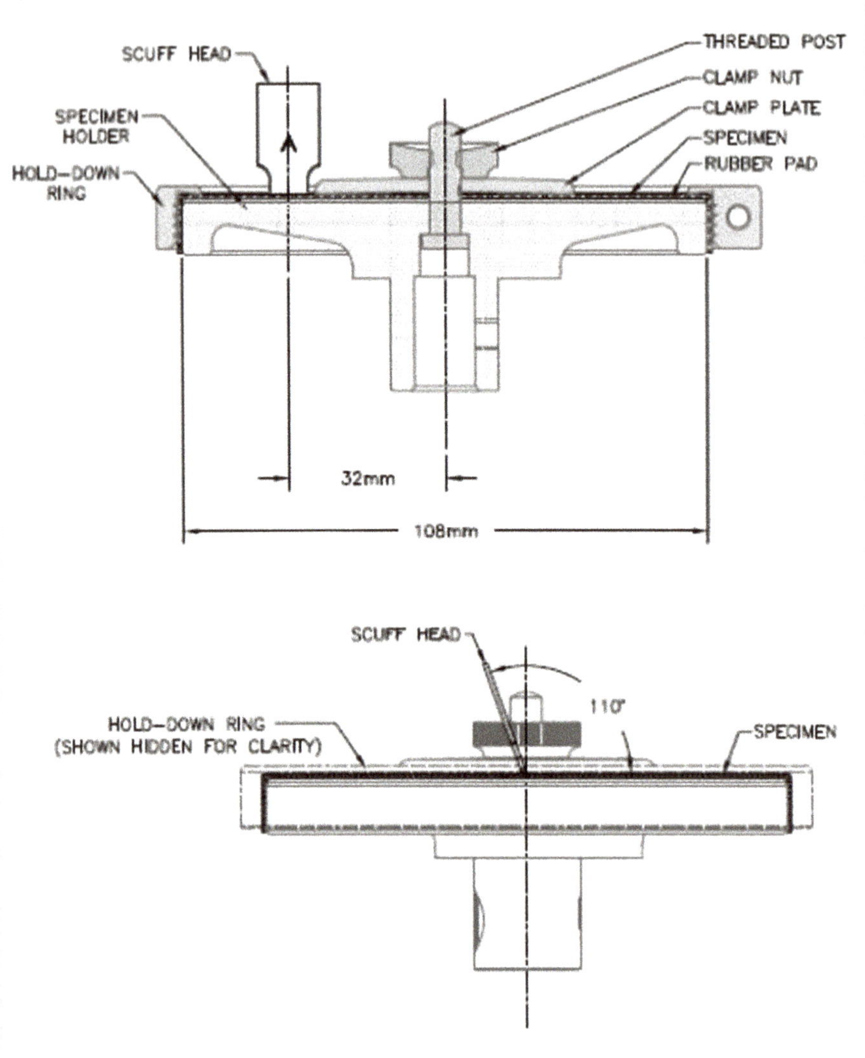

Some examples of trim tests that are commonly performed in the automotive industry include the following:

1. Abrasion Resistance Testing: This measures the ability of a trim component to resist wear and abrasion from normal use, such as rubbing against clothing or other surfaces.
2. Scratch Resistance Testing: This measures the ability of a trim component to resist scratches from objects that it may come into contact with, such as keys or other sharp objects.

3. UV Resistance Testing: This measures the ability of a trim component to resist fading, discoloration, and other damage caused by exposure to ultraviolet (UV) radiation from sunlight or other sources.
4. Temperature and Humidity Testing: This measures the ability of a trim component to withstand extreme temperatures and humidity levels, which can affect the component's performance, durability, and appearance.

By performing trim testing, automotive engineers and quality assurance professionals can ensure that interior components are safe, durable, and able to withstand the rigors of normal use over the vehicle's life. This helps to improve the overall quality and safety of automotive products and reduces the risk of costly recalls or warranty claims.

U

"Statistics are used much like a drunk uses a lamppost: for support, not illumination."

—Vin Scully

Ultraviolet Testing

The accelerated exposure of automotive exterior materials using a fluorescent UV and condensation apparatus is a testing method commonly employed in the automotive industry to assess the durability and performance of materials exposed to harsh environmental conditions (see Figure U.1). This testing apparatus combines two key elements: fluorescent UV radiation and condensation.

Fluorescent UV lamps emit high-intensity ultraviolet (UV) radiation that simulates the damaging effects of sunlight, particularly the short-wavelength UV rays. Condensation is generated by exposing the test samples to high humidity and elevated temperature cycles, simulating the moisture and temperature fluctuations experienced by automotive materials in real-world conditions.

During the testing process, automotive exterior materials, such as paints, coatings, plastics, rubbers, and textiles, are subjected to continuous exposure to fluorescent UV radiation and alternating periods of condensation and drying. This testing aims to simulate and accelerate the degradation mechanisms caused by ultraviolet radiation, moisture, and temperature, which can lead to color fading, gloss reduction, cracking, peeling, embrittlement, and other forms of material deterioration.

The test specimens are typically evaluated periodically to assess changes in visual appearance, mechanical properties, adhesion, and other performance attributes. This allows automotive manufacturers and suppliers to predict the long-term behavior of materials and make informed decisions regarding their selection, formulation, and design.

The results obtained from accelerated exposure testing using a fluorescent UV and condensation apparatus help automotive industry professionals develop more durable and reliable exterior materials, optimize coating and surface protection systems, meet quality standards, comply with regulatory requirements, and enhance the overall longevity and aesthetics of vehicles exposed to various weathering conditions.

FIGURE U.1 Parts of a vehicle can be exposed to a wide range of light wavelength exposure.

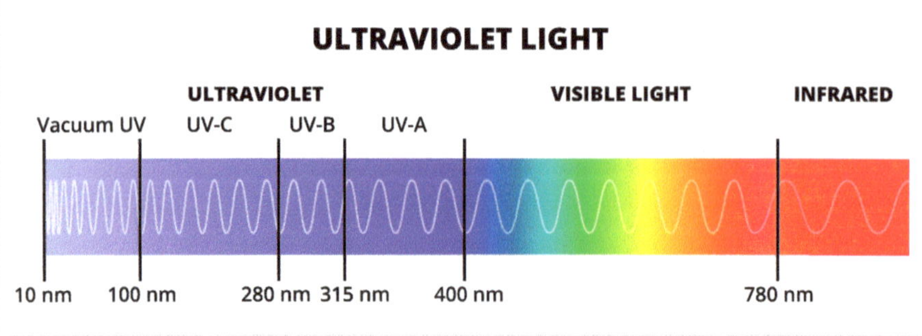

petroudny43/Shutterstock.com.

Unit Testing

Unit testing involves testing individual units or components of a software system in isolation from the rest of the system. It is designed to ensure that each unit or component of the system functions correctly and meets specified requirements.

The development team typically does unit testing during the implementation phase of the software development process. It is an important step in the development process, as it helps identify defects or issues in the code early on, before the system is integrated and tested.

Some common activities involved in unit testing include the following:

1. Writing Test Cases: Developing a set of test cases that cover the functionality and requirements of the unit or component being tested
2. Executing Tests: Executing the test cases and evaluating the results to ensure that the unit or component meets the specified requirements
3. Documenting Findings: Documenting any defects or issues that are identified during the unit testing process, including the location and severity of the issue

Unit testing is an important step in the development process, as it allows the development team to ensure that each unit or component of the system functions correctly and meets all specified requirements. It is typically done in conjunction with other types of testing, such as integration testing and system testing, to ensure the overall quality and reliability of the software system.

Unit under Test

The thing being tested is referred to as a "unit under test" (UUT). Other names for it include "device under test" (DUT). The test points of the test system are connected to the connecting points (such as plugs) via the adapter. See Device Under Test.

Unreachable Code

Unreachable code is a section of code that can never be executed during the runtime of a program. This can occur when a condition or loop control variable prevents the execution of the code, or when the code is located after a return statement, which will cause the function to exit before reaching the code.

Unreachable code can indicate a logic error in the program and lead to inefficiencies in the code. It is important to identify and remove unavailable code during the testing phase of software development to ensure that the program is running efficiently and without errors.

There are several ways to identify unreachable code:

1. Code Analysis Tools: Many software development tools, such as integrated development environments (IDEs) and code analysis tools, can detect unreachable code and flag it as an error or warning.
2. Code Reviews: Code reviews by other developers can help identify potential instances of unreachable code and suggest modifications to resolve the issue.
3. Automated Testing: Automated tests can help identify instances of unreachable code by executing the program and flagging any code that is not executed during the test.

Removing unreachable code can improve the overall quality and efficiency of the program and make the code easier to maintain and modify in the future.

Usability Testing

Usability testing evaluates the ease of use and user-friendliness of a software system or application. It is designed to ensure that the system is intuitive, easy to use, and meets the users' needs and expectations.

In the automotive industry, usability testing may be conducted on a variety of software systems, including vehicle control systems, diagnostic tools, and dealership management systems. It is an important step in the development process to ensure that these systems are easy to use and understand and provide a positive user experience.

Some common activities involved in usability testing include the following:

1. Defining Usability Goals: Identifying the software's specific usability requirements or goals
2. Recruiting Test Participants: Selecting a group of representative users to participate in the usability testing

3. Conducting User Interviews: Interviewing users to gather feedback on their experience with the software, including any issues or frustrations they encountered
4. Observing User Interactions: Observing users as they interact with the software to identify any usability issues or areas for improvement
5. Analyzing Data: Assessing the data collected during the usability testing process to identify trends and issues and determine improvement areas.

Use Case

A use case is a detailed description of a specific interaction or scenario that outlines how a system, product, or service is used to achieve a particular goal or outcome. It provides a step-by-step representation of the interactions between different actors (users, systems, entities) and the system itself. Use cases are commonly employed in software development, system design, and project management to define the functional requirements and behavior of a system from the user's perspective.

User Acceptance Testing

User acceptance testing (UAT) evaluates a software system's functionality and usability from the end users' perspective. It is designed to ensure that the system meets the needs and expectations of the users and that it is easy to use and understand. UAT is typically conducted near the end of the development process, after the software has undergone other types of testing.

UAT is typically conducted by a group of representative users, rather than by the development team or testers. This allows for a more realistic evaluation of the software, as the users better understand how they will employ the system in their daily work.

Some common activities involved in UAT include the following:

1. Defining Acceptance Criteria: Identifying the requirements or goals the software must meet to succeed
2. Creating Test Cases: Developing a set of test cases that fully cover the software's functionality and usability
3. Executing Tests: Running the test cases and evaluating the results to ensure that the software meets the acceptance criteria
4. Documenting Findings: Recording any defects or issues that are identified during the UAT process, including the location and severity of the issue.

V

"The man of science has learned to believe in justification, not by faith, but by verification."

—Thomas Huxley

Valid Partitions

Valid partitions are used in the testing of automotive systems, such as advanced driver assistance systems (ADAS) and autonomous vehicles. In this context, the input domain would include various environmental conditions, such as different types of roads, weather conditions, lighting conditions, and different vehicle states, such as accelerating or braking.

Partitioning the input domain into valid partitions can help automotive engineers design and execute effective tests that evaluate the system's behavior under various conditions. For example, engineers might partition the input domain into subsets based on road type (e.g., highway, city streets, off-road), weather conditions (e.g., rain, snow, fog), or vehicle speed (e.g., low speed, high speed).

By selecting a representative set of inputs from each partition and designing test cases that cover the behavior of the system for each input, automotive engineers can ensure that the system behaves correctly under a wide range of conditions. Testing with valid partitions can also help identify potential edge cases or corner cases that might be missed in less systematic testing approaches.

Validation

In the automotive industry, validation refers to verifying that a product or system meets the specified requirements and performs as intended (see Figure V.1). This can include verifying a product's functionality and specific customer application, performance, and reliability, and ensuring that it meets regulatory standards and customer expectations [39, 174].

FIGURE V.1 An overview of the steps to validation [174].

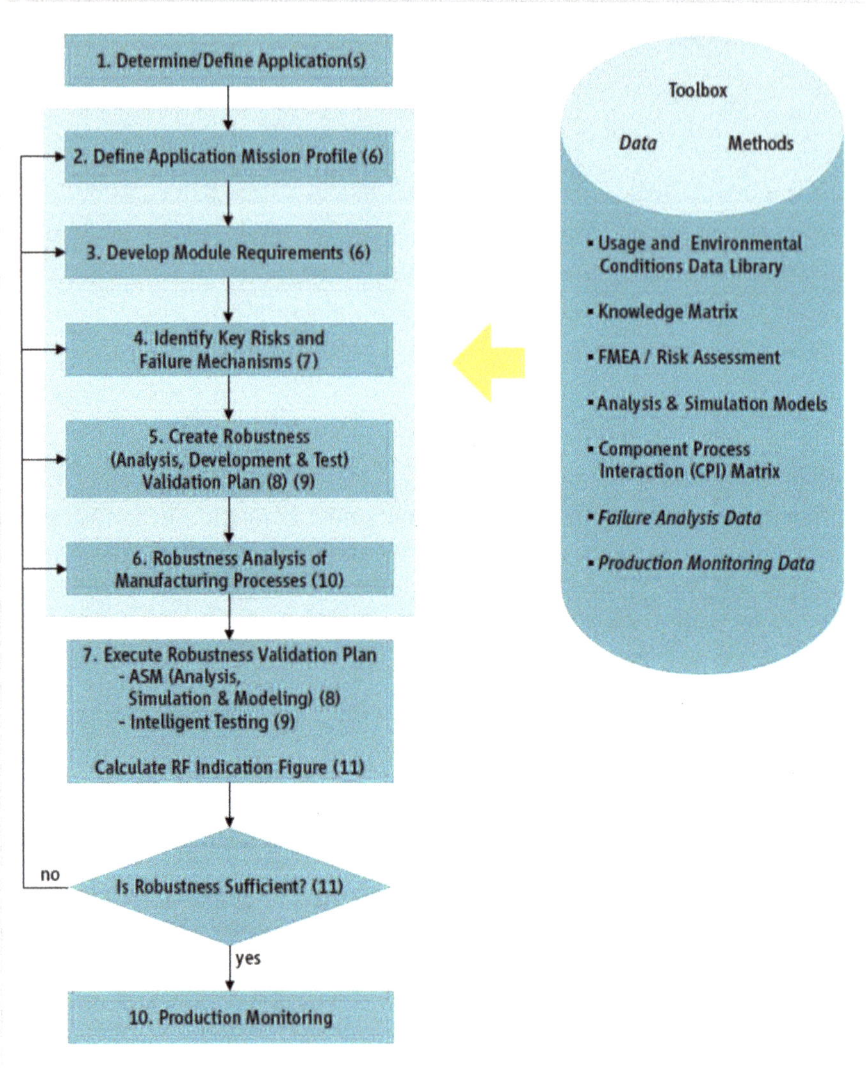

Several types of validation are conducted in the automotive industry:

1. Design Validation: Verifying that the design of a product or system meets the specified requirements and performs as intended. Customer exploration and testing of prototype parts is part of adjusting the design and learning
2. Process Validation: Verifying that the manufacturing process for a product or system can consistently produce products that meet the specified requirements

3. System Validation: Verifying that a system, such as a vehicle or a component, functions as needed by the customer and intended when integrated into the final product
4. Product Validation: Verifying that a finished product meets the specified requirements and performs as intended

Validation is essential for ensuring that products and systems are safe, reliable, and meet the required standards. It is typically conducted throughout the development process, with various types of validation being undertaken at different stages (see Figure V.2).

FIGURE V.2 A juxtaposition of the requirements development to validation [174].

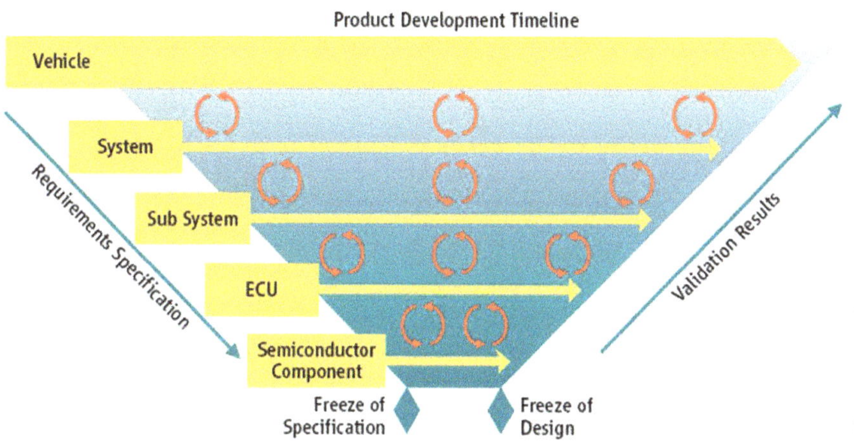

Variance

The expectation of a random variable's squared divergence from its population mean or sample mean is known as its variance. Variance serves as a proxy for dispersion, or the degree to which a set of numbers deviates from their mean.

$$\sigma^2 = \sum \frac{(x_i - \mu)^2}{n}$$

Where:

σ^2 is the sample variance.
Σ is the summation symbol.
x_i is each individual data point in the set.
μ is the mean value of all observations.
n is the number of observations.

Vehicle-in-the-Loop Simulation

A vehicle-in-the-loop (VIL) simulator is a testing tool that allows developers to assess a vehicle's or automotive system's performance and behavior in a virtual environment. It is used to evaluate the performance and reliability of the vehicle or system under different driving scenarios or environmental factors.

A VIL simulator typically consists of a computer-based model of the vehicle or system, which is integrated with sensors and actuators to simulate the inputs and outputs of the real-world system. It also includes a simulation environment, which can be used to simulate different driving scenarios or environmental conditions.

VIL simulators are commonly used in the automotive industry to test and evaluate the performance and reliability of vehicles and systems before they are built or tested. They allow for more efficient and cost-effective testing and greater control over the testing conditions. However, VIL simulators are not a replacement for physical testing, and they are typically used in conjunction with physical testing to ensure the reliability and performance of automotive products.

Verification

In the automotive industry, verification refers to confirming that a product or system has been designed and implemented according to specifications and standards. It checks that a product or system meets the specified requirements and performs as intended.

Several types of verification may be conducted in the automotive industry:

1. Design Verification: This confirms that the design of a product or system meets the specified requirements and performs as intended.
2. Process Verification: This confirms that the manufacturing process for a product or system can consistently produce products that meet the specified requirements.
3. System Verification: This confirms that a system, such as a vehicle or a component, functions as intended when integrated into the final product.
4. Product Verification: This confirms that a finished product meets the specified requirements and performs as intended.

Verification is typically conducted throughout the development process, with various types of verification being conducted at different stages, to ensure that products and systems are of high quality and meet all specifications.

Verification Plan

A verification plan outlines the approach and activities for certifying that a product or system meets specified requirements and performs as intended. The verification plan includes details on the specific requirements that need to be verified, the methods and tools that will be used for verification, the schedule and timeline for verification activities, and the roles and responsibilities of the team members involved in the verification process.

Some key elements that may be included in a verification plan for an automotive product or system include the following [39]:

1. Verification Objectives: These are the specific goals and objectives of the verification process, such as ensuring that the product meets regulatory standards or customer expectations.
2. Verification Methods: These are the methods and tools used to verify the product or system, such as testing, inspection, or analysis.
3. Verification Schedule: This is the time frame for conducting verification activities, including the order in which different approaches to the verification are conducted and scheduling milestones.
4. Verification Budget: The budget notes any special equipment and costs associated with internal and external verification and test activities.
5. Verification Criteria: These are measures that will be used to determine if the product or system meets the specified requirements and performs as intended.
6. Verification Reporting: This is the process for documenting and reporting on the verification process's results, including any identified issues or defects.

The verification plan is essential for ensuring that the verification process is thorough, consistent, and well organized and that the product or system meets the required standards.

Version Control

Version control is a process used to track and manage changes to a software system or application, particularly in the automotive industry. It allows developers to track and undo changes to the product, collaborate with other team members, and maintain a history of the development process. Equally important is that the testers can prepare for testing in advance and are testing the content in each version of the product (see Figure V.3) [57, 122].

FIGURE V.3 Version control ensures the product that goes to testing is the product version that should go to testing.

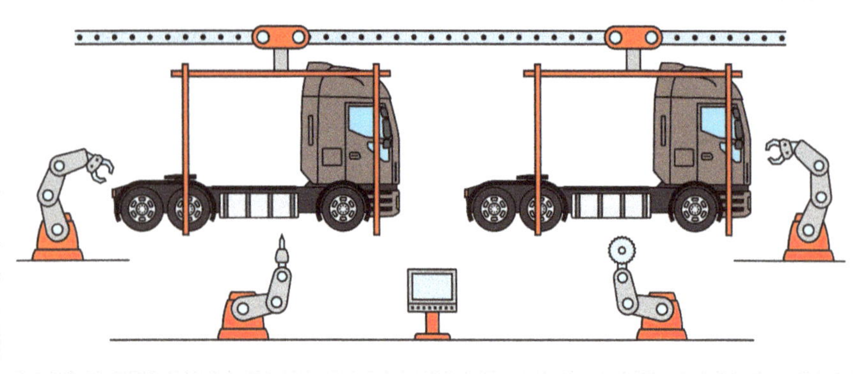

all_is_magic/Shutterstock.com.

In the automotive industry, version control develops and maintains complex systems, such as vehicle control systems, diagnostic tools, or dealership management systems. It helps to ensure that changes to software are properly documented and that there is a clear record of the development process. Version control tools allow software developers to track changes to the codebase, create branches for new features or fixes, and merge changes back into the main codebase. They also allow team members to collaborate and work on the same codebase without conflicts or overlaps.

Vertical Traceability

Vertical traceability refers to tracking the relationship between the requirements of a product or system and the design, implementation, and testing activities used to develop and validate those requirements. In the example in Figure V.4, there are component requirements but this could easily be mapped to system-level requirements, and often are. Vertical traceability is a way to ensure that all needs are being adequately addressed and that there is a clear link between the requirements and the final product [57, 175].

FIGURE V.4 Vertical traceability map of requirements to test cases.

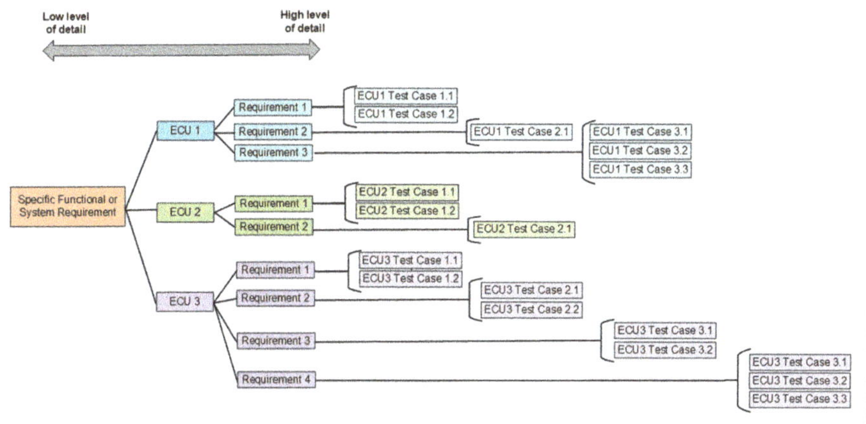

Courtesy of Jon Quigley.

Vertical traceability ensures that the requirements have mapped test cases defining what is tested and how. Additionally, test results are traced to those requirements, all of which have been considered for testing. This can involve creating a traceability matrix that maps the relationship between the requirements, corresponding test cases, and results. Lastly, this traceability makes it possible to predict the consequences and predict the severity of a single test case failure on the entire system.

Vertical traceability ensures the quality and reliability of the product or system, allows for more thorough testing, and helps to identify any gaps or issues in the development process. Perhaps more importantly, it provides traceability of consequences for any test case failure.

Virtual Homologation

Virtual homologation is a process in which the safety and performance of a vehicle or automotive system is evaluated through virtual simulations and tests rather than through physical testing. It is a way to reduce the cost and time associated with traditional homologation processes, which typically involve testing prototypes of the vehicle or system under various conditions.

Virtual homologation employs computer simulations and virtual testing tools to evaluate the performance and safety of a vehicle or system. This includes simulations of crash tests, durability testing, and other types of performance evaluations.

Virtual homologation is a growing trend in the automotive industry as it allows for more efficient and cost-effective testing and evaluation of vehicles and systems. It can also help reduce the environmental impact of traditional testing methods, as it eliminates the need for physical prototypes and testing.

However, virtual homologation is not a replacement for physical testing, and it is typically used in conjunction with physical testing to ensure the reliability and safety of automotive products.

Virtual Reality

Virtual reality (VR) technology is used for a variety of testing purposes in the automotive industry. Here are some examples:

1. Design Validation: Automotive manufacturers use VR technology to create virtual prototypes of cars, allowing designers to explore and test various design options and configurations before building physical prototypes. This can help streamline the design process and reduce development costs.
2. Ergonomics Testing: VR technology is used to simulate the interior of a car, allowing designers to evaluate the ergonomics and comfort of the seating, controls, and other features. This can help ensure that the car's design is optimized for driver and passenger comfort and safety.
3. Safety Testing: VR technology is used to simulate driving scenarios and test a car's safety features, such as collision detection systems and airbags. This can help identify and address safety issues before the car is released to the public.
4. Training: Automotive manufacturers also use VR technology for training purposes, allowing mechanics and technicians to practice working on cars in a simulated environment. This can help improve the quality and efficiency of repairs and maintenance.
5. User Experience Testing: VR technology is used to simulate a driver's interface and test its usability and functionality. This can help identify and address issues with the car's controls and displays before it is released to the public.

VR technology represents a valuable testing tool for automotive manufacturers, allowing them to identify and address issues early in the development process, streamline design and testing processes, and improve vehicle safety and user experience.

Virtual Validation

Virtual validation is a process in which the performance and reliability of an automotive product or system is evaluated through simulations (computer-aided design, CAD, and computer-aided engineering, CAE) and tests, rather than through physical testing (see Figure V.5). It is a way to reduce the cost and time associated with traditional validation processes, which typically involve testing prototypes of the product or system under various conditions.

FIGURE V.5 Virtual validation is used with the smallest components to the entirety of a vehicle or system.

Gorodenkoff/Shutterstock.com.

Virtual validation uses computer simulations and virtual testing tools to evaluate the performance and reliability of a product or system. This includes simulations of real-world conditions, such as driving scenarios or environmental factors, as well as simulations of component failures or malfunctions.

Virtual validation is a growing trend in the automotive industry, allowing for more efficient and cost-effective evaluation of products and systems. It also helps reduce the environmental impact of traditional testing methods, as it eliminates the need for physical prototypes and testing. However, virtual validation is not a replacement for physical testing, and it is typically used in conjunction with physical testing to ensure the reliability and performance of automotive products.

V-Model

The V-model is a software development approach that combines the linear structure of the waterfall model with the iterative nature of Agile development. It is called the "V-model" because it visually represents the process as a V shape, with the bottom of the V representing the development phase and the top of the V representing the testing phase (see Figure V.6).

FIGURE V.6 An example of the V-model juxtaposing development artifacts and testing levels.

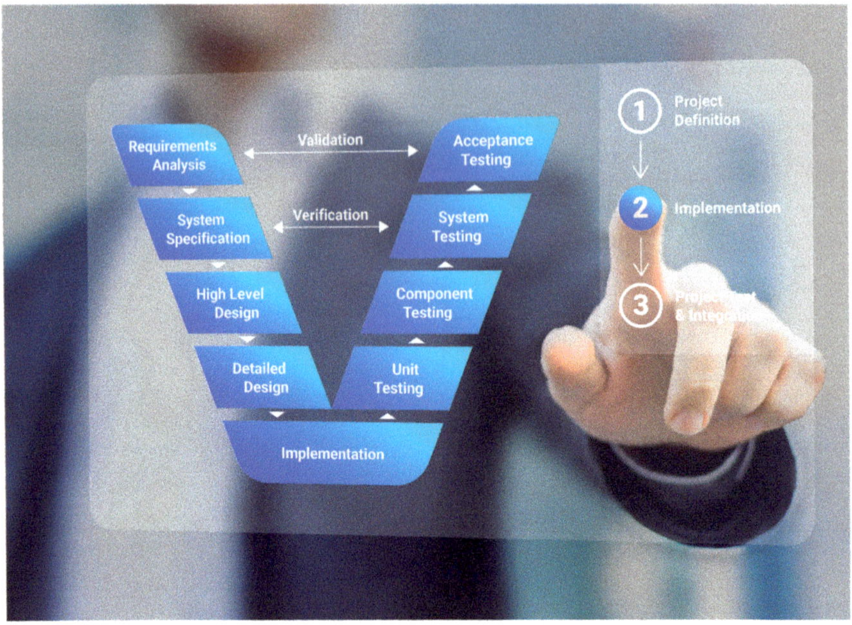

NicoElNino/Shutterstock.com.

In the V-model, testing is integrated into each phase of the development process rather than treated as a separate phase at the end. This means that testing is performed throughout the development process, starting with unit testing at the beginning and progressing to system testing at the end.

The V-model is often used in industries with strict regulatory or quality requirements, such as aviation and healthcare, as it allows for more thorough and comprehensive testing. It is also useful for projects with well-defined requirements and a fixed scope, as it allows for more predictability and control over the development process.

The V-model does not always proceed in a single loop. It can move through a loop multiple times, thus making the model look more like a W or zigzag (see Figure V.7) [122, 123].

FIGURE V.7 A representation of the cyclic nature of an incarnation of the V-model [123].

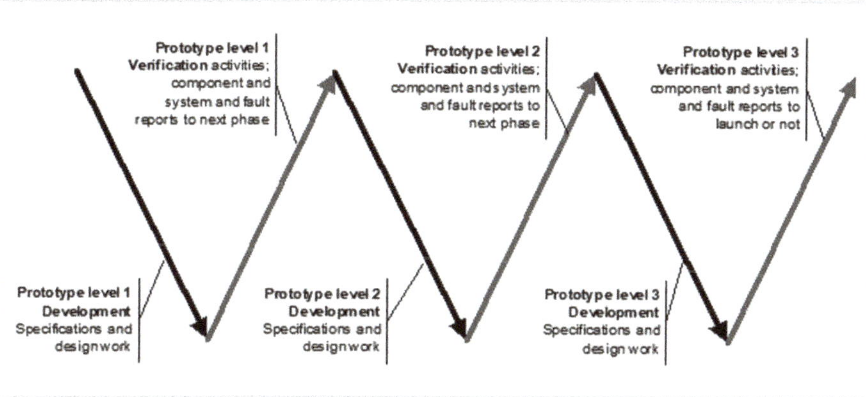

Courtesy of Jon Quigley.

Volume Testing (Audio)

Audio volume testing in the automotive industry typically involves evaluating the sound system's performance for both entertainment and safety purposes (see Figure V.8). The goal is to provide a satisfactory listening experience to vehicle occupants while maintaining clarity and avoiding distortion. Here are some key aspects of automotive audio volume testing:

FIGURE V.8 Vehicle in-cabin audio testing covers safety and entertainment systems.

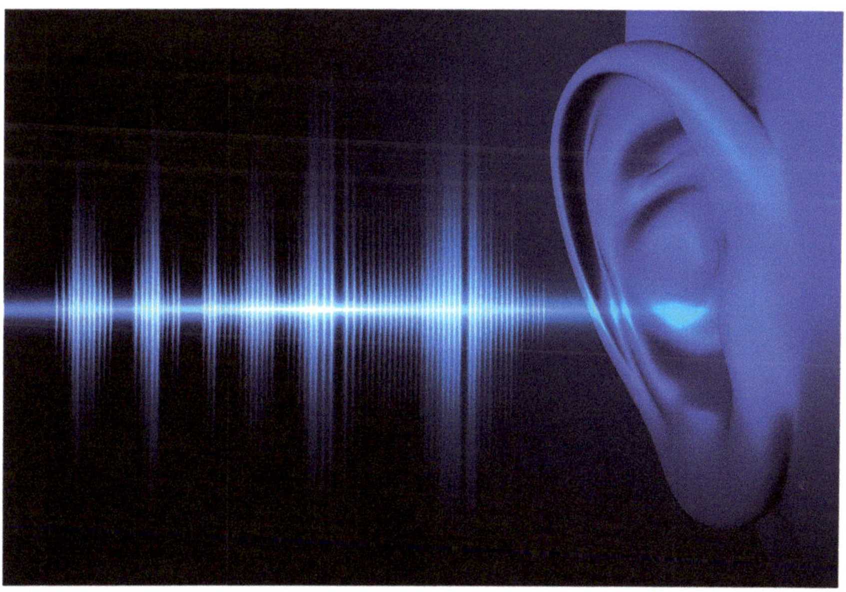

Tatiana Shepeleva/Shutterstock.com.

1. Decibel (dB) Measurement: Decibels are used to quantify sound levels. In some vehicle safety systems (e.g., commercial vehicles), minimum dB is required to ensure it is heard and prompts immediate action. The audio system's volume is adjusted during testing, and the sound pressure level (SPL) is measured at various frequencies and positions within the vehicle cabin. This helps determine the system's maximum volume capability and ensures it complies with safety regulations.
2. Frequency Response Analysis: Automotive audio systems (radios and CD players) should accurately reproduce sound across a wide range of frequencies. Testing involves measuring the infotainment system's performance at different volumes to ensure no significant peaks or dips in the frequency response curve. This provides a balanced and enjoyable listening experience for vehicle occupants.
3. Distortion Analysis: Distortion can occur when an audio system is pushed to its limits. Testing involves gradually increasing the volume until distortion becomes audible. Ideally the level at which distortion occurs should not compromise the listening experience or harm the system components.
4. Speaker Performance: The performance of individual speakers, including their power-handling capabilities, frequency response, and distortion levels, is evaluated. This involves testing speakers at different volumes and frequencies to ensure consistent and clear sound reproduction across the entire range.
5. Equalization (EQ) Tuning: Automotive audio systems often incorporate equalization to adjust the sound output to compensate for variations in the vehicle's acoustics. Testing involves tuning the EQ settings to achieve a balanced sound well-suited to the vehicle's cabin environment.
6. Sound Localization: In some advanced audio systems, sound localization techniques are employed to create a more immersive listening experience. Testing involves evaluating the accuracy of sound positioning to ensure alignment with the intended spatial audio effects.
7. Noise and Vibration Testing: Automotive audio systems are designed to minimize external and interior noise and vibration. Testing involves subjecting the system to various road conditions and measuring the cabin audio performance. This ensures the audio system remains clear and audible even in challenging environments.

Audio volume testing in the automotive industry is typically conducted in specialized testing facilities equipped with calibrated audio measurement instruments. These tests help ensure that the audio systems in vehicles meet quality standards and deliver an optimal listening experience to passengers.

Volume Testing (Throughput)

Volume testing evaluates the performance and stability of a software system under heavy load conditions. It is designed to simulate high-data bus traffic or processor load to determine how the system handles increased usage and identify potential bottlenecks or issues.

Some common focus areas for volume testing include the following:

1. Response Time: Measuring the time it takes for the system to respond to requests or queries under heavy load conditions
2. Throughput: Measuring the amount of data or transactions the system can handle over a given period
3. Scalability: Evaluating the system's ability to handle an increase in user or data volume without negatively affecting performance
4. Resource Utilization: Monitoring the system's use of resources such as memory, central processing unit (CPU), and disk space to identify any potential issues

Volume testing is often used to verify the performance and stability of a software system in real-world conditions, such as during peak usage periods or when handling large amounts of data. It is an essential step in the testing process to ensure that systems can handle the expected loads and provide a satisfactory user experience.

Vulnerability

A vulnerability refers to a weakness or flaw in a software system that attackers can exploit to compromise the system's security or integrity. Testing is essential in identifying and mitigating vulnerabilities, as it helps uncover potential weaknesses before attackers can exploit them.

Several types of testing can be used to identify vulnerabilities:

1. Penetration Testing: This involves simulating an attack on the system to identify vulnerabilities that could be exploited. Penetration testing can be conducted manually or using automated tools.
2. Vulnerability Scanning: This involves using automated tools to scan the system for known vulnerabilities. Vulnerability scanning can be conducted on a regular basis to identify new vulnerabilities.
3. Code Review: This involves reviewing the code of the system to identify potential vulnerabilities that could be exploited by attackers. Code review can be conducted manually or using automated tools.

4. Fuzz Testing: This involves testing the system with random or malformed inputs to identify potential vulnerabilities that could be exploited by attackers.
5. User Acceptance Testing: This involves testing the system from the perspective of end users to identify potential vulnerabilities that could be exploited by attackers.

By conducting these types of testing, developers and security professionals can identify and address vulnerabilities in the system before attackers can exploit them. This can help improve the security and reliability of the system and protect it against potential attacks.

W

"The beginning is the most important part of the work."

—Plato

Walkthrough

A walkthrough involves a group of people, usually developers, testers, and other stakeholders, reviewing a software system or piece of code to identify potential issues, defects, or areas for improvement [12].

During a walkthrough, the group typically follows a structured process that involves the following steps:

1. Planning: The group agrees on the scope and objectives of the walkthrough, as well as the roles and responsibilities of each participant.
2. Preparation: The developer or team responsible for the code or system prepares the necessary documentation, including design documents, test plans, and code.
3. Presentation: The developer or team responsible for the code or system presents the documentation and code to the group, explaining its purpose, design, and implementation.
4. Review: The group examines the documentation and code in detail, asking questions and providing feedback on potential issues, defects, or areas for improvement.
5. Follow-up: The developer or team responsible for the code or system addresses any issues or defects identified during the walkthrough and updates the documentation and code as necessary.

Walkthroughs can be conducted during various software development life cycle stages, including requirements gathering, design, coding, and testing. They can help improve the quality and reliability of the system by identifying potential issues early in the development process, reducing the likelihood of defects and rework later on. Walkthroughs can also help improve collaboration and communication among team members and provide knowledge sharing and learning opportunities.

Waterfall Development

Waterfall development is a software methodology that follows a sequential, linear process. It is called "waterfall" because the process flows steadily downward, with each stage following the previous one in a cascade-like manner.

The waterfall development process typically includes the following phases:

1. Requirements Gathering: The first stage involves collecting and documenting the project's requirements, including functional and nonfunctional requirements, and determining the scope of the project.
2. Design: In this stage, the design of the software system is planned and documented, including the system architecture, data model, and user interface.
3. Implementation: This stage involves the actual coding and development of the software system based on the design specifications.
4. Testing: In this phase, the software is tested for defects, errors, and compliance with the requirements.
5. Deployment: Once the software has been tested and approved, it is deployed to the production environment.
6. Maintenance: After the software has been deployed, it is monitored and maintained to ensure it meets the users' requirements and functions properly.

One of the benefits of the waterfall model is that it provides a clear and structured approach to software development, with each stage building upon the previous one. It also emphasizes the importance of thorough planning and documentation. Theoretically this model does have loops to update the prior steps in the list above, it is often represented as a single pass through the steps. This is not always true, however even when the team pass through these steps repeatedly, the durations between these steps are of longer duration than in an Agile sprint.

Wear-Out Parts

Wear-out parts are components or elements of a system that are prone to deteriorating or failing over time due to normal usage or wear and tear. They are typically subject to higher levels of stress or wear than other parts of the system, and may require frequent replacement or maintenance to ensure that the system continues to function correctly (see Figure W.1).

FIGURE W.1 Parts have a defined life expectancy and they can require periodic servicing based on time or mileage. Testing confirms their life expectations and serviceability.

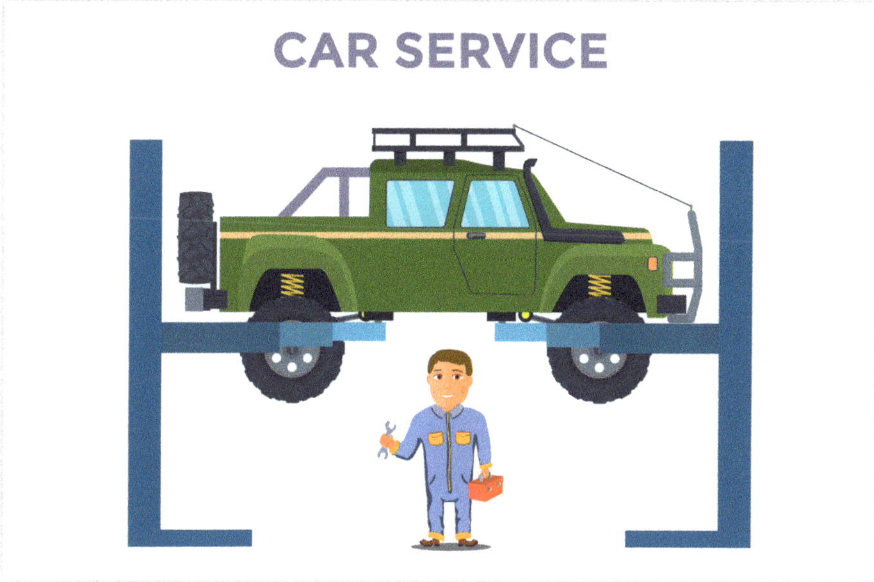

dimair/Shutterstock.com.

Testing focuses on wear-out parts in order to evaluate their performance and identify any issues or weaknesses that may arise over time. This involves subjecting the parts to simulated wear or abuse in order to evaluate their durability and reliability.

Understanding the performance and durability of wear-out parts is important for ensuring that a system is able to function correctly and reliably over time. Testing wear-out parts can reveal potential issues or weaknesses and help to optimize the maintenance and replacement schedule for those parts.

Web Applications Testing

Web application testing verifies the functionality and performance of off-vehicle web interfaces. An example of an automotive web application is the telemetry system that transmits hard braking events via cellular or satellite mechanisms to a "back office" for aggregation and analysis (mainly applies to commercial vehicles). Web application testing includes evaluating compatibility with different car models and systems, assessing the user experience for car dealers and mechanics, and verifying the accuracy and reliability of the application's data and information (see Figure W.2).

FIGURE W.2 Modern vehicles are connected to external systems for diagnosing and prediction.

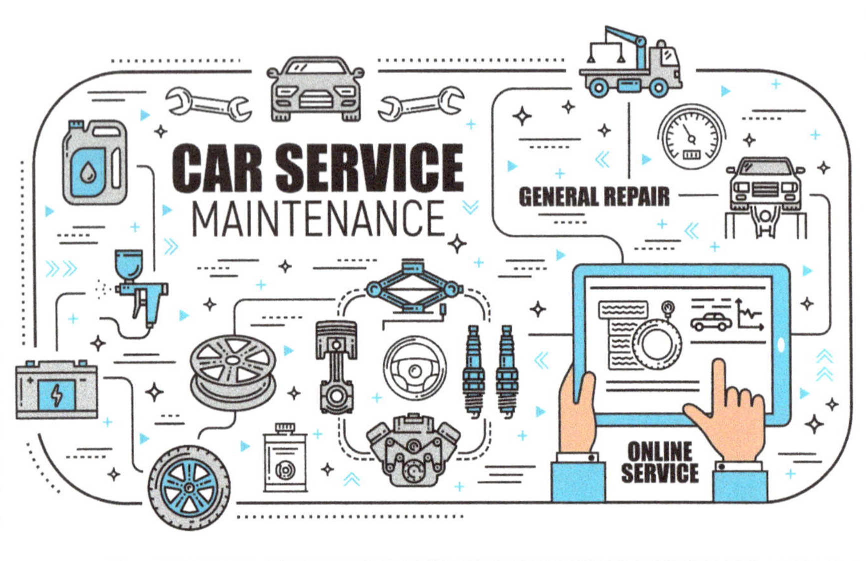

Vector Tradition/Shutterstock.com.

Some key areas to focus on during web application testing for automotive applications include the following:

1. Compatibility Testing: Ensuring the web application works seamlessly with different car models and systems, including operating systems and hardware configurations
2. Performance Testing: Testing the application's speed, stability, and ability to handle large amounts of data and traffic
3. Usability Testing: Verifying that the user interface is intuitive and easy to use for car dealers and mechanics and that the application can meet the needs of different user groups
4. Data and Information Accuracy: Ensuring that the data and information displayed in the application is accurate and reliable and sourced from trusted sources
5. Security Testing: Ensuring that the application is secure and that user data is protected from unauthorized access or data breaches

White-Box Testing

A subsystem with viewable but typically unchangeable internals is referred to as a "white box." Both systems engineering and software engineering make use of the phrase.

White-box testing focuses on the internal structure and functionality of a system, component, or software program (see Figure W.3). Also known as glass box testing or structural testing, it typically tests the individual components and functions of a system rather than the system as a whole.

FIGURE W.3 In white-box testing, we know the product's interior electrically and algorithmically.

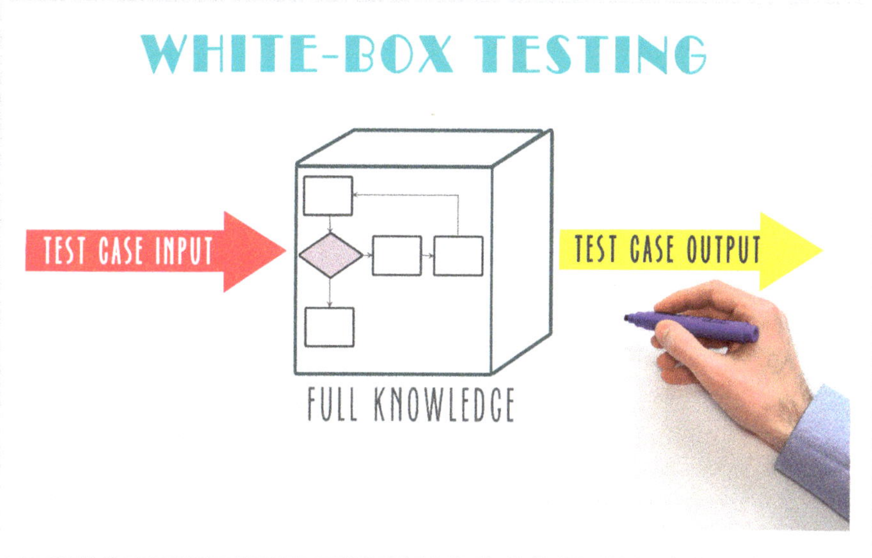

astel design/Shutterstock.com.

In white-box testing, the tester has full knowledge of the system's internal structure and can thoroughly examine the behavior and performance of individual components or functions. This involves testing the code or algorithms used in the system, the data structures or interfaces between components, or the system's error handling and recovery capabilities.

White-box testing is an important step in the testing process, as it helps to ensure that a system's individual components and functions are functioning correctly and meeting the desired performance standards. It is typically conducted early in the testing process before higher-level tests, such as integration or acceptance tests, are performed. However, it can be time-consuming and resource-intensive and may require specialized expertise to set up and run effectively.

Wideband Delphi

The wideband Delphi estimate method is a strategy for estimating work that relies on consensus. It derives from the Delphi technique, which was created as a forecasting tool at the RAND Corporation in the 1950s and 1960s.

The wideband Delphi (WBD) method is a technique for gathering and aggregating expert opinions and knowledge to make informed decisions or predictions. It is commonly used in software development, project management, and engineering to identify and prioritize potential risks or issues.

The WBD method is used to gather opinions and insights on potential issues or challenges that may arise during the testing process. This information can help testers to identify and prioritize areas of concern and to develop strategies for addressing those issues [146].

The WBD method involves multiple rounds of anonymous feedback and discussion among the expert group, in which each member provides their perspective and insights on the issues being considered. This process is repeated until a consensus is reached or the group reaches a stable state of understanding.

The WBD method is useful for gathering and synthesizing the knowledge and experience of a diverse group of experts, as well as for identifying potential issues that may not be apparent to individuals. However, it can be time-consuming and require careful planning and coordination to ensure that it is conducted effectively.

Wire Harness Test

A wire harness is a bundle of wires that are organized and wrapped together, often with connectors or terminals attached to each end. The wire harness connects various electrical components within a vehicle, such as the engine control module, sensors, and actuators (see Figure W.4). Wire harnesses are subjected to many environmental stimuli that can cause failure [176]. Wire harness testing is therefore essential for ensuring the safe and reliable operation of automotive electrical systems.

FIGURE W.4 An example of an electrical distribution center with wire harness interconnects.

Wire harness failures can lead to electrical malfunctions and pose risks to vehicle performance and safety. The testing suite is defined for a specific harness and includes testing immediate functionality as well as reliability over the life of the vehicle [176]. There are a number of wire harness failures and tests:

1. Fretting: This is the repetitive motion or vibration between two surfaces in contact, resulting in damage over time. In automotive wire harnesses, fretting can occur at connection points, such as terminals and connectors. The constant movement can wear away protective coatings, leading to exposed metal surfaces. This can cause intermittent electrical connections, signal loss, or short circuits.
2. Chafing: This refers to the friction or rubbing of wires against abrasive or sharp surfaces. In automotive wire harnesses, chafing can happen when wires come into contact with rough edges, vehicle components, or improperly secured brackets. The constant rubbing can wear away insulation or shielding, potentially exposing the wires. Chafing can lead to short circuits, open circuits, or electrical arcing.
3. Galvanic Corrosion: This occurs when dissimilar metals come into contact with an electrolyte, such as moisture or salts. In automotive wire harnesses, galvanic corrosion can arise when different metals, such as copper and aluminum, are used in connectors or terminals. The contact between the metals and the presence of moisture or other contaminants creates an electrochemical reaction that corrodes the metals. This electrochemical reaction can result in poor electrical conductivity, signal degradation, or complete connection failure. See Salt Fog (Spray).
4. Continuity Testing: See Cross-Continuity Test.
5. Insulation Resistance Testing: This involves testing the resistance of the wire insulation to ensure that it is sufficient to prevent electrical shorts or leakage. Insulation resistance testing can help identify damaged, worn, or otherwise compromised wires.
6. Dielectric Strength Testing: This involves testing the ability of the wire insulation to withstand high voltages without breaking down or arcing. Dielectric strength testing can help identify improperly insulated wires that pose a risk of electrical shock or fire.
7. Voltage Drop Testing: This involves testing to ensure that the voltage drop across each wire in the harness is within acceptable limits. Voltage drop testing can identify wires that are undersized, damaged, or otherwise not capable of handling the electrical load. Starter motors are susceptible to voltage drops, specifically, the lower voltage does not engage the starter; sensors that require a well-defined voltage source are similarly susceptible [177]. For more information, see **J3226_202212** [178].

8. Connector Testing: This involves examining the connectors or terminals on the ends of the wires to ensure that they are correctly crimped or soldered and making a secure electrical connection (risk of pull out; see Figures W.5 and W.6). For more information, check out AS39029/64C Contacts, Electrical Connector, Pin, Crimp Removeable (for MIL-DTL-24308-Connectors).

FIGURE W.5 An example of pull-out force requirements.

Approx. (mm^2)	AWG	Minimum Pull-out Force (N)	Reference MinimumCable Break Strength (N)
0.8	18	90	136
1.0	16	120	202
1.5		150	
2.0	14	180	331
2.5		210	
3.0	12	240	521
4.0		265	
5.0	10	290	833
6.0		320	
8.0	8	350	

FIGURE W.6 Example of a prototype interface to wire harness tester connector test.

9. Short Circuit Testing: This involves shorting heavy current–carrying wire elements to the ground and viewing the thermal impacts or implications to assess the possibility of fire during these events. There are three general classes of circuit breakers: Type 1, automatic reset; Type 2, modified reset; and Type 3, manual reset with a few variations. For more information, see **J1533_202205** [180].

By conducting these types of wire harness testing, automotive manufacturers can help ensure the safe and reliable operation of their vehicle's electrical systems, reducing the risk of electrical failures, fires, or other hazards.

Work Breakdown Structure

In project management and systems engineering, a work breakdown structure (WBS) divides a project team's work into digestible chunks [122]. It is a hierarchical representation of the tasks and activities involved in a project or process. It is typically used to break down a complex project or process into smaller, more manageable components and define their relationships and dependencies.

A WBS can be used to organize and plan the testing activities for a project or system. It identifies the specific tasks and activities that need to be completed as part of the testing process and defines the relationships and dependencies between those tasks. An example of a work breakdown structure for testing might look like this:

1. Product Testing
    - 1.1. Test Planning
        - 1.1.1. Define Test Objectives and Scope
        - 1.1.2. Identify Test Requirements and Success Criteria
        - 1.1.3. Develop Test Strategy and Approach
        - 1.1.4. Create Test Plan
    - 1.2. Test Design
        - 1.2.1. Identify Test Cases and Scenarios
        - 1.2.2. Define Test Data and Environment
        - 1.2.3. Design Test Procedures
        - 1.2.4. Develop Test Scripts or Automation Framework
    - 1.3. Test Execution
        - 1.3.1. Set up Test Environment
        - 1.3.2. Execute Test Cases and Scenarios
        - 1.3.3. Record Test Results and Defects
        - 1.3.4. Monitor and Report Test Progress

- 1.4. Test Evaluation
    - 1.4.1. Analyze Test Results and Defects
    - 1.4.2. Assess Test Coverage and Traceability
    - 1.4.3. Validate Test Results against Acceptance Criteria
    - 1.4.4. Prepare Test Summary Report
- 1.5. Test Closure
    - 1.5.1. Complete Test Documentation
    - 1.5.2. Perform Lessons Learned and Knowledge Transfer
    - 1.5.3. Obtain Stakeholder Sign-off on Test Completion

Workflow Testing

Workflow testing entails simulating the production environment throughout the testing phase to assess it from an end user's perspective. The test database must contain enough data to test each workflow adequately.

Workflow testing evaluates the performance and functionality of a system or process as it moves through a series of steps or stages. It is used to test systems involving a series of interdependent tasks or actions, such as business, manufacturing, or software development.

In workflow testing, the tester typically follows a specific set of steps or scenarios to examine the system's behavior and performance as it progresses. This exploration may involve simulating different conditions or scenarios to test the system's response, ensuring that it functions correctly and meets the desired performance standards.

Workflow testing can help identify bottlenecks or other issues that may impact the system's performance or efficiency and identify opportunities for improvement. However, it can be resource-intensive and may require careful planning and coordination to ensure that it is conducted effectively.

References

1. B. L. S. H. Dodson, *R-518 Accelerated Testing: A Practitioner's Guide to Accelerated and Reliabilty Testing 2nd Edition*, Warrendale, PA: SAE Publishing, 2021.
2. Society of Automotive Engineers, *J2017-01-0389 Accelerated Testing of Brake Hoses for Durability Assessment*, Warrendale, PA: SAE Publishing, 2017.
3. Society of Automotive Engineers, *J2020_202210 Accelerated Exposure of Automotive Exterior Materials Using a Fluorescent UV and Condensation Apparatus*, Warrendale, PA: SAE Publishing, 2022.
4. Society of Automotive Engineers, *J2100_202101 Accelerated Environmental Testing for Bonded Automotive Assemblies*, Warrendale, PA: SAE Publishing, 2021.
5. Society of Automotive Engineers, *J3014_201805 Select Highly Accelerated Failure Test (HAFT) for Automotive Lamps with LED Assembly*, Warrendale, PA: SAE Publishing, 2018.
6. L. Klyatis, "*Basic Negative and Positive Trends in Development Accelerated Testing*," SAE Publishing, Warrendale, PA, 2019.
7. IEEE Standards, "EEE Std 610.12-1990 IEEE Standard of Software Engineering Terminology," in *Software Engineering Volume One Customer and Terminology Standards*, New York, NY, The Institute of Electrical and Electronics Engineers, Inc, 1999.
8. Society of Automotive Engineers, "*AMS2442 Magnetic Particle Acceptance Criteria for Parts*," Society of Automotive Engineers Publishing, Warrendale, PA, 2022.
9. A. Thiyagaraj, *Parasitic Battery Drain Problems and AUTOSAR Acceptance Testing*, Warrendale, PA: SAE Publishing, 2018.
10. Society of Automotive Engineers, *J2364_201506 Navigation and Route Guidance Function Accessibility While Driving*, Warrendale, PA: SAE Publishing, 2015.

11. Society of Automotive Engineers, *J2678_201609 Navigation and Route Guidance Function Accessibility While Driving Rationale*, Warrendale, PA: SAE Publishing, 2016.

12. IEEE Standards, "IEEE Std 1028-1997 IEEE Standard for Software Reviews," in *Software Engineering Volume Two Process Standards*, New York, NY, The Institute of Electrical and Electronic Engineers, Inc., 1999.

13. Defense for Research and Engineering, *Systems Engineering Guidebook*, Washington, DC: Office of the Under Secretary of Defense for Research and Engineering, 2022.

14. Office of the Deputy Director for Engineering, *Engineering of Defense Systems Guidebook*, Washington, DC: Office of the Under Secretary of Defense for Research and Engineering, 2022.

15. ISTQB, "ISTQB Ad Hoc Testing," International Software Test Qualification Board, [Online]. Available: https://glossary.istqb.org/en_US/term/ad-hoc-testing-3-2. [Accessed 2 5 2023].

16. Society of Automotive Engineers, *J1211_201211 Handbook for Robustness Validation of Automotive Electrical/Electronic Modules*, Warrendale, PA: SAE Publishing, 2012.

17. International Software Testing Qualification Board, "ISTQB Glossary," 23 4 2023. [Online]. Available: https://glossary.istqb.org/en_US/term/anomaly. [Accessed 2 5 2023].

18. IEEE Standards, "IEEE Std 1044-1993 IEEE Standard Classification for Software Anomalies," in *Software Engineering Volume Four Resource and Technique Standards*, New York, NY, The Institute of Electrical and Electronics Engineers, Inc, 1999.

19. IEEE Standards, "IEEE std 1044.1-1995 IEEE Guide to Classification for Software Anomalies," in *Software Engineering Volume Four Resource and Technique Standards*, New York, NY, The Institute of Electrical and Electronics Engineers, Inc, 1999.

20. Society of Automotive Engineers, *J1455 Recommended Environmental Practices for Electronic Equipment Design in Heavy-Duty Vehicle Applications*, Warrendale, PA: SAE Publishing, 2017.

21. Society of Automotive Engineers, *J377_200712 Vehicular Traffic Sound Signaling Devices (Horns)*, Warrendale, PA: SAE Publishing, 2007.

22. Society of Automotive Engineers, *J2883_202003 Laboratory Measurement of Random Incidence Sound Absorption Tests Using a Small Reverberation Room*, Warrendale, PA: SAE Publishing, 2020.

23. Society of Automotive Engineers, *J1503_200409 Performance Test for Air-Conditioned, Heated, and Ventilated Off-Road, Self-Propelled Work Machines*, Warrendale, PA: SAE Publishing, 2004.

24. Society of Automotive Engineers, *J1400_201707 Laboratory Measurement of the Airborne Sound Barrier Performance of Flat Materials and Assemblies*, Warrendale, PA: SAE Publishing, 2017.

25. ISTQB, "API Testing Glossary," International Software Testing Qualification Board, [Online]. Available: https://glossary.istqb.org/en_US/term/api-testing. [Accessed 2 5 2023].

26. Society of Automotive Engineers, "*J2534 Pass-Thru Interface - Alternate Platforms for API Version 04.04*," Society of Automotive Engineers Publishing, Warrendale, PA, 2022.

27. Society of Automotive Engineers, *J2057/4_202212 Class A Multiplexing Architecture Strategies*, Warrendale, PA: SAE Publishing, 2022.

28. Society of Automotive Engineers, *J3131_202203 Definitions for Terms Related to Automated Driving Systems Reference Architecture*, Warrendale, PA: SAE Publishing, 2022.

29. A. A. B. C. Ametller, "A Coupling Architecture for Remotely Validating Powertrain Assemblies," *SAE International: J Electric Vehicle*, vol. 12, no. 2, 2023.

30. ISTQB, "ISTQB Glossary Audit," International Software Testing Qualification Board, [Online]. Available: https://glossary.istqb.org/en_US/term/audit-2-2. [Accessed 2 5 2023].

31. DIN - Society of Automotive Engineer, *Terms and Definitions Related to Testing of Automated Vehicle Technologies*, Warrendale, PA: SAE Publishing, 2023.

32. J. Tao, F. Klueck, H. Felbinger, M. Nica, F. Zieher, C. Wolf and C. Wang, *Automated Test Case Generation and Virtual Assessment Framework for UN Regulation on Automated Lane Keeping Systems*, Warrendale, PA: SAE Publishing, 2021.

33. S. Aly, *Consolidating AUTOSAR with Complex Operating Systems (AUTOSAR on Linux)*, Warrendale, PA: SAE Publishing, 2017.

34. S. Mirheidari, A. Fallahi, D. Zhang and K. Kuppam, "*AUTOSAR Model-Based Software Component Integration of Supplier Software*," SAE Publishing, Warrendale, PA, 2015.

35. R. Mader, A. Graf and G. Winkler, "*AUTOSAR Based Multicore Software Implementation for Powertrain Applications*," SAE Publishing, Warrendale, PA, 2015.

36. Society of Automotive Engineers, "*J2933 Verification of Brake Rotor and Drum Modal Frequencies*," SAE Publishing, Warrendale, PA, 2022.

37. Society of Automotive Engineers, "*J1113-4 Immunity to Radiated Electromagnetic Fields - Bulk Current Injection (BCI) Method*," Society of Automotive Engineers Publishing, Warrandale, PA, 2020.

38. CMMI Product Team, *CMMI For Development, Vesion 1.3*, Pitsburgh, PA: Carnegie Mellon University, 2010.

39. K. H. A. Q. J. M. Pries, *Testing Complex and Embedded Systems*, Boca Raton, FL: CRC Press, Taylor & Francis Group, 2011.

40. K. Ishikawa, *Introduction to Quality Control*, White Planes, NY: 3A Corporation, 1993.

41. IEEE Standard, "IEEE Std 1042-1987 (reaff 1993) IEEE Guide to Software Configuration Management," in *Software Engineering Volume Two Process Standard*, New York, NY, The Institute of Electrical and Electronic Engineers, Inc., 1999.

42. IEEE Standards, "IEEE Std 1490-1998 IEEE Guide - Adoption of PMI Standard - A Guide to the Project Management Body of Knowledge," in *Software Engineering Volume Two Process Standards*, New York, NY, The Institute of Electrical and Electronics Engineering, Inc., 1999.

43. Society of Automotive Engineers, *J1976 Outdoor Weathering of Exterior Materials*, Warrendale, PA: SAE Publishing, 2022.

44. K. Nagarajan, A. Ranga, K. M. Kalkura, R. Anegundi and A. Ariharan, "*Virtual Software-In-Loop (Closed Loop) Simulation Setup during Software Development*," SAE Publishing, Warrendale, PA, 2022.

45. S. Klein, R. Savelsberg, F. Xia, D. Guse, J. Andert, T. Blochwitz, C. Bellanger, S. Walter, S. Beringer, J. Jochheim and N. Amringer, *Engine in the Loop: Closed Loop Test Bench Control with Real-Time Simulation*, Warrendale, PA: SAE Publishing, 2017.

46. D. R. Bothe, "*Reducing Process Variation*," Landmark Publishing Co., Cedarburg, WI, 2007.

47. Society of Automotive Engineers, "*Requirements for a COTS Assembly Management Plan*," SAE Publishing, Warrendale, PA, 2020.

48. A. Himmler, K. Lamberg, T. Schulze and J.-E. Stavesand, "*Testing of Real-Time Criteria in ISO 26262 Related Projects - Maximizing Productivity Using a Certified COTS Test Automation Tool*," SAE Publishing, Warrendale, PA, 2016.

49. National Highway Traffic Safety Administration, "*FMVSS No. 101, Controls and Displays*," US Department of Transportation, Washington, 2020.

50. IEEE Standards, "IEEE Std 1233, 1998 Edition, IEEE Guide for Developing Systems Requirements Specifications," in *Software Engineering Volume One Customer and Terminology Standards*, New York, NY, The Institute of Electrical and Electronics Engineering, Inc., 1999.

51. IEEE Standards, "IEEE Std 830-1998 IEEE Recommended Practice for Software Requirements Specification," in *Software Engineering Volume Four*

Resource and Technique Standards, New York, NY, The Institute of Electrical and Electronics Engineering, Inc, 1999.

52. Q. Zhou, T. Lucchini, G. D'Errico, R. Novella, J. M. Garcia-Oliver and X. Lu, "*CFD Modeling of Reacting Diesel Sprays with Primary Reference Fuel*," SAE Publishing, Warrendale, PA, 2021.

53. K. J. Link and N. A. Pohlman, "*CFD Windshield Deicing Simulations for Commercial Vehicle Applications*," SAE Publishing, Warrendale, PA, 2018.

54. IEEE Standards, "IEEE Std 1348-1995, IEEE Recommended Practice for Adoption of Computer Aided Software Engineering (CASE) Tools," in *Software Engineering Volume Four Resource and Technique Standards*, New York, NY, The Institute of Electrical and Electronics Engineers, Inc, 1999.

55. J. Alfonso, J. M. Rodriguez, J. C. Salazar, J. Orús, V. Schreiber, V. Ivanov, K. Augsburg, J. V. Molina, M. Al Sakka and J. A. Castellanos, *Distributed Simulation and Testing for the Design of a Smart Suspension*, Warrendale, PA: SAE Publishing, 2020.

56. S. R. Nehe, A. Ghogare, S. Vatsa, L. Tekade, S. Thakare, P. Yadav, P. Shool, P. Wankhade, V. Dhakane and S. Charjan, *Design, Analysis, Simulation and Validation of Automobile Suspension System Using Drive-Shaft as a Suspension Link*, Warrendale, PA: SAE Publishing, 2018.

57. J. M. Quigley and K. L. Robertson, *Configuration Management: Theory and Application for Engineers, Managers, and Practitioners*, Boca Raton, FL: CRC Press, 2020.

58. IEEE Standards, "IEEE Std 828-1998, IEEE Standard for Software Configuration Management Plans," in *Software Engineering Volume Two Process Standards*, New York, NY, The Institute of Electrical and Electronic Engineers, Inc., 1999.

59. Society of Automotive Engineers, *J2721 Recommended Corrosion Test Methods for Commercial Vehicle Components*, Warrendale, PA: SAE Publishing, 2011.

60. Society of Automotive Engineers, *J1128 Low Voltage Primary Cable*, Warrendale, PA: SAE Publishing, 2020.

61. Society of Automotive Engineers, *J1292 Automobile and Motor Coach Wiring*, Warrendale, PA: SAE Publishing, 2016.

62. Society of Automotive Engineers, *J2202 Heavy-Duty Wiring Systems for On-Highway Trucks*, Warrendale, PA: SAE Publishing, 2019.

63. Society of Automotive Engineers, *J1673 High Voltage Automotive Wiring Assembly Design*, Warrendale, PA: SAE Publishing, 2012.

64. Society of Automotive Engineers, "*SAE/USCAR-7-2 Deembrittlement Verification Test*," SAE Publishing, Warrandale, PA, 2020.

65. Society of Automotive Engineers, "*J3215 Hydrogen Embrittlement Testing of Ultra High Strength Steels and Stampings by Acid Immersion*," Society of Automotive Engineers Publishing, Warrandale, PA, 2023.

66. N. Okui, "*A Study on Alternative Test Method of Real Driving Emissions for Heavy-duty Vehicle by Using Engine In the Loop Simulation*," SAE Publishing, Warrendale, PA, 2021.

67. Society of Automotive Engineers, " *J1113-4 Immunity to Radiated Electromagnetic Fields - Bulk Current Injection (BCI) Method*," SAE Publishing, Warrendale, PA, 2014.

68. Society of Automotive Engineers, "*J1113-1 Electromagnetic Compatibility Measurement Procedures and Limits for Components of Vehicles, Boats (up to 15 m), and Machines (Except Aircraft) (16.6 Hz to 18 GHz)*," SAE Publishing, Warrendale, PA, 2013.

69. Society of Automotive Engineers, "*J1812 Function Performance Status Classification for EMC Immunity Testing*," SAE Publishing, Warrendale, PA, 2018.

70. Society of Automotive Engineers, "*J1113-12 Electrical Interference by Conduction and Coupling - Capacitive and Inductive Coupling via Lines Other than Supply Lines*," SAE Publishing, Warrendale, PA, 2022.

71. Society of Automotive Engineers, "*J1113-21 Electronmagnetic Compatibility Measurement Procedure for Vehicle Components -Part 21: Immunity to Electromagnetic Fields, 30 MHz to 18 GHz, Absorber-Lined Chamber*," SAE Publishing, Warrendale, PA, 2013.

72. Society of Automotive Engineers, "*J1113-13 Electromagnetic Compatibility Measurement Procedure for Vehicle Components - Part 13: Immunity to Electrostatic Discharge*," SAE Publishing, Warrandale, PA, 2015.

73. Society of Automotive Engineers, "*J551 Vehicle Electromagnetic Immunity - Electrostatic Discharge (ESD)*," SAE Publishing, Warrendale, PA, 2020.

74. Society of Automotive Engineers, "*J113/13 Electromagnetic Compatibility Measurement Procedure for Vehicle Components - Part 13: Immunity to Electrostatic Discharge*," SAE Publishing, Warrendale, PA, 2015.

75. L. Navarenho de Souza Fino and R. Navarenho de Souza, "*Improving the ESD Performance and Its Effects in CMOS – SOI/BULK Technologies and Automotive Electronic Components*," SAE Publishing, Warrendale, PA, 2014.

76. Society of Automotive Engineers, *J1739_202101 Potential Failure Mode and Effects Analysis (FMEA) Including Design FMEA, Supplemental FMEA-MSR, and Process FMEA*, Warrendale, PA: SAE Publishing, 2021.

77. Automotive Industry Action Group, *Potential Failure Mode and Effects Analysis for Tooling and Equipment*, Detroit, MI: Automotive Industry Action Group, 2001.

78. Automotive Industry Action Group, *Potential Failure Mode and Effects Analysis Third Edition*, Detroit, MI: Automotive Industry Action Group, 2001.

79. Society of Automotive Engineers, "*J2562 Biaxial Wheel Fatigue Test*," SAE Publishing, Warrendale, PA, 2021.

80. Society of Automotive Engineers, "*J2649 Strain-Life Fatigue Data File Format*," SAE Publishing, Warrendale, PA, 2018.

81. Society of Automotive Engineers, "*J2409 Strain-Life Fatigue Data File Format*," SAE Publishing, Warrendale, PA, 2018.

82. IEEE Standards, "IEEE Std 829-1998, IEEE Standard for Software Test Documentation," in *Software Engineering Volume Four Resource and Technique Standards*, New York, NY, The Institute of Electrical and Electronics Engineers, Inc, 1999.

83. Society of Automotive Engineers, "*J369 Flammability of Polymeric Interior Materials - Horizontal Test Method*," Society of Automotive Engineers Publishing, Warrendale, PA, 2019.

84. G. Schagerl, D. Brameshuber, K. Rom and M. Hammer, "*21SIAT-0638 - Fleet Analytics - A Data-Driven and Synergetic Fleet Validation Approach*," SAE Publishing, Warrendale, PA, 2021.

85. G. Gafencu, "*Aspects of Aircraft Certification Concerning Fungus Testing and Analysis*," Society of Automotive Engineers Publishing, Warrendale, PA, 2020.

86. D. Gulati and M. Sain, "*Effect of Fungal Modification on Fiber-Matrix Adhesion in Natural Fiber Reinforced Polymer Composites*," Society of Automotive Engineers Publishing, Warrendale, PA, 2006.

87. Automotive Industry Action Group, *Measurement Systems Analysis Third Edition*, Detroit, MI: Automotive Industry Action Group Publishing, 2002.

88. Automotive Industry Action Group, *Statistical Process Control (SPC) Second Edition*, Detroit, MI: Automotive Industry Action Group, 2005.

89. Transport Canada Motor Vehicle Safety, "*TECHNICAL STANDARDS DOCUMENT No. 101, Revision 0 Controls, Tell-tales, Indicators and Sources of Illumination*," Minister Motor Vehicle Safety, Ottawa, 2019.

90. Society of Automotive Engineers, "*J417 Hardness Tests and Hardness Number Conversions*," Society of Automotive Engineers Publishing, Warrandale, PA, 2018.

91. Automotive Industry Action Group, *Statistical Process Control (SPC) Second Edition*, Detroit, MI: Automotive Industry Action Group Publishing, 2005.

92. Automotive Industry Action Group, *Production Part Approval Process (PPAP) Fourth Edition*, Detroit, MI: Automotive Industry Action Group, 2006.

93. S. Solmaz, M. Rudigier, M. Mischinger and J. Reckenzaun, "Hybrid Testing: A Vehicle-in-the Loop Testing Method for the Development of Automated Driving Functions," *SAE Int. J. of CAV,* vol. 4, no. 1, pp. 133-148, 2021.

94. R. Dekate, S. V. P. Sharma and A. Reddi, "Model Based Design, Simulation and Experimental Validation of SCR Efficiency Model," *SAE Int. J. Advances & Curr. Prac. in Mobility,* vol. 4, no. 3, pp. 870-875, 2021.

95. Society of Automotive Engineers, "*J2812 Road Load Tire Model Validation Procedures for Dynamic Behavior,*" Society of Automotive Engineers Publishing, Warrendale, PA, 2023.

96. M. Norouzi and E. Nikolaidis, "Separable and Standard Monte Carlo Simulation of Linear Dynamic Systems Using Combined Approximations," *SAE Int. J. Commer. Veh.,* vol. 12, no. 2, pp. 103-114, 2019.

97. S. Chowdhury, S. Ravuri, N. Roy and Y. Mehta, "*Differential Case Imbalance Calculation Using Monte Carlo Simulation,*" Society of Automotive Engineers Publishing, Warrendale, PA, 2023.

98. Society of Automotive Engineers, *J2602 LIN Network for Vehicle Applications Conformance Test,* Warrendale, PA: Society of Automotive Engineers Publishing, 2021.

99. L. Zhang and Y. Xuke, "Accelerating In-Vehicle Network Intrusion Detection System Using Binarized Neural Network," *SAE Int. J. Advances & Curr. Prac. in Mobility,* vol. 4, no. 6, pp. 2037-2050, 2022.

100. Z. Yu and W. Klier, *Efficiency of Safety-Related Non-Functional Software Unit Test,* Warrendale, PA: Society of Automotive Engineers Publishing, 2013.

101. P. Yadav and H. Nalin, *Framework for Expressing Non-functional Requirements in System Engineering,* Warrendale, PA: Society of Automotive Engineers Publishing, 2022.

102. Society of Automotive Engineers, "*J1850 Verification Test Procedures,*" Society of Automotive Engineers Publishing, Warrendale, PA, 2021.

103. Society of Automotive Engineers, "*J1699-2 Test Cases for OBD-II Scan Tools and I/M Test Equipment,*" Society of Automotive Enginees Publishing, Warrendale, PA, 2017.

104. Society of Automotive Engineers, "*J1699-3 Vehicle OBD II Compliance Test Cases,*" Society of Automotive Engineers Publishing, Warrendale, PA, 2021.

105. Society of Automotive Engineers, "*J1699-4 OBD-II Communications Anomaly List,*" Society of Automotive Engineers Publishing, Warrendale, PA, 2021.

106. A. Koenig, M. Gutbrod, S. Hohmann and J. Ludwig, "*Bridging the Gap between Open Loop Tests and Statistical Validation for Highly Automated Driving,*" Society of Automotive Engineers, Warrendale, PA, 2017.

107. A. Taleb-Bendiab, "*EPR2020012 Unsettled Topics Concerning User Experience and Acceptance of Automated Vehicles*," Society of Automotive Engineers Publishing, Warrendale, PA, 2020.

108. Society of Automotive Engineers, "*J2944 Operational Definitions of Driving Performance Measures and Statistics*," Society of Automotive Engineers Publishing, Warrandale, PA, 2023.

109. Society of Automotive Engineers, "*Avoiding Electrical Overstress for Automotive Semiconductors by New Connecting Concepts*," Society of Automotive Engineers Publishing, Warrendale, PA, 2009.

110. A. A. Jauhri, "*Determine Thermal Fatigue Requirements for PEPS Determine Thermal Fatigue Requirements for PEPS Determine Thermal Fatigue Requirements for PEPS Determine Thermal Fatigue Requirements for PEPS Defined Reliability Requirements*," Society of Automotive Engineers Publishing, Warrendale, PA, 2019.

111. Society of Automotive Engineers, "*SAE/USCAR-2 Revision 8 Performance Specification for Automotive Electrical Connector Systems*," Society of Automotive Engineers Publishing, Warrendale, PA, 2022.

112. D.R. Bothe, *Reducing Process Variation Using the DOT Star Problem Solving Strategy*, Cedarburg, WI: Landmark Pub, 2002.

113. J. Dürrwang, J. Braun, M. Rumez, R. Kriesten and A. Pretschner, "Enhancement of Automotive Penetration Testing with Threat Analyses Results," *Transp. Cyber. & Privacy*, vol. 1, no. 2, pp. 92-95, 2018.

114. Society of Automotive Engineers, "*J3220 Lithium-Ion Cell Performance Testing*," Society of Automotive Engineers Publishing, Warrendale, PA, 2022.

115. Society of Automotive Engineers, "*J994 202306 Alarm - Backup - Electric Laboratory Performance Testing*," Society of Automotive Engineers Publishing, Warrendale, PA, 2023.

116. Society of Automotive Engineers, "*J2432 202111 Performance Testing of PK Section V-Ribbed Belts*," Society of Automotive Engineers Publishing, Warrendale, PA, 2021.

117. Automotive Industry Action Group, *Potential Failure Mode and Effects Analysis*, Southfield, MI: Automotive Industry Action Group Publishing, 2001.

118. International Software Testing Qualification Board, "International Software Testing Qualification Board," 12 5 2023. [Online]. Available: https://glossary.istqb.org/en_US/term/postcondition-4-2. [Accessed 5 7 2023].

119. International Software Testing Qualification Board, "International Software Testing Qualification Board," 10 5 2023. [Online]. Available: https://glossary.istqb.org/en_US/term/probe-effect-2-2. [Accessed 5 7 2023].

120. IEEE Standards, "IEEE Std 1008-1987 IEEE Standard for Software Unit Testing," in *Software Engineering Volume Two Process Standards*, New York, NY, The Institute of Electrical and Electronic Engineers, Inc., 1999.

121. International Software Testing Qualification Board, "International Software Testing Qualification Board," 2011. [Online]. Available: https://glossary.istqb.org/en_US/term/process-compliant-test-strateg. [Accessed 5 7 2023].

122. K. Pries and J. M. Quigley, "*Project Management of Complex and Embedded Systems, Ensuring Product Integrity and Program Quality*," Boca Raton, CRC Press, 2008, p. 4.

123. K. Pries and J. M. Quigley, "*Project Management of Complex and Embedded Systems, Ensuring Product Integrity and Program Quality*," Boca Raton, CRC Press, 2008, p. 256.

124. IEEE Standards, "IEEE Std 1058-1998 IEEE Standard for Software Project Management Plans," in *Software Engineering Volume Two Process Standards*, New York, NY, The Institute of Electrical and Electronic Engineers, Inc, 1999.

125. R. J. Shenoy and J. M. Quigley, *Project Management for Automotive Engineers: A Field Guide*, Warrendale, PA: Society of Automotive Engineers Publishing, 2016.

126. Automotive Industry Action Group, *Quality System Requirements QS-9000*, Detroit, MI: Automotive Industry Action Group Publishing, 1995.

127. Society of Automotive Engineers, "*J661 Brake Lining Quality Test Procedure*," Society of Automotive Engineers Publishing, Warrendale, PA, 2021.

128. Society of Automotive Engineers, "*J3119 Reliability, Maintainability, and Sustainability Terms and Definitions*," Society of Automotive Engineers Publishing, Warrandale, PA, 2020.

129. Society of Automotive Engineers, "*J3083 Reliability Prediction for Automotive Electronics Based on Field Return Data*," Society of Automotive Engineers Publishing, Warrendale, PA, 2017.

130. Society of Automotive Engineers, "*J2958_202002 Report on Unmanned Ground Vehicle Reliability*," Society of Automotive Engineers Publishing, Warrendale, PA, 2020.

131. Society of Automotive Engineers, "*J2816 Guide for Reliability Analysis Using the Physics-of-Failure Process*," Society of Automotive Engineers Publishing, Warrendale, PA, 2018.

132. Society of Automotive Engineers, "*J2940_202002 Use of Model Verification and Validation in Product Reliability and Confidence Assessments*," Society of Automotive Engineers Publishing, Warrendale, PA, 2020.

133. International Software Testing Qualification Board, "International Software Testing Qualification Board," 23 3 2023. [Online]. Available: https://glossary.istqb.org/en_US/term/requirements-based-testing-1-2. [Accessed 5 7 2023].

134. Society of Automotive Engineers, "*J1739 Potential Failure Mode and Effects Analysis (FMEA) Including Design FMEA, Supplemental FMEA-MSR, and Process FMEA*," SAE Publishing, Warrendale, PA, 2021.

135. International Software Testing Qualification Board, "International Software Testing Qualification Board," 10 5 2023. [Online]. Available: https://glossary.istqb.org/en_US/term/test-data-1-3. [Accessed 9 7 2023].

136. International Software Qualification Board, "International Software Qualification Board," 10 5 2023. [Online]. Available: https://glossary.istqb.org/en_US/term/risk-based-testing-2. [Accessed 6 7 2023].

137. C. Young, D. King and G. Siegmun, "Rollover and Near-Rollover Kinematics During," *Advances & Curr. Prac. in Mobility* , vol. 5, no. 1, pp. 84-96, 2022.

138. Society of Automotive Engineers, "*J2926_202110 Rollover Testing Methods*," Society of Automotive Engineers Publishing, Warrendale, PA, 2021.

139. Society of Automotive Engineers, "J2114," Society of Automotive Engineers Publishing, Warrendale, PA, 2014 Dolly Rollover Recommended Test Procedure.

140. D. E. Verbitsky, "Systemic Root Cause Early Failure Analysis during Accelerated Reliability Testing of Mass Produced Mobility Electronics," *SAE Int. J. Mater. Manf,* vol. 9, no. 3, pp. 534-544, 2016.

141. Society of Automotive Engineers, "*J3063 Active Safety Systems Terms and Definitions*," Society of Automotive Engineers Publishing, Warrendale, PA, 2023.

142. Society of Automotive Engineers, "*USCAR1-3 Salt Spray Testing and Evaluation Of Fastener Finishes*," Society of Automotive Engineers Publishing, Warrendale, PA, 2022.

143. International Software Testing Qualification Board, "International Software Testing Qualification Board," 2018. [Online]. Available: https://glossary.istqb.org/en_US/term/scalability-testing. [Accessed 6 7 2023].

144. International Software Testing Qualifications Board, "International Software Testing Qualifications Board," 2016. [Online]. Available: https://glossary.istqb.org/en_US/term/security-vulnerability. [Accessed 6 7 2023].

145. Society of Automotive Engineers, "*J2057 Class A Multiplexing Sensors*," Society of Automotive Engineers Publishing, Warrandale, PA, 2022.

146. J. M. Quigley and A. Gulve, *Modernizing Product Development Processes: Guide for Engineers*, Warrendale, PA: Society of Automotive Engineers Publishing, 2023.

147. P. Keller, *Six Sigma Demystified*, New York: McGraw Hill, 2005.

148. IEEE Standards, "IEEE std 1062-1998 Edition, IEEE Recommended Practice for Software Acquisition," in *Software Engineering Volume One Customer*

and Terminology Standards, New York, NY, The Institute of Electrical and Electronics Engineers, Inc, 1999.

149. IEEE Standards, "IEEE Std 12207.0-1996 Software Life Cycle Processes," in *Software Engineering Volume One Customer and Terminology Standards*, New York, NY, The Institute of Electrical and Electronics Engineers, Inc, 1999.

150. IEEE Standards, "IEEE Std 1074-1997, IEEE Standard for Developing Software Life Cycle Processes," in *Software Engineering Volume Two Process Standards*, Ney York, NY, The Institute of Electrical and Electronics Engineers, Inc., 1999.

151. IEEE Standards, "IEEE Std 1012-1998 IEEE Standard for Software Verification and Validation," in *Software Engineering Volume Two Process Standards*, New York, NY, The Institute of Electrical and Electronic Engineers, Inc., 1999.

152. IEEE Standards, "IEEE Std 730.1-1995 IEEE Guide for Softwar Quality Assurance Planning," in *Software Engineering Volume Two Process Standards*, New York, NY, The Institute of Electrical and Electronic Engineers, Inc., 1999.

153. IEEE Standards, "IEEE std 1061-1998, IEEE Standard for a Software Quality Metrics Methodology," in *Software Engineering Volume Three Product Standards*, New York, NY, The Institute of Electrical and Electronics Engineers, Inc., 1999.

154. International Software Testing Qualification Board, "International Software Testing Qualification Board," 2019. [Online]. Available: https://glossary.istqb.org/en_US/term/state-transition-testing-1-3. [Accessed 7 7 2023].

155. International Software Testing Qualification Board, "International Software Testing Qualification Board," 2019. [Online]. Available: https://glossary.istqb.org/en_US/term/state-transition-testing-1-3. [Accessed 7 7 2023].

156. International Software Testing Qualification Board, "International Software Testing Qualification Board," 10 10 2023. [Online]. Available: https://glossary.istqb.org/en_US/term/test-object-3-2. [Accessed 9 7 2023].

157. International Software Testing Qualifications Board, "International Software Testing Qualifications Board," 10 5 2023. [Online]. Available: https://glossary.istqb.org/en_US/term/statement-testing-2. [Accessed 8 7 2023].

158. Interrnational Software Testing Qualifications Board, "Interrnational Software Testing Qualifications Board," 10 5 2023. [Online]. Available: https://glossary.istqb.org/en_US/term/test-suite-1-3. [Accessed 9 7 2023].

159. International Software Testing Qualification Board, "International Software Testing Qualification Board," 10 5 2023. [Online]. Available: https://glossary.istqb.org/en_US/term/static-testing-6-4. [Accessed 9 7 2023].

160. International Software Testing Qualification Board, "International Software Testing Qualification Board," 10 5 2023. [Online]. Available: https://glossary.istqb.org/en_US/term/test-procedure. [Accessed 9 7 2023].

161. International Software Test Qualification Board, "International Software Test Qualification Board," 10 5 2023. [Online]. Available: https://glossary.istqb.org/en_US/term/test-case-2. [Accessed 9 7 2023].

162. International Software Testing Qualification Board, "International Software Testing Qualification Board," 10 5 2023. [Online]. Available: https://glossary.istqb.org/en_US/term/static-analysis-2. [Accessed 9 7 2023].

163. Society of Automotive Engineers, "*J2130 Identification of Self-Propelled Sweepers and Cleaning Equipment Part 1 and - Machines with a Gross Vehicle Mass Greater than 5000 kg*," Society of Automotive Engineers Publishing, Warrendale, PA, 2021.

164. International Software Testing Qualification Board, "International Software Testing Qualification Board," 10 5 2023. [Online]. Available: https://glossary.istqb.org/en_US/term/stub. [Accessed 9 7 2023].

165. International Software Testing Qualification Board, "International Software Testing Qualification Board," 23 4 2023. [Online]. Available: https://glossary.istqb.org/en_US/term/test-driven-development-2-1. [Accessed 9 7 2023].

166. International Software Testing Qualification Board, "International Software Testing Qualification Board," 10 5 2023. [Online]. Available: https://glossary.istqb.org/en_US/term/test-harness-3-2. [Accessed 9 7 2023].

167. International Software Testing Qualification Board, "International Software Testing Qualification Board," 10 5 2023. [Online]. Available: https://glossary.istqb.org/en_US/term/test-item-2. [Accessed 9 7 2023].

168. International Software Testing Qualification Board, "International Software Testing Qualification Board," 10 5 2023. [Online]. Available: https://glossary.istqb.org/en_US/term/test-monitoring-3-1. [Accessed 9 7 2023].

169. International Software Testing Qualification Board, "International Software Testing Qualification Board," 23 3 2023. [Online]. Available: https://glossary.istqb.org/en_US/term/test-point-analysis-3. [Accessed 9 7 2023].

170. The TMMi Foundation, "*Test Maturity Model integration (TMMi®) Guidelines for Test Process Improvement Release 1.3*," TMMi Foundation, United Kingdom, 2022.

171. Society of Automotive Engineers, "*J2723 Engine Power Test Code - Engine Power and Torque Certification*," Society of Automotive Engineers Publishing, Warrendale, PA, 2021.

172. Society of Automotive Engineers, "*J1349 Engine Power Test Code - Spark Ignition and Compression Ignition - As Installed Net Power Rating*," Society of Automotive Engineers Publishing, Warrendale, PA, 2011.

173. Society of Automotive Engineers, "*J365 Method of Testing Resistance to Scuffing of Trim Materials*," Society of Automotive Engineers Publishing, Warrendale, PA, 2020.

174. Society of Automotive Engineers, "*J1211 Handbook for Robustness Validation of Automotive Electrical/Electronic Modules*," Society of Automotive Engineers Publishing, Warrendale, PA, 2012.

175. Department of Defense, *Department of Defense Handbook Configuration Management Guidance*, Washington: Department of Defense, 2020.

176. Society of Automotive Engineers, "*J1742 Connections for High Voltage On-Board Vehicle Electrical Wiring Harnesses - Test Methods and General Performance Requirements*," Society of Automotive Engineers Publishing, Warrendale, PA, 2022.

177. Society of Automotive Engineers, "*J541 Voltage Drop for Starting Motor Circuits*," Society of Automoive Engineers Publishing, Warrendale, PA, 2013.

178. Society of Automotive Engineers, "*J3226 Voltage Regulation and Limits*," Society of Automotive Engineers Publishing, Warrendale, PA, 2022.

179. "*MIL-DTL-24308H Connectors, Electric, Rectangular, Nonenvironmental, Miniature, Polarized Shell, Rack and Panel,General Specification For*," DLA Land and Maritime, Columbus, OH, 2017.

180. Society of Automotive Engineers, "*J1533 Circuit Breakers*," Societiy of Automotive Engineers Publishing, Warrandale, PA, 2022.

181. A. Jauhri, "*Determine Thermal Fatigue Requirements for PEPS Antenna Copper Wire over Vehicle Lifetime with Defined Reliability Requirements*," Society of Automotive Engineers Publishing, Warrendale, PA, 2019.

Appendix A: Units and Numbers

Use of SI (Metric Units of Measure in SAE Technical Papers)

The long-term goal for SAE is international communication with minimal effort and confusion. Therefore, the use of SI units in all technical publications and presentations is preferred. The Society will strive toward universal usage of SI units and will encourage their use whenever appropriate.

However, the Society also recognizes that sectors of the mobility market do not yet use SI units because of tradition, regulatory language, or other reasons. Mandating the use of SI units in these cases will impede rather than facilitate technical communication. Therefore, it is the policy to allow non-SI units and dual dimensioning where communication will be enhanced. This shall not be viewed as an avenue to circumvent the long-term goal of 100% SI usage.

Instructions on SAE-approved techniques for conversion of units are contained in "SAE Recommended Practices, Rules for SAE Use of SI (METRIC) Units—TSB003." Copies of TSB003 can be obtained from SAE headquarters (Table A.1).

Although what follows represents a change to the current policy, it is not a change to the SAE Board of Directors' Policy since it falls within the scope of the words, "where a conflicting industry practice exists." Dual (metric/US customary) units for the following vehicle characteristics may be considered where communication will be enhanced.

TABLE A.1 Metric and US customary units.

Vehicle characteristic	Metric units	US customary units
Volume, engine displacement	Liters, L, or cubic cm, cm^3	Cubic inches, $in.^3$
Liquid volume	Liters, L	Pints/quarts/gallons
Engine power	Kilowatts, kW	Brake horsepower, bhp
Engine torque	Newton meters, N - m	Foot-pounds, lb-ft
Mass	Kilograms, kg	Slugs, lb-s^2/ft
Pressure, stress	Kilopascal, kPa	Pounds per square inch, psi
Temperature	Degrees Celsius, °C	Degrees Fahrenheit, °F
Area	Square cm, cm^2	Square inches, $in.^2$
Linear dimensions	Millimeters, mm, meters, m, or kilometers, km	Inches, in., feet, ft, miles, mi
Spring rates	Newtons per millimeter, N/mm	Pounds per inch, lb/in.
Speed	Kilometers per hour, km/h or kph	Miles per hour, mph
Fuel economy	Kilometers per liter, km/L or kmpL	Miles per gallon, mpg
Force	Newtons, N	Pounds, lb
Acceleration	Kilometers per second per second, km/s^2, g	Feet per second per second, ft/s^2, g

Numbers, Significant Figures, and Rounding

Significant Figures

In all branches of science and technology, numbers are used to express values, that is, levels or amounts of physical quantities. It is important to state numbers appropriately so that they properly convey the intended information. The number of significant figures contained in a stated number reflects the accuracy to which that quantity is known. For example, suppose the speed of a vehicle is reported as 21 m/s (69 ft/s). Is 21 m/s different from 21.0 m/s? According to the rules of significant figures, yes, but in practice, it may or may not. Could the number 21 m/s imply 20.9 m/s or less, or could it imply 21.1 m/s or greater? It could, but such implications or interpretations must be determined from context, not the number 21 itself. Answers to some of these questions are related to the topic of uncertainty (covered in Chapter 1). To properly quantify and communicate a physical measurement or property, it should be stated as a reference value plus and minus an uncertainty. For example, a speed stated as $v = 21.0 \pm 0.6$ m/s clearly is meant to be between 20.4 m/s and 21.6 m/s. This is one of the ways of estimating and revealing the uncertainty of results. But the basic rules of using significant figures and rounding must be understood before uncertainty can be expressed. Some of the rules for handling and interpreting the significance of numbers are covered in this Appendix. Note that the terms *significant figures* and *significant digits* are used synonymously.

The number of significant figures in a number is defined in the following way [A.1, A.2]:

1. The leftmost nonzero digit of a number is the most significant digit.
2. If there is no decimal point, the rightmost nonzero digit is the least significant digit.
3. If there is a decimal point, the rightmost digit is the least significant digit, even if it is a zero.
4. All digits, from the least to the most significant, are counted as significant.

So, for example, 2.610. and 2,498 have four significant digits each, whereas 0.125 and 728,000 have three significant digits. The following numbers each have five significant digits: 1000.0, 1206.5, 12,065,000, and 0.00012065. Unless it is stated to be exact, the speed of 21 m/s has two significant figures. If it is *exact*, then 21 is equivalent to 21.0000 …, with an unlimited number of zeros. Each of the speeds 20.4 and 21.6 has three significant figures.

When numbers are very large or very small, it is convenient to express them in *scientific notation*. To use scientific notation, a decimal point is placed immediately after the leftmost significant digit and the number is given a suffix of 10 raised to a power n. The value of n is positive or negative. If the magnitude (disregarding the sign) of the stated number is less than 1, then $n < 0$. If the stated number is greater than 10, $n > 0$. If the stated number is between 1 and 10, $n = 0$. The value of n is the power of 10 that returns the number in scientific notation to its original value. For example, 0.0000687 becomes 6.87×10^{-5} and 12,360,000 becomes 1.236×10^{7}. Note that the number of significant digits does not change when converting to or from scientific notation.

Rounding of Numbers

After completing calculations or when listing the results of measurements, it usually is necessary to round numbers to a lesser number of significant figures by discarding digits. Three possibilities can arise; these are:

1. *The leftmost discarded digit is less than 5.* When rounding such numbers, the last digit retained should remain unchanged. For example, if 3.46325 is to be rounded to four digits, the digits 2 and 5 would be discarded and 3.463 remains.
2. *The leftmost discarded digit is greater than 5 or it is a 5 followed by at least one digit other than 0.* In such cases, the last figure retained should

be increased by one. For example, if rounded to four digits, 8.37652 would become 8.377; if rounded to three digits, it would be 8.38.

3. *The leftmost discarded digit is a 5, followed only by zeros or no other numbers.* Here, the last digit retained should be rounded up if it is an odd number, but no adjustment made if it is an even number. For example, 21.165, when rounded to four significant digits, becomes 21.16. The number 21.155 would likewise round to the same value, 21.16.

A reason for this last rule [A.2] is to avoid systematic errors that otherwise would be introduced into the average of a group of such numbers. Not all computer software follows this rule, however,[1] and when rounding for purposes of reporting results of measurements and/or calculations, the even-odd rule is not critical.

Consistency of Significant Figures When Adding and Subtracting

When adding and subtracting numbers, proper determination of the number of significant figures is stated as a rule [A.1]. The rule is *the answer shall contain no significant digits farther to the right than occurs in the number with the least significant digits.* The simplest way of following this rule is first to add or subtract the numbers using all of the stated significant figures[2] followed by rounding of the final answer. For example, consider the addition of the three numbers, 964,532, 317,880, and 563,000. These have six, five, and three significant figures, respectively. The sum by direct addition is 1,845,412. The answer then is adjusted, or rounded, to conform to the number with the least significant figures (563,000 with three), giving the final result, 1,845,000. This number has no more zero digits to the right of the comma than does 563,000. Now consider the sum of the three numbers, 964,532, –317,880, and –563,000; the direct result is 83,652. As above, this must be made to conform with the significant figures of 563,000 by using the rounding rule and is 84,000.

In the last example, the concept being conveyed is that the number 563,000 is "indefinite" to the right of the "3" digit. It is not known if 563,000 could really mean 562,684 or 563,121 or other values because 563,000, itself, may have been obtained by rounding. If it had been stated as 563,000.0, then everything would be different (since 563,000.0 would have seven significant figures and 317,880 would then have the least significant digits of the three numbers to be added in the above example).

---

1 The reader may wish to try such an example in their favorite software.

2 ASTM SI-10 suggests first rounding each individual number to one significant figure greater than the least *before* adding or subtracting and then rounding the final answer. Though this may be better, it is not the way most computer software operates. Rounding after summing typically gives the same result.

Consistency of Significant Figures When Multiplying and Dividing

ASTM SI-10 [A.1] states a rule for multiplying and dividing as the product or quotient shall contain no more significant digits than are contained in the number with the fewest significant digits. For example, consider the product, $125.64 \times 829.4 \times 1.25$, of the three numbers with five, four, and three significant digits, respectively. The answer from straightforward multiplication is 130,257.27. After rounding to three significant figures, the proper end result of the multiplication is 130,000. Note that the answer, 130,000, by itself appears to have only two significant figures. This illustrates that ambiguities sometimes can arise when determining significant figures and that the amount of significant figures of a number may need to be found from context. A way of resolving such ambiguities is to express results of rounding in scientific notation. In this case the result would be 1.30×10^5.

Other Forms of Number Manipulation

Not all calculations are done with addition, subtraction, multiplication, and division. There is the taking of roots, logarithms, trigonometric functions, etc. In addition, sometimes strict adherence to rounding rules can produce paradoxical or impractical results (see the following example). So more general rules are needed. In summary, two very general but some *practical* rules are recommended:

1. In rounding of numbers and conversion of units, retain a number of significant digits such that accuracy and precision are neither sacrificed nor exaggerated.
2. When making and reporting calculations, continually carry all of the significant figures of a calculating device without rounding intermediate values, and round only the final answer.
3. Unit conversion should precede rounding.
4. Whenever possible, explicitly state the uncertainty of the results of measurements and calculations.

Example A.1

Suppose a vehicle skids to a stop over a distance of $d = 33.9$ m from an initial speed, v, on a pavement with a uniform frictional drag coefficient of $f = 0.7 \pm 0.1$. Use the minimum and maximum values of f and Equation (1.1) to calculate bounds on the initial speed. Convert the results to US customary units of feet per second (ft/s).

Solution The lower value of speed for $f = 0.6$ is

$$v = \sqrt{2fgd} = \sqrt{2 \times 06 \times 9.806650 \times 33.9} = 19.973345\ldots$$

Similarly, the initial speed for $f = 0.8$ is

$$v = \sqrt{2 \times 08 \times 9.806650 \times 33.9} = 23.063233\ldots$$

The frictional drag coefficient and its uncertainty have the fewest number of significant figures of the input values. According to the rules, the final results should be rounded to one significant figure. Rounding 19.973345... to a single significant digit gives a speed of $v = 20$ m/s. Rounding 23.063235... to a single significant digit also gives a speed of $\nu = 20$ m/s. Both upper and lower bounds result in the same speed, $v = 20$ m/s. Clearly, the result is an exaggeration of precision. Consider now another approach.

The variation of $f = \pm\ 0.1$ is another way of saying that because of uncertainty, f can take on any value between 0.6 and 0.8.[3] From the above discussion of significant figures and rounding, a point of view can be taken that the lower value, 0.6, for example, could be the result of rounding to one significant figure of any number from 0.55+ to 0.65− (such as 0.551, 0.642, etc.). Similarly, the upper value, 0.8, could be viewed as the result of rounding of any number from 0.75+ to 0.85− (such as 0.751, 0.842, etc.). So the full range of values of the frictional drag coefficient corresponding to the stated uncertainty and from the concepts of significant figures is $0.55 \le f \le 0.85$. At this point the calculations are performed *as if* all numbers are exactly giving a speed range of $19.123022\ldots \le v \le 23.773036\ldots$m/s. Since rounding to one significant figure here produces an exaggeration of precision (as above), rounding is done to an additional significant figure. Consequently, the final result is stated as $19 \le v \le 24$ m/s, or $v = 19.5 \pm 2.5$ m/s. Precision no longer is exaggerated. An initial ±14% variation (0.7 ± 0.1) becomes a 12% variation of v(19.5 ± 2.5) through the use of Equation (1.1).

Finally, the speed is to be converted to the US customary units of feet per second (ft/s). The proper conversion factor is 1 ft = 0.3048 m (this is an *exact* conversion; see the following unit conversion table). Unit conversions should be done before rounding, so $19.123022\ldots \le v \le 23.773036\ldots$m/s becomes $62.739573\ldots \le v \le 77.995525\ldots$ft/s. Rounding again to one significant figure gives the same result, 70 ft, so another significant figure is acceptable, giving $63 \le v \le 78$ ft/s, or $v = 70.5 \pm 7.5$ ft/s.

[3] Note that there is no implication of the likelihood of any of the values within this range.

Another consideration that must be kept in mind when rounding is the use or purpose of the results, for example, if the speed calculated in the last example is to be compared to a speed limit, say, 25 m/s. Rounding to a number of significant digits to the right of the decimal point is superfluous. The result $19 \le v \le 24$ m/s is satisfactory to conclude that the calculated speed is less than the speed limit. Instead, suppose that the calculated speed is a measure of vehicle braking performance and is to be compared to a governmental regulation stated to three significant figures. Rounding to an additional significant figure leads to an exaggeration of accuracy. To compare the speed to such a regulation requires a more accurate value of friction, stated at least to two significant figures.

Unit Conversions for Common Units

Factors in boldface are exact. When options exist, units in the first column printed in *italics* are preferred by the National Institute for Standards and Technology [A.3].

To convert from	To	Multiply by	
Acre (based on US survey foot)	Square meter (m^2)	4.046 873	E+03
Acre foot (based on US survey foot)	Cubic meter (m^3)	1.233 489	E+03
Ampere hour (A·h)	Coulomb (C)	**3.6**	**E+03**
Atmosphere, standard (atm)	Pascal (Pa)	**1.013 25**	**E+05**
Atmosphere, standard (atm)	Kilopascal (kPa)	**1.013 25**	**E+02**
Atmosphere, technical (at)	Pascal (Pa)	**9.806 65**	**E+04**
Atmosphere, technical (at)	Kilopascal (kPa)	**9.806 65**	**E+01**
Bar (bar)	Pascal (Pa)	**1.0**	**E+05**
Bar (bar)	Kilopascal (kPa)	**1.0**	**E+02**
Barn (b)	Square meter (m^2)	**1.0**	**E−28**
Barrel [for petroleum, 42 gallons (US)] (bbl)	Cubic meter (m^3)	1.589 873	E−01
Barrel [for petroleum, 42 gallons (US)] (bbl)	Liter (L)	1.589 873	E+02
British thermal unit (mean) (Btu)	Joule (J)	1.055 87	E+03
Bushel (US) (bu)	Cubic meter (m^3)	3.523 907	E−02
Bushel (US) (bu)	Liter (L)	3.523 907	E+01
Calorie (cal) (mean)	Joule (J)	4.190 02	E+00
Candela per square inch (cd/in.^{2})	Candela per square meter (cd/m^2)	1.550 003	E+03
Carat, metric	Kilogram (kg)	**2.0**	**E−04**
Carat, metric	Gram (g)	**2.0**	**E−01**

To convert from	To	Multiply by	
Centimeter of mercury (0°C)	Pascal (Pa)	1.333 22	E+03
Centimeter of water (4°C)	Pascal (Pa)	9.806 38	E+01
Centimeter of water, conventional (cm H_2O)	Pascal (Pa)	**9.806 65**	**E+01**
Centipoise (cP)	Pascal second (Pa·s)	**1.0**	**E−03**
Centistokes (cSt)	Meter squared per second (m^2/s)	**1.0**	**E−06**
Chain (based on US survey foot) (ch)	Meter (m)	2.011 684	E+01
Circular mil	Square meter (m^2)	5.067 075	E−10
Cord (128 ft^3)	Cubic meter (m^3)	3.624 556	E+00
Cubic foot (ft^3)	Cubic meter (m^3)	2.831 685	E−02
Cubic inch ($in.^3$)	Cubic meter (m^3)	1.638 706	E−05
Cubic mile (mi^3)	Cubic meter (m^3)	4.168 182	E+09
Cubic yard (yd^3)	Cubic meter (m^3)	7.645 549	E−01
Cup (US)	Cubic meter (m^3)	2.365 882	E−04
Cup (US)	Liter (L)	2.365 882	E−01
Day (d)	Second (s)	**8.64**	**E+04**
Day (sidereal)	Second (s)	8.616 409	E+04
Degree (angle) (°)	Radian (rad)	1.745 329	E−02
Degree Celsius (temperature) (°C)	Kelvin (K)	K = °C + **273.15**	
Degree Celsius (temperature interval) (°C)	Kelvin (K)	**1.0**	**E+00**
Degree centigrade (temperature)	Degree Celsius (°C)	°C = deg. cent.	
Degree centigrade (temperature interval)	Degree Celsius (°C)	1.0	E+00
Degree Fahrenheit (temperature) (°F)	Degree Celsius (°C)	°C = (°F − **32**)/**1.8**	
Degree Fahrenheit (temperature) (°F)	Kelvin (K)	K = (°F + **459.67)/1.8**	
Degree Fahrenheit (temperature interval) (°F)	Degree Celsius (°C)	5.555 556	E−01
Degree Fahrenheit (temperature interval) (°F)	Kelvin (K)	5.555 556	E−01
Degree Rankine (°R)	Kelvin (K)	K = (°R)/**1.8**	
Degree Rankine (temperature interval) (°R)	Kelvin (K)	5.555 556	E−01
Denier	Kilogram per meter (kg/m)	1.111 111	E−07
Dyne (dyn)	Newton (N)	**1.0**	**E−05**
Dyne centimeter (dyn·cm)	Newton meter (N·m)	**1.0**	**E−07**

To convert from	To	Multiply by	
Dyne per square centimeter (dyn/cm^2)	Pascal (Pa)	**1.0**	**E−01**
Erg (erg)	Joule (J)	**1.0**	**E−07**
Erg per second (erg/s)	Watt (W)	**1.0**	**E−07**
Fathom (based on US survey foot)	Meter (m)	1.828 804	E+00
Fluid ounce (US) (fl oz)	Cubic meter (m^3)	2.957 353	E−05
Fluid ounce (US) (fl oz)	Milliliter (mL)	2.957 353	E+01
Foot (ft)	Meter (m)	**3.048**	**E−01**
Foot (US survey) (ft)	Meter (m)	3.048 006	E−01
Footcandle	Lux (lx)	1.076 391	E+01
Footlambert	Candela per square meter (cd/m^2)	3.426 259	E+00
Foot of water, conventional (ftH_2O)	Pascal (Pa)	2.989 067	E+03
Foot of water, conventional (ftH_2O)	Kilopascal (kPa)	2.989 067	E+00
Foot per hour (ft/h)	Meter per second (m/s)	8.466 667	E−05
Foot per minute (ft/min)	Meter per second (m/s)	**5.08**	**E−03**
Foot per second (ft/s)	Meter per second (m/s)	**3.048**	**E−01**
Foot per second squared (ft/s^2)[4]	Meter per second squared (m/s^2)[2]	**3.048**	**E−01**
Foot-poundal	Joule (J)	4.214 011	E−02
Foot pound-force (ft · lbf)	Joule (J)	1.355 818	E+00
Foot pound-force per hour (ft · lbf/h)	Watt (W)	3.766 161	E−04
Foot pound-force per minute (ft · lbf/min)	Watt (W)	2.259 697	E−02
Foot pound-force per second (ft · lbf/s)	Watt (W)	1.355 818	E+00
Gal (Gal)	Meter per second squared (m/s^2)	**1.0**	**E−02**
Gallon [Canadian and UK (Imperial)] (gal)	Cubic meter (m^3)	**4.546 09**	**E−03**
Gallon [Canadian and UK (Imperial)] (gal)	Liter (L)	**4.546 09**	**E+00**
Gallon (US) (gal)	Cubic meter (m^3)	3.785 412	E−03
Gallon (US) (gal)	Liter (L)	3.785 412	E+00
Gallon (US) per day (gal/d)	Cubic meter per second (m^3/s)	4.381 264	E−08

[4] Standard value of free-fall acceleration is $g = 9.80665\ m/s^2$.

To convert from	To	Multiply by	
Gallon (US) per day (gal/d)	Liter per second (L/s)	4.381 264	E−05
Gallon (US) per horsepower-hour [gal/(hp·h)]	Cubic meter per joule (m^3/J)	1.410 089	E−09
Gallon (US) per horsepower-hour [gal/(hp·h)]	Liter per joule (L/J)	1.410 089	E−06
Gallon (US) per minute (gpm) (gal/min)	Cubic meter per second (m^3/s)	6.309 020	E−05
Gallon (US) per minute (gpm) (gal/min)	Liter per second (L/s)	6.309 020	E−02
Grain (gr)	Kilogram (kg)	**6.479 891**	**E−05**
Grain (gr)	Milligram (mg)	**6.479 891**	**E+01**
Grain per gallon (US) (gr/gal)	Kilogram per cubic meter (kg/m^3)	1.711 806	E−02
Grain per gallon (US) (gr/gal)	Milligram per liter (mg/L)	1.711 806	E+01
Gram-force per square centimeter (gf/cm^2)	Pascal (Pa)	**9.806 65**	**E+01**
Gram per cubic centimeter (g/cm^3)	Kilogram per cubic meter (kg/m^3)	**1.0**	**E+03**
Hectare (ha)	Square meter (m^2)	**1.0**	**E+04**
Horsepower (550 ft·lbf/s) (hp)	Watt (W)	7.456 999	E+02
Horsepower (boiler)	Watt (W)	9.809 50	E+03
Horsepower (electric)	Watt (W)	**7.46**	**E+02**
Horsepower (metric)	Watt (W)	7.354 988	E+02
Horsepower (UK)	Watt (W)	7.4570	E+02
Horsepower (water)	Watt (W)	7.460 43	E+02
Hour (h)	Second (s)	**3.6**	**E+03**
Hour (sidereal)	Second (s)	3.590 170	E+03
Hundredweight (long, 112 lb)	Kilogram (kg)	5.080 235	E+01
Hundredweight (short, 100 lb)	Kilogram (kg)	4.535 924	E+01
Inch (in.)	Meter (m)	**2.54**	**E−02**
Inch (in.)	Centimeter (cm)	**2.54**	**E+00**
Inch of mercury, conventional (in. Hg)	Pascal (Pa)	3.386 389	E+03
Inch of mercury, conventional (in. Hg)	Kilopascal (kPa)	3.386 389	E+00
Inch of water, conventional (in. H_2O)	Pascal (Pa)	2.490 889	E+02
Kelvin (K)	Degree Celsius (°C)	t/°C = T/K − **273.15**	
Kilocalorie (mean) (kcal)	Joule (J)	4.190 02	E+03
Kilogram-force (kgf)	Newton (N)	**9.806 65**	**E+00**

To convert from	To	Multiply by	
Kilogram-force meter (kgf·m)	Newton meter (N·m)	**9.806 65**	**E+00**
Kilogram-force per square centimeter (kgf/cm^2)	Kilopascal (kPa)	**9.806 65**	**E+01**
Kilogram-force per square meter (kgf/m^2)	Pascal (Pa)	**9.806 65**	**E+00**
Kilometer per hour (km/h)	Meter per second (m/s)	2.777 778	E−01
Kilopond (kilogram-force) (kp)	Newton (N)	**9.806 65**	**E+00**
Kilowatt hour (kW·h)	Joule (J)	**3.6**	**E+06**
Kilowatt hour (kW·h)	Megajoule (MJ)	**3.6**	**E+00**
Kip (1 kip = 1000 lbf)	Newton (N)	4.448 222	E+03
Kip (1 kip = 1000 lbf)	Kilonewton (kN)	4.448 222	E+00
Kip per square inch (ksi) (kip/in.^{2})	Pascal (Pa)	6.894 757	E+06
Kip per square inch (ksi) (kip/in.^{2})	Kilopascal (kPa)	6.894 757	E+03
Knot (nautical mile per hour)	Meter per second (m/s)	5.144 444	E−01
Lambert	Candela per square meter (cd/m^2)	3.183 099	E+03
Light-year (l.y.)	Meter (m)	9.460 73	E+15
Liter (L)	Cubic meter (m^3)	**1.0**	**E−03**
Lumen per square foot (lm/ft^2)	Lux (lx)	1.076 391	E+01
Microinch	Meter (m)	**2.54**	**E−08**
Microinch	Micrometer (μm)	**2.54**	**E−02**
Micron (μ)	Meter (m)	**1.0**	**E−06**
Micron (μ)	Micrometer (μm)	**1.0**	**E+00**
Mil (0.001 in.)	Meter (m)	**2.54**	**E−05**
Mil (0.001 in.)	Millimeter (mm)	**2.54**	**E−02**
Mil (angle)	Radian (rad)	9.817 477	E−04
Mil (angle)	Degree (°)	**5.625**	**E−02**
Mile (mi)	Meter (m)	**1.609 344**	**E+03**
Mile (mi)	Kilometer (km)	**1.609 344**	**E+00**
Mile (based on US survey foot) (mi)	Meter (m)	1.609 347	E+03
Mile (based on US survey foot) (mi)	Kilometer (km)	1.609 347	E+00
Mile, nautical	Meter (m)	**1.852**	**E+03**
Mile per gallon (US) (mpg) (mi/gal)	Meter per cubic meter (m/m^3)	4.251 437	E+05

To convert from	To	Multiply by	
Mile per gallon (US) (mpg) (mi/gal)	Kilometer per liter (km/L)	4.251 437	E−01
Mile per gallon (US) (mpg) (mi/gal)	Liter per 100 kilometers (L/100 km)	Divide 235.215 by number of miles per gallon	
Mile per hour (mi/h)	Meter per second (m/s)	**4.4704**	**E−01**
Mile per hour (mi/h)	Kilometer per hour (km/h)	**1.609 344**	**E+00**
Mile per minute (mi/min)	Meter per second (m/s)	**2.682 24**	**E+01**
Mile per second (mi/s)	Meter per second (m/s)	**1.609 344**	**E+03**
Millibar (mbar)	Pascal (Pa)	**1.0**	**E+02**
Millibar (mbar)	Kilopascal (kPa)	**1.0**	**E−01**
Millimeter of mercury, conventional (mmHg)	Pascal (Pa)	1.333 224	E+02
Millimeter of water, conventional (mm H_2O)	Pascal (Pa)	**9.806 65**	**E+00**
Minute (angle) (N)	Radian (rad)	2.908 882	E−04
Minute (min)	Second (s)	**6.0**	**E+01**
Minute (sidereal)	Second (s)	5.983 617	E+01
Ounce (avoirdupois) (oz)	Kilogram (kg)	2.834 952	E−02
Ounce (avoirdupois) (oz)	Gram (g)	2.834 952	E+01
Ounce (troy or apothecary) (oz)	Kilogram (kg)	3.110 348	E−02
Ounce (troy or apothecary) (oz)	Gram (g)	3.110 348	E+01
ounce [Canadian and UK fluid (Imperial)] (fl oz)	Cubic meter (m^3)	2.841 306	E−05
Ounce [Canadian and UK fluid (Imperial)] (fl oz)	Milliliter (mL)	2.841 306	E+01
Ounce (US fluid) (fl oz)	Cubic meter (m^3)	2.957 353	E−05
Ounce (US fluid) (fl oz)	Milliliter (mL)	2.957 353	E+01
Ounce (avoirdupois)-force (ozf)	Newton (N)	2.780 139	E−01
Ounce (avoirdupois)-force inch (ozf·in.)	Newton meter (N·m)	7.061 552	E−03
Ounce (avoirdupois)-force inch (ozf·in.)	Millinewton meter (mN·m)	7.061 552	E+00
Ounce (avoirdupois) per cubic inch (oz/in.^{3})	Kilogram per cubic meter (kg/m^3)	1.729 994	E+03
Peck (US) (pk)	Cubic meter (m^3)	8.809 768	E−03
Peck (US) (pk)	Liter (L)	8.809 768	E+00
Pennyweight (dwt)	Kilogram (kg)	1.555 174	E−03
Pennyweight (dwt)	Gram (g)	1.555 174	E+00

To convert from	To	Multiply by	
Pica (computer) (1/6 in.)	Meter (m)	4.233 333	E−03
Pica (computer) (1/6 in.)	Millimeter (mm)	4.233 333	E+00
Pica (printer's)	Meter (m)	4.217 518	E−03
Pica (printer's)	Millimeter (mm)	4.217 518	E+00
Pint (US dry) (dry pt)	Cubic meter (m^3)	5.506 105	E−04
Pint (US dry) (dry pt)	Liter (L)	5.506 105	E−01
Pint (US liquid) (liq pt)	Cubic meter (m^3)	4.731 765	E−04
Pint (US liquid) (liq pt)	Liter (L)	4.731 765	E−01
Point (computer) (1/72 in.)	Meter (m)	3.527 778	E−04
Point (computer) (1/72 in.)	Millimeter (mm)	3.527 778	E−01
Point (printer's)	Meter (m)	3.514 598	E−04
Point (printer's)	Millimeter (mm)	3.514 598	E−01
Poise (P)	Pascal second (Pa · s)	**1.0**	**E−01**
Pound (avoirdupois) (lb)	Kilogram (kg)	4.535 924	E−01
Pound (troy or apothecary) (lb)	Kilogram (kg)	3.732 417	E−01
Poundal	Newton (N)	1.382 550	E−01
Poundal per square foot	Pascal (Pa)	1.488 164	E+00
Poundal second per square foot	Pascal second (Pa · s)	1.488 164	E+00
Pound foot squared (lb · ft^2)	Kilogram meter squared (kg · m^2)	4.214 011	E−02
Pound-force (lbf)[5]	Newton (N)	4.448 222	E+00
Pound-force foot (lbf · ft)	Newton meter (N · m)	1.355 818	E+00
Pound-force foot per inch (lbf · ft/in.)	Newton meter per meter (N · m/m)	5.337 866	E+01
Pound-force inch (lbf · in.)	Newton meter (N · m)	1.129 848	E−01
Pound-force inch per inch (lbf · in./in.)	Newton meter per meter (N · m/m)	4.448 222	E+00
Pound-force per foot (lbf/ft)	Newton per meter (N/m)	1.459 390	E+01
Pound-force per inch (lbf/in.)	Newton per meter (N/m)	1.751 268	E+02
Pound-force per pound (lbf/lb) (thrust-to-mass ratio)	Newton per kilogram (N/kg)	**9.806 65**	**E+00**
Pound-force per square foot (lbf/ft^2)	Pascal (Pa)	4.788 026	E+01
Pound-force per square inch (psi) (lbf/$in.^2$)	Pascal (Pa)	6.894 757	E+03

[5] If the local value of the acceleration of free fall is taken as the standard value g = 9.80665 m/s^2, then the exact conversion factor is 4.448 221 615 260 5 E + 00.

To convert from	To	Multiply by	
Pound-force per square inch (psi) ($lbf/in.^2$)	Kilopascal (kPa)	6.894 757	E+00
Pound-force second per square foot ($lbf \cdot s/ft^2$)	Pascal second ($Pa \cdot s$)	4.788 026	E+01
Pound-force second per square inch ($lbf \cdot s/in.^2$)	Pascal second ($Pa \cdot s$)	6.894 757	E+03
Pound inch squared ($lb \cdot in.^2$)	Kilogram meter squared ($kg \cdot m^2$)	2.926 397	E−04
Pound per cubic foot (lb/ft^3)	Kilogram per cubic meter (kg/m^3)	1.601 846	E+01
Pound per cubic inch ($lb/in.^3$)	Kilogram per cubic meter (kg/m^3)	2.767 990	E+04
Pound per cubic yard (lb/yd^3)	Kilogram per cubic meter (kg/m^3)	5.932 764	E−01
Pound per foot (lb/ft)	Kilogram per meter (kg/m)	1.488 164	E+00
Pound per foot hour [$lb/(ft \cdot h)$]	Pascal second ($Pa \cdot s$)	4.133 789	E−04
Pound per foot second [$lb/(ft \cdot s)$]	Pascal second ($Pa \cdot s$)	1.488 164	E+00
Pound per gallon [Canadian and UK (Imperial)] (lb/gal)	Kilogram per cubic meter (kg/m^3)	9.977 637	E+01
Pound per gallon [Canadian and UK (Imperial)] (lb/gal)	Kilogram per liter (kg/L)	9.977 637	E−02
Pound per gallon (US) (lb/gal)	Kilogram per cubic meter (kg/m^3)	1.198 264	E+02
Pound per gallon (US) (lb/gal)	Kilogram per liter (kg/L)	1.198 264	E−01
Pound per horsepower-hour [$lb/(hp \cdot h)$]	Kilogram per joule (kg/J)	1.689 659	E−07
Psi (pound-force per square inch) ($lbf/in.^2$)	Pascal (Pa)	6.894 757	E+03
Psi (pound-force per square inch) ($lbf/in.^2$)	Kilopascal (kPa)	6.894 757	E+00
Quad (10^{15} Btu_{IT})	Joule (J)	1.055 056	E+18
Quart (US dry) (dry qt)	Cubic meter (m^3)	1.101 221	E−03
Quart (US dry) (dry qt)	Liter (L)	1.101 221	E+00
Quart (US liquid) (liq qt)	Cubic meter (m^3)	9.463 529	E−04
Quart (US liquid) (liq qt)	Liter (L)	9.463 529	E−01
Rad (absorbed dose) (rad)	Gray (Gy)	**1.0**	**E−02**
Revolution (r)	Radian (rad)	6.283 185	E+00
Revolution per minute (rpm) (r/min)	Radian per second (rad/s)	1.047 198	E−01
Rod (based on US survey foot) (rd)	Meter (m)	5.029 210	E+00

To convert from	To	Multiply by	
Rpm (revolution per minute) (r/min)	Radian per second (rad/s)	1.047 198	E−01
Second (angle) (°)	Radian (rad)	4.848 137	E−06
Second (sidereal)	Second (s)	9.972 696	E−01
Shake	Second (s)	**1.0**	**E−08**
Shake	Nanosecond (ns)	**1.0**	**E+01**
Slug (slug)	Kilogram (kg)	1.459 390	E+01
Slug per cubic foot (slug/ft^3)	Kilogram per cubic meter (kg/m^3)	5.153 788	E+02
Slug per foot second [slug/(ft·s)]	Pascal second (Pa·s)	4.788 026	E+01
Square foot (ft^2)	Square meter (m^2)	**9.290 304**	**E−02**
Square foot per hour (ft^2/h)	Square meter per second (m^2/s)	**2.580 64**	**E−05**
Square foot per second (ft^2/s)	Square meter per second (m^2/s)	**9.290 304**	**E−02**
Square inch (in.^{2})	Square meter (m^2)	**6.4516**	**E−04**
Square inch (in.^{2})	Square centimeter (cm^2)	**6.4516**	**E+00**
Square mile (mi^2)	Square meter (m^2)	2.589 988	E+06
Square mile (mi^2)	Square kilometer (km^2)	2.589 988	E+00
Square mile (based on US survey foot) (mi^2)	Square meter (m^2)	2.589 998	E+06
Square mile (based on US survey foot) (mi^2)	Square kilometer (km^2)	2.589 998	E+00
Square yard (yd^2)	Square meter (m^2)	8.361 274	E−01
Stokes (St)	Meter squared per second (m^2/s)	**1.0**	**E−04**
Tablespoon	Cubic meter (m^3)	1.478 676	E−05
Tablespoon	Milliliter (mL)	1.478 676	E+01
Teaspoon	Cubic meter (m^3)	4.928 922	E−06
Teaspoon	Milliliter (mL)	4.928 922	E+00
Therm (EC)	Joule (J)	**1.055 06**	**E+08**
Therm (US)	Joule (J)	**1.054 804**	**E+08**
Ton, assay (AT)	Kilogram (kg)	2.916 667	E−02
Ton, assay (AT)	Gram (g)	2.916 667	E+01
Ton-force (2000 lbf)	Newton (N)	8.896 443	E+03
Ton-force (2000 lbf)	Kilonewton (kN)	8.896 443	E+00
Ton, long (2240 lb)	Kilogram (kg)	1.016 047	E+03
Ton, long, per cubic yard	Kilogram per cubic meter (kg/m^3)	1.328 939	E+03
Ton, metric (t)	Kilogram (kg)	**1.0**	**E+03**
Tonne (called "metric ton" in the US) (t)	Kilogram (kg)	**1.0**	**E+03**

To convert from	To	Multiply by	
Ton of refrigeration (12 000 Btu_{IT}/h)	Watt (W)	3.516 853	E+03
Ton of TNT (energy equivalent)	Joule (J)	**4.184**	**E+09**
Ton, register	Cubic meter (m^3)	2.831 685	E+00
Ton, short (2000 lb)	Kilogram (kg)	9.071 847	E+02
Ton, short, per cubic yard	Kilogram per cubic meter (kg/m^3)	1.186 553	E+03
Ton, short, per hour	Kilogram per second (kg/s)	2.519 958	E−01
Torr (Torr)	Pascal (Pa)	1.333 224	E+02
Watt-hour (W · h)	Joule (J)	**3.6**	**E+03**
Yard (yd)	Meter (m)	**9.144**	**E−01**
Year (365 days)	Second (s)	**3.1536**	**E+07**
Year (sidereal)	Second (s)	3.155 815	E+07
Year (tropical)	Second (s)	3.155 693	E+07

About the Author

Jon M. Quigley, PMP, CTFL, is a principal and founding member of Value Transformation, a product development—from idea to product retirement—and cost improvement organization established in 2009. Jon has an engineering degree from the University of North Carolina at Charlotte, two master's degrees from the City University of Seattle, and two globally recognized certifications. Jon has over thirty years of product development and manufacturing experience, ranging from embedded hardware and software to verification, process, and project management.

Jon won the Volvo-3P Technical Award in 2005 and went on to win the 2006 Volvo Technology Award. Jon has secured seven US patents and several international patents, ranging from multiplexing systems and human-machine interfaces to telemetry systems and driver's aides. Jon has been on the Western Carolina University Master of Project Management Advisory Board and the Forsyth Technical Community College Advisory Board. He has also been a guest lecturer at Wake Forest University's Charlotte, NC, campus and Eindhoven Technical University (The Netherlands).

Jon has authored or contributed to numerous books on a range of product development and project management topics. The books he writes are used in bachelor- and master-level classes at universities around the globe, including the Eindhoven Technical University, Manchester Metropolitan University, San Beda College Manila in the Philippines, and Tecnológico de Monterrey. Among his titles are the following:

- Co-author of *Project Management of Complex and Embedded Systems: Ensuring Product Integrity and Program Quality* (Taylor and Francis, 2008, ISBN 1420070256)
- Co-author of *Scrum Project Management* (Taylor and Francis, 2010, ISBN 1439825157)
- Co-author of *Testing of Complex and Embedded Systems* (Taylor and Francis, 2010, ISBN 1439821402)
- Co-author of *Saving Software with Six Sigma*, Redwood Collaborative Media Professional Development Series (Software Test Professionals, ISBN 978-0-9831220-0-5)
- Co-author of *Aggressive Testing for Real Pros*, Redwood Collaborative Media Professional Development Series (Software Test Professionals, ISBN 978-0-9831220-1-2)
- Co-author of *Total Quality Management for Project Managers* (Taylor and Francis, 2012, ISBN 978-1-4398-85055-5)
- Co-author of *Reducing Process Costs with Lean, Six Sigma, and Value Engineering Techniques* (Taylor and Francis, 2012, ISBN 978-1-4398-8725-7)
- Co-author of *Project Management for Automotive Engineers: A Field Guide* (Society of Automotive Engineering, 2016, eISBN PDF 978-0-7680-8315-6; eISBN prc 978-0-7680-8316-3; eISBN epub 978-0-7680-8317-0)
- Contributor to *Opening the Door: 10 Predictions on the Future of Project Management in the Professional Services Industry* (eBook with MavenLink and ProjectManagers.net)
- Co-author of *Configuration Management: Theory and Application for Engineers and Managers*, two editions (Taylor and Francis, 2015 and 2019, ISBN 978-1482229356)
- Co-author of *Continuous and Embedded Learning for Organizations* (Taylor and Francis, 2020)
- Co-author of *Principles, Processes and Practices of Risk Management* (under contract with Taylor and Francis)
- Co-contributor to "Scrum Project Management" for the *Encyclopedia of Software Engineering* (ISBN – 1-4200-5977-7 and e-ISBN – 1-4200-5978-5)

- Contributor to *The Project Manager Who Smiled* by Peter Taylor (eBook, 2023, ISBN 978-0-9576689-0-4)
- Co-author of *Modernizing Product Development Processes: Guide for Engineers* (Society of Automotive Engineers, forthcoming 2023)

Jon has co-authored or been interviewed for hundreds of magazine articles on various product development teams and manufacturing topics, appearing in more than 60 magazines, e-zines, and other outlets. In addition, he writes three recurring columns:

- PMTips Quigley and Lauck's Expert Column
- *Assembly* magazine, P's and Q's on Project Management and Quality
- Automotive Industries, Quigley's Corner on automotive product development

Jon also contributes to Thomas Cagley's Software Process and Measurement CAST (SPaMCAST) podcast. He also regularly presents at technical conferences on various domains of product development, including product testing, learning, Agile, and project management.

Jon lives in Lexington, North Carolina, where he enjoys the beauty of nature, hiking in the woods, and playing the bass. Jon. Email Jon at jon.quigley@valuetransform.com.

www.ingramcontent.com/pod-product-compliance
Lightning Source LLC
LaVergne TN
LVHW050906210926
848482LV00012B/362
* 9 7 8 1 4 6 8 6 0 5 9 0 7 *